Algebra

Dedicated to the
Memory of
Mikhail Ivanovich Kargapolov

Preface

This volume contains a selection of the talks delivered at the IIIrd International Conference on Algebra. The conference, dedicated to the memory of Mikhail Ivanovich KARGAPOLOV (1928–1976), was held on the 23–28 of August, 1993, in Krasnoyarsk (Russia), basing on the Krasnoyarsk State University.

The conference was organized by the Krasnoyarsk State University and the Institute of Mathematics of the Siberian Branch of the Russian Academy of Sciences.

The collection of the abstracts, presented by all of those who applied for participation, was published before the opening, to be available at the conference. The total number of participants was about 240, including scientists from 13 countries all over the world, the majority being, naturally, from Russia. The most numerous groups arrived from Krasnoyarsk (38), Moscow (27), Novosibirsk (19), Ekaterinburg (15), Omsk (15).

There were 18 morning plenary one-hour lectures (see Appendices).

On the 25th of August, the evening plenary session was held, dedicated to the memory of M. I. Kargapolov, where many of his pupils and colleagues spoke on his life and work, and on the results of M. I. Kargapolov's school in Group Theory.

Five sections were organized:

1. Finite Groups.
2. Infinite Groups.
3. Ring Theory.
4. Algebraic Systems and Model Theory.
5. Applied and Computer Algebra.

There were 7 one-hour lectures (see Appendices), 50 half-hour reports, and 42 short communications during the sectional sessions altogether.

The cultural program included excursions around the city of Krasnoyarsk, along the Eniseĭ river, to the Krasnoyarsk hydro-power station, and to the reserve "Stolby" (= "Pillars" — the rocky mountains near Krasnoyarsk).

The main bulk of expenses of the Conference was taken by the Krasnoyarsk State University. The Conference was also supported by the Siberian Fund of Algebra and Logic and by the Krasnoyarsk Regional Science Foundation. The International Mathematical Union provided partial support for the travel expenses of the participants.

These Proceedings appear due to a very constructive and encouraging attitude of the Walter de Gruyter & Co, and especially that of Dr. M. Karbe.

The volume was set at the Novosibirsk Institute of Mathematics, after the papers, submitted in Russian, were translated into English, and some of those, contributed in English, were polished.

We are grateful to Tamara Rozhkovskaya and her team who set the manuscripts in TEX and produced a camera-ready copy.

We thank also our colleagues at the Institute of Mathematics, headed by Alexander Ryaskin, who contributed by reading the proofs and language polishing.

Yuriĭ L. ERSHOV
Evgeniĭ I. KHUKHRO
Vladimir M. LEVCHUK
Nikolaĭ D. PODUFALOV

Organizing Committee

Chairman	ERSHOV Yuriĭ L.
Vice-Chairmen	MAZUROV Victor D. PODUFALOV Nikolaĭ D. SHESTAKOV Ivan P.
Secretaries	MOROZOV Andreĭ S. LEVCHUK Vladimir M.
Committee Members	BOKUT' Leonid A. BOREVICH Zenon I. BUSARKIN Viktor M. COHN Paul GONCHAROV Sergeĭ S. GUPTA Chander-Kanta GUPTA Narain D. DROZD Yuriĭ A. HARTLEY Brian KOSTRIKIN Alekseĭ I. MACINTYRE Angus MIKHALEV Aleksandr A. NERODE Anil NEUMANN Bernard OL'SHANSKII Aleksandr Yu. PLOTKIN Boris I. POIZAT Bruno REMESLENNIKOV Vladimir N. RYBAKOV Vladimir V. SENASHOV Vladimir I. SHEMETKOV Leonid A. SHMEL'KIN Alfred L. SHUNKOV Vladimir P. STAROSTIN Albert I. ZEL'MANOV Efim I.

Table of Contents

Lattices of centralizers in groups

Vladimir A. Antonov

Abstract. Some questions about groups with various restrictions on the lattice of centralizers (for instance, distributive lattice, modular lattice and so on) are discussed. It is shown that there is a connection between the lattice of centralizers and that of subgroups. The results are given without proofs.

1991 Mathematics Subject Classification: 20F14, 20E15, 20F50, 06B05.

Let G be a group and let $M \subseteq G$. Denote the centralizer of M in G by $C(M)$. A subgroup H of G is called a *centralizer* if $H = C(M)$ for some subset M of G. Let $C^2(M) = C(C(M))$. It is clear that $C^2(M)$ is the minimal centralizer containing M. Moreover, C^2 is a closure operation in the set of all subgroups of G.

The set of all centralizers of any group G, with the operations

$$A \wedge B = A \cap B,\ A \vee B = C^2(\langle A, B\rangle) = C(C(A) \cap C(B)),$$

forms a complete lattice, called the *lattice of centralizers of* G and denoted by $\mathfrak{C}(G)$. The first result on $\mathfrak{C}(G)$ has been obtained by R. Schmidt [1].

In this paper we consider groups with various restrictions on the lattice of centralizers. In particular, we discuss decomposition of $\mathfrak{C}(G)$ into a direct sum, groups with the distributive lattice $\mathfrak{C}(G)$, groups with the modular lattice $\mathfrak{C}(G)$, a characterization of groups by their lattice of centralizers, and connection between the lattice of centralizers of a group and that of its subgroups. Finally, we indicate what lattice can be the lattice of centralizers of some group.

1. Decomposition of $\mathfrak{C}(G)$ into a direct sum. Suppose that $\mathfrak{L}_1$, $\mathfrak{L}_2$ are lattices and L_1, L_2 are their units. The notation $\mathfrak{L}_1 \oplus \mathfrak{L}_2$ ($L_1 \oplus L_2$) means the direct sum of $\mathfrak{L}_1$ and $\mathfrak{L}_2$. It is obvious that if $G = G_1 \times G_2$ then $\mathfrak{C}(G) = \mathfrak{C}(G_1) \oplus \mathfrak{C}(G_2)$. The converse assertion is not true.

The following theorem presents a criterion for decomposability of the lattice of centralizers into a direct sum of its sublattices.

Theorem 1 [2, 3]. *Let G be a group with atomic lattice of centralizers $\mathfrak{C}(G)$. Then $\mathfrak{C}(G)$ is the direct sum of its sublattices $\mathfrak{L}_1$ and $\mathfrak{L}_2$ if the following conditions are valid:*

— *there exist elementwise commuting subgroups H_1 and H_2 of G such that $\mathfrak{C}(G) \cong \mathfrak{C}(H_1H_2)$, $\mathfrak{C}(H_i) \cong \mathfrak{L}_i$,*

— *the sets of commutators of subgroups H_1 and H_2 have the trivial intersection.*

Recall that a lattice $\mathfrak{L}$ is called *atomic* if any non-zero element $A \in \mathfrak{L}$ contains an atom of $\mathfrak{L}$. Note that G does not have to be equal to H_1H_2. Of most interest is the modular lattice $\mathfrak{C}(G)$.

In the sequel, $\omega(A)$ denotes the sum of atoms of the lattice $\mathfrak{C}(G)$ contained in A, the symbol $l(A)$ stands for the length of the interval $[Z(G), A]$ of $\mathfrak{C}(G)$, and the notation $A \oplus B$ means that $[Z(G), A \vee B]$ is the direct sum of $[Z(G), A]$ and $[Z(G), B]$.

Theorem 2 [4, 5]. *Let G be a group with modular lattice of centralizers of finite length.*

(i) *If $\omega(G) = \omega_1 \oplus \omega_2$ then $G = G_1 \oplus G_2$ where $\omega(G_i) = \omega_i$.*

(ii) *If $\mathfrak{C}(G)$ is indecomposable into a direct sum of its sublattices and $A \in \mathfrak{C}(G)$, then $\mathfrak{C}(A)$ is also indecomposable into a direct sum.*

2. Groups with distributive lattice $\mathfrak{C}(G)$. R. Schmidt [1] showed that a group with distributive lattice of centralizers of finite length is abelian. The following theorem can be viewed as a generalization of [1].

Theorem 3 [6]. *Let G be a non-abelian group with distributive lattice $\mathfrak{C}(G)$. Then $\mathfrak{C}(G)$ is dense, i.e., $A, B \in \mathfrak{C}(G)$ and $A > B$ imply the existence of $C \in \mathfrak{C}(G)$ such that $A > C > B$.*

The problem of existence of a group with dense lattice of centralizers is far from being solved.

3. Groups with modular lattice $\mathfrak{C}(G)$. R. Schmidt [1] studied finite non-primary groups with modular lattice of centralizers of length 2 ($\mathfrak{M}$-groups). This description is easily transferred to the class of locally finite groups. Finite primary $\mathfrak{M}$-groups were examined by D. M. Rocke [7].

Suppose that G is a group with the modular lattice $\mathfrak{C}(G)$ of a finite length. By Theorem 1 it suffices to consider the lattice $\mathfrak{C}(G)$ indecomposable into a direct sum. By Theorem 2(i) the interval $[Z(G), \omega(G)]$ is also indecomposable into a direct sum but it is a projective geometry.

Denote by $\mathfrak{U}$ the class of locally finite groups with modular lattice of centralizers of finite length which satisfies the following conditions:

(i) if $G \in \mathfrak{U}$ and $A \in \mathfrak{C}(G)$ then $A \in \mathfrak{U}$,

(ii) if $G \in \mathfrak{U}$ and $\mathfrak{C}(G)$ is a projective geometry then the group G is an $\mathfrak{M}$-group or the factor-group $G/Z(G)$ is primary.

Theorem 4 [2,6]. *Let G be a locally finite group with modular lattice $\mathfrak{C}(G)$ of finite length. Then $G \in \mathfrak{U}$ if G satisfies one of the following conditions:*

(a) *G is finite,*

(b) *the lattice $\mathfrak{C}(G)$ is a sublattice of the lattice of subgroups,*

(c) *for any $x \in G$, the interval $[Z(G), C^2(x)]$ of the lattice $\mathfrak{C}(G)$ is a chain.*

The problem of existence of locally finite groups with modular lattice of centralizers of finite length not belonging to $\mathfrak{U}$ is far from being solved.

Theorem 5 [6]. *Let G be a $\mathfrak{U}$-group, $\mathfrak{C}(G)$ indecomposable into direct sum. If $\omega(G)$ is a non-abelian subgroup then G is an $\mathfrak{M}$-group, i.e. $l(G) = 2$, or $G/Z(G)$ is a p-group for some prime p.*

Corollary [2]. *Let G be a finite simple group with modular lattice of centralizers. Then $G \cong SL(2, 2^n)$, $n \geqslant 2$.*

For $\omega(G)$ abelian, the factor-group $G/Z(G)$ may not be a p-group. However, the following statement is true.

Theorem 6 [6]. *Let G be a $\mathfrak{U}$-group, $\mathfrak{C}(G)$ indecomposable into a direct sum. If $l(\omega(G)) > 2$ and, for any $x, y \in G$, the subgroup $C^2(x) \vee C^2(y)$ does not contain all the atoms of $\mathfrak{C}(G)$ then $G/Z(G)$ is a p-group for some prime p.*

In the case where $l(\omega(G)) = 2$, even the following more strong condition is not sufficient for $G/Z(G)$ to be a p-group: for any $x \in G$, the interval $[Z(G), C^2(x)]$ of $\mathfrak{C}(G)$ is a chain. For such groups the following theorem holds.

Theorem 7 [6]. *Let $G \in \mathfrak{U}$. Suppose that $l(\omega(G)) = 2$ and, for any $x \in G$, the interval $[Z(G), C^2(x)]$ of the lattice $\mathfrak{C}(G)$ is a chain. Then G is an $\mathfrak{M}$-group or $C(\omega(G))/Z(G)$ is a p-group for some prime p and either $G/C(\omega(G))$ is a p-group or $G/C(\omega(G))$ is a partitioned group and any component of this partition is a p-group or a locally cyclic p'-group.*

In the case of finite group G, the factor-group $G/C(\omega(G))$ can be one of the following types (see [4]):

a p-group,

a Frobenius group whose kernel is a p-group and the complement is a cyclic p'-group,

$PSL(2,p^n)$ or $PGL(2,p^n)$ where p is an odd prime.

4. A characterization of a group by its lattice of centralizers. In the general case, a group G is not characterized by its lattice of centralizers. Indeed, for an abelian group A we have $\mathfrak{C}(G) \cong \mathfrak{C}(G \times A)$. If $G_1 = PSL(2,p^n)$, $p^n \geqslant 4$, and $G^2 = D_{4(p^{2n}+p^n)}$ then $\mathfrak{C}(G^1) \cong \mathfrak{C}(G_2)$.

R. Schmidt [1, 8, 9] and A. V. Vasil'eva [10–12] showed that the finite simple groups with $l(G) \leqslant 4$ are characterized by their lattices of centralizers in the class of finite groups with trivial centre.

A finite simple group with $l(G) \leqslant 4$ is one of the following types [13]: $PSL(2,q)$, $q \geqslant 4$; $Sz(2^{2n+1})$, $n \geqslant 1$; $PSL(3,q)$, $q > 2$; $PSU(3,q^2)$, $q > 2$; J_1; A_7; M_{11}.

5. Connection between the lattice of centralizers of a group and the lattice of its subgroups. We write $\mathfrak{L}(G)$ for the lattice of subgroups of group G.

(a) Finite groups with $\mathfrak{C}(G) = \mathfrak{L}(G)$ were studied by W. Gaschütz [14]. The result of [14] is easily generalized to the case of locally finite groups.

(b) Groups in which $\mathfrak{C}(G)$ coincides with the interval $[Z(G), G]$ of the lattice $\mathfrak{L}(G)$ were described in the case of locally finite $G/Z(G)$ (see [15]). Finite groups with $\mathfrak{C}(G) = [Z(G), G]$ were studied independently by M. Reuther [16] and Y. Cheng [17, 18].

(c) Groups in which the map $C^2 : H \Rightarrow C^2(H)$ is a homomorphism of the lattice $\mathfrak{L}(G)$ (in the general case, that of the interval $[Z(G), G]$) into the lattice $\mathfrak{C}(G)$ are also described in the case where $G/Z(G)$ is locally finite [15]. By their structure, these groups are similar to the groups of type (b).

(d) Of most interest is the question of investigation of groups whose lattice of centralizers is a sublattice of the lattice of subgroups ($\mathfrak{C}(G) \geqslant \mathfrak{L}(G)$). The following results are obtained.

Theorem 8 [3]. *Let G be a group with the normalizer condition (i.e., $N(H) > H$ for every proper subgroup H of G). Suppose that $\mathfrak{C}(G) \leqslant \mathfrak{L}(G)$ and $\mathfrak{C}(G)$ is of finite length. Then $\mathfrak{C}(G)$ is modular and every maximal series of centralizers is subnormal.*

Theorem 9 [3, 19, 20]. *Let G be a locally finite group. If $\mathfrak{C}(G)$ has a finite length and $\mathfrak{C}(G) \leqslant \mathfrak{L}(G)$ then $\mathfrak{C}(G)$ is modular. If the lattice $\mathfrak{C}(G)$ is indecomposable into a direct sum of sublattices then the following assertions hold:*

(i) *if G is locally nilpotent then its every maximal series of centralizers is subnormal and all its factors are isomorphic elementary abelian groups,*

(ii) *if G is not locally nilpotent then G has a subnormal series of centralizers such that all its factors, except the last one, are isomorphic elementary abelian p-groups and the last factor is a locally cyclic p'-group acting irreducibly on the first factor.*

6. What lattice can be the lattice of centralizers of some group? An element A of a lattice $\mathfrak{L}$ is said to be *join irreducible* if $A = B \vee C$ implies $A = B$ or $A = C$.

A lattice $\mathfrak{L}$ with involutory anti-automorphism φ will be called a *C-lattice* (see [5]) provided that

(i) if $A \in \mathfrak{L}$ and A is join irreducible then $A \leqslant \varphi(A)$,

(ii) if $A \in \mathfrak{L}$ and $A \not\subset \varphi(A)$ then there are more than two elements of B with $\varphi(B) \wedge A = B$ under A and if $l(A) < \infty$ there are more than two maximal elements under A,

(iii) if $A \in \mathfrak{L}$ then A coincides with the sum of $B \in \mathfrak{L}$ with $\varphi(B) \wedge A = B$.

Theorem 10 [21]. *The lattice of centralizers of any group is a C-lattice.*

The converse assertion is not true. However, the following statement is valid.

Theorem 11 [5]. *Let $\mathfrak{L}$ be a finite modular complemented C-lattice. Then there exists a finite group G such that $\mathfrak{C}(G) \cong \mathfrak{L}$.*

Some propositions, which generalize Theorem 11, have been also obtained [21]. For the complete proofs of the above statements we refer to [22].

References

[1] Schmidt, R., Zentralisatorverbände endlicher Gruppen. Rend. Semin. Mat. Univ. Padova 44 (1970), 97–131.

[2] Antonov, V. A., Finite groups with modular lattice of centralizers. Algebra Logika 26:6 (1987), 653–683. English translation: Algebra Logic 26:6 (1987), 403–422.

[3] — On the connection between the lattice of centralizers of a group and the lattice of its subgroups. Mat. Zap. 14:4 (1989), 4–13.

[4] — Finite groups with modular lattice of centralizers, 2. Deposited in VINITI, No 4822–B90. Moscow 1990.

[5] — On a class of modular lattices. Algebra Logika 30:1 (1991), 3–14. English translation: Algebra Logic 30:1 (1991), 1–8.

[6] — Locally finite groups with the modular lattice of centralizers (to appear).

[7] Rocke, D. M., P-groups with Abelian centralizers. Proc. Lond. Math. Soc., III. Ser. 30:1 (1975), 55–75.

[8] Schmidt, R., Characterisierung der einfachen Gruppe der Ordnung 175560 durch ihren Zentralisatorverbände. Math. Z. 116:4 (1970), 299–306.

[9] — A characterization of $PSL(2,p^n)$ by its centralizer lattice. J. Algebra 21:2 (1972), 280–291.

[10] Vasil'eva, A. V., A characterization of the group $PSL_3(q)$ by its centralizer lattice. Algebra Logika 15:5 (1976), 509–534. English translation: Algebra Logic 15:5 (1976), 320–334.

[11] — Determinability of some simple groups by its centralizer lattice. In: 5th All–Union Symp. on Theory of Groups (Krasnodar, October 1–3, 1976), Abstracts of reports, 18. Institut Matematiki SO AN SSSR, Novosibisk 1976.

[12] — A characterization of $PSU_3(q)$ by its centralizer lattice. Deposited in VINITI, No 4806–85. Moscow 1985.

[13] — Centralizer lattices of finite simple groups. Sib. Mat. Zh. 18:2 (1977), 263–273. English translation: Sib. Math. J. 18:2 (1977), 188–196.

[14] Gaschütz, W., Gruppen, deren sämtliche Untergruppen Zentralisatoren sind. Arch. Math. 6:1 (1954), 5–8.

[15] Antonov, V. A., Groups of Gaschütz type and groups that are close to them. Mat. Zametki 27:6 (1980), 839–857. English translation: Math. Notes 27:5–6 (1980), 405–414.

[16] Reuther, M., Endliche Gruppen, in deren alle das Zentrum enthaltenden Untergruppen Zentralisatoren sind. Arch. Math. 29:1 (1977), 45–54.

[17] Cheng, Y., On double centralizer subgroups of some finite p-groups. Proc. Am. Math. Soc. 86:2 (1982), 205–208.

[18] — On finite p-groups with cyclic commutator subgroup. Arch. Math. 39:4 (1982), 295–298.

[19] Antonov, V. A., On finite groups with modular lattice of centralizers. In: Int. Conf. on Algebra in Honor of A. I. Mal'cev (Novosibirsk, August 21–26, 1989), 6. Abstracts of Reports on Group Theory. Institut Matematiki SO AN SSSR, Novosibirsk 1989.

[20] — Finite groups in which the lattice of centralizers is a sublattice of the lattice of subgroups. Izv. Vyssh. Uchebn. Zaved., Mat. 3 (1991), 5–14. English translation: Sov. Math. 35:3 (1991), 4–12.

[21] — Right C-lattices and lattices of centralizers (to appear).

[22] — Groups with restrictions on centralizers. Ph. D. Thesis. Chelyabinsk 1993.

Varieties of discrete representations of hyperbolic 3-lattices

Boris Apanasov *

Abstract. We survey global properties of the varieties of conjugacy classes of discrete faithful representations of the fundamental groups of closed (or finite volume) hyperbolic 3-manifolds M in linear groups $O(1, m)$, $m \geq 3$, especially for $m = 4$ and $m = 5$ and their relation to Teichmüller spaces of geometric structures. Of special interest are questions of connectedness of these spaces, cohomology groups $H^1(\pi_1(M), o(1, m))$ (infinitesimal deformations), smooth paths in the varieties and a link with Platonov's conjecture on arithmeticity and super-rigidity. For the last problem, we find a Coxeter group which is nonarithmetic but "hyperbolically super-rigid." Also, in contrast to Teichmüller spaces of hyperbolic surfaces, we show disconnectedness of varieties $\mathcal{T}(\pi_1(M), m), m = 4, 5$. For $m = 4$, this effect is related to an equivariant and everywhere wild embedding of a closed 3-ball into 3-sphere S^3. For $m = 5$, such phenomenon has been discovered by a recent author's construction of "twisted" 3-lattice representations in $O(1, 5)$. It uses the so-called block-building method and a topological background related to 2-knots in the 4-sphere. This has also a relation to first constructions of non-trivial circle and 2-plane bundles (with geometric structures) over a closed hyperbolic 3-manifold and wild (even locally wild) quasi-symmetric 2-knots in the 4-sphere.

1991 Mathematics Subject Classification: 20G05, 22E40, 57M30, 57M60, 30C65, 55R10.

1. Varieties of group representations

Let $\Gamma = \langle g_1, \dots, g_d \mid s_\alpha, \alpha \in A\rangle$ be a finitely generated torsion-free group, $\mathbf{H}$ be a classical simple algebraic group defined over $\mathbb{R}$ with real points H and $\rho_l : \Gamma \to H$ be a fixed embedding of Γ as a lattice in H. If H is not locally isomorphic to $PSL_2(\mathbb{R})$ then Γ is rigid by the Mostow–Margulis rigidity theorem [1, 2]. In addition, we suppose that $\mathbf{H}$ is represented in another algebraic group $\mathbf{G}$, ρ_a : $\mathbf{H} \to \mathbf{G}$, also defined over $\mathbb{R}$ where the image of $\mathbf{H}$ is not contained in a proper parabolic subgroup of $\mathbf{G}$. Then the composition of embeddings $\rho_0 = \rho_a\rho_l$ embeds Γ into G as a discrete subgroup (usually no longer a lattice). The main problem is to study the deformation space $\mathcal{R}_G(\Gamma) = \mathrm{Hom}(\Gamma, G)$ (with the algebraic convergence

* The author gratefully acknowledges partial support by National Science Foundation (grants DMS-920719 and DMS-9306311).

topology) which has the structure of an affine algebraic variety over $\mathbb{R}$ (obtained by embedding it in a product of d copies of G by sending the representation ρ to the d-tuple $(\rho(g_i)_{i=1,\dots,d})$. Its geometric properties reflect the relationships among the same degree representations of Γ. Of most interest is the quotient $\mathcal{S}_G(\Gamma) = R_G(\Gamma)/G$ of this variety by the conjugation action of G. The points of $\mathcal{S}_G(\Gamma)$ correspond to isomorphism classes of simple representations of Γ. A curve in $\mathcal{S}_G(\Gamma)$ through the (isomorphism class $[\rho]$ of the) simple representation ρ can be thought of as a deformation of ρ. For a general introduction to representation varieties, see [3].

As applications of representations varieties (see [4]), it is possible to consider a space of locally homogeneous structures on the original locally symmetric space $M = \Gamma\backslash H/K$ where K is a maximal compact subgroup of H. Of most interest are the groups either $SO(1,n)$ or $SU(1,n)$ or $PGL_n(\mathbb{R})$ because of the following observation [5]:

Theorem 1.1 *The representation ρ is locally rigid in G, i.e., its orbit is open in* $\mathrm{Hom}(\Gamma, G)$ *unless H is locally isomorphic to $SO(1,n)$ or $SU(1,n)$.*

It makes sense to consider Γ as a lattice in n-dimensional hyperbolic space (either real or complex). Due to the fact that the case of complex hyperbolic structures is much more rigid than the real one [6], our basic object will be the fundamental group $\Gamma = \pi_1(M)$ of a closed (or finite volume) hyperbolic n-manifold $M = \mathbb{H}^n/\Gamma$ or a hyperbolic n-lattice $\Gamma \subset O^\circ(1,n) = H$, $n \geq 3$.

2. Representations of hyperbolic lattices and geometric structures

Now we assume that Γ is a hyperbolic lattice (in a semisimple Lie group $H = O^\circ(1,n)$ of real rank 1, $n \geq 3$) and we have its representations in another semisimple Lie group G (especially, $G = O^\circ(1,m)$, $m \geq 4$).

Inside of $\mathcal{R}_m(\Gamma) = \mathcal{R}_{O^\circ(1,m)}(\Gamma)$ we have two sequential subspaces of faithful representations with discrete (correspondingly, discontinuous somewhere at $\partial\mathbb{H}^m = S^{m-1}$) images. The group $O^\circ(1,m)$ acts on $\mathcal{R}_m(\Gamma)$ by conjugation leaving both subspaces invariant. We denote their quotient spaces as

$$\mathcal{H}_m(\Gamma), \quad \mathcal{T}_{m-1}(\Gamma) \subset \mathcal{R}_m(\Gamma)/O^\circ(1,m). \tag{2.1}$$

There is a natural identification of the biggest space $\mathcal{H}_m(\Gamma)$ with the space of n-dimensional hyperbolic structures on Γ. An element of this space is determined by an equivalence class of pairs $\{N, \phi\}$ where N is an m-dimensional hyperbolic manifold and $\phi : \Gamma \to \pi_1(N)$ is an isomorphism (up to isometries of manifolds N and inner automorphisms of Γ).

As an example, if $\Gamma = \pi_1(S_g)$ for a closed surface of genus $g > 1$, then $\mathcal{H}_2(\Gamma)$ is the Teichmüller space of Γ and homeomorphic to $\mathbb{R}^{6g-\epsilon}$. Another example of $\Gamma = \pi_1(M)$ for a hyperbolic closed n-manifold M provides the space $\mathcal{H}_n(\Gamma)$ which degenerates to a point due to the Mostow–Margulis rigidity [1, 2].

Theorem 2.1. *Let M_1, M_2 be closed hyperbolic n-manifolds with isomorphic fundamental groups. Then M_1 and M_2 are isometric.*

Similarly, via holonomy, we have a natural identification of the space $\mathcal{T}_{m-1}(\Gamma)$ with the (Teichmüller) space of uniformizable $(m-1)$-dimensional conformal structures on Γ (cf. [7, 8]). An element of this space is determined by a pair $\{N, \phi\}$, where N is a conformal (=conformally flat) $(m-1)$-manifold with a non-surjective developing map $d : \tilde{N} \to S^{m-1}$ of its universal covering space $\tilde{N}$ into $d(\tilde{N}) = \Omega \subset S^{m-1}$ and $\phi : \Gamma \to \pi_1(N)$ is a monomorphism corresponding to the short exact sequence

$$0 \to \pi_1(\Omega) \to \pi_1(N) \to \Gamma \to 0, \tag{2.2}$$

By [9, 10], the developing map d and the natural projection $\pi : \Omega \to \Omega/G \cong N$ (where $\Gamma \cong G = d_*(\pi_1(N)) \subset O^\circ(1, m)$) are covering maps which factor the universal projection $\tilde{N} \to N$. Two pairs $\{N_0, \phi_0\}$ and $\{N_1, \phi_1\}$ determine the same conformal structure on Γ if there is an orientation preserving conformal homeomorphism $f : N_0 \to N_1$ such that $f_* \circ \phi_0$ and ϕ_1 differ (up to the isotropy subgroup $Z(\rho_0)$ of the inclusion $\rho_0 : \Gamma \to O^\circ(1, m)$) by an inner automorphism of Γ. (Notice that $f_* : \pi_1(N_0) \to \pi_1(N_1)$ is only well defined up to inner automorphism of $\pi_1(N_0)$.)

3. Deformations of representations

The basic object of this survey is the varieties $\mathcal{T}_{m-1}(\Gamma) \subset \mathcal{H}_m(\Gamma)$. The first surprising result (beyond Mostow–Margulis rigidity) is that there are large classes of groups Γ for which these varieties are positive-dimensional. So the Margulis super-rigidity theorem [11] fails for such lattices, and a question of A. Borel and N. Wallach [12] on triviality of the first cohomology group $H^1(\Gamma, o(1, m))$ (i.e., of the tangent space for the variety $\mathcal{T}_{m-1}(\Gamma)$, consisting of infinitesimal deformations) has a negative answer.

Namely, in [13] the author gave a method for constructing nontrivial deformations under some assumptions on lattices Γ. Referring to more later results [5, 7, 14, 15], some of these non-rigid lattices $\Gamma \subset O^\circ(1, n)$ might be described as the fundamental groups of hyperbolic n-manifolds M with at least one embedded totally geodesic hypersurface $N \subset M = \mathbb{H}^n/\Gamma$. For any such hypersurface, the construction assigns a real analytic curve in $\mathcal{T}_{m-1}(\Gamma)$ through the representation

ρ_0, consisting of the so-called bending representations of Γ. We shall outline this construction from an algebraic point of view.

Let $\Gamma_N \subset \Gamma$ be the fundamental group of the hypersurface $N \subset M$. Then Γ_N is a lattice in $O^\circ(1, n-1) \subset O^\circ(1,n)$ because N is totally geodesic. It shows that the center $\mathbb{Z}(\Gamma_N, O^\circ(1,m))$, $m > n$, is not trivial. Now, due to Seifert–Van Kampen theorem, Γ is represented as either a free product with Γ_N as the amalgamated subgroup or a HNN-extension, correspondingly to the property of the complement $M \backslash N$ either to have two connected components M_0 and M_1 or to be connected ($M \backslash N = M_0$). Let us denote corresponding subgroups of Γ (the fundamental groups of M_i) by Γ_i, $i = 0, 1$.

In the first case of a free amalgamated product, $\Gamma = \Gamma_0 *_{\Gamma_N} \Gamma_1$, we have constructed a non-trivial smooth curve through $[\rho_0] \in \mathcal{T}_{m-1}(\Gamma)$ so that its points correspond to conjugacy classes of the following representations:

$$\rho_\alpha \Gamma = \Gamma_0 *_{\Gamma_N} g_\alpha \Gamma_1 g_\alpha^{-1}, \tag{3.1}$$

where α, $-\epsilon < \alpha < \epsilon$, is the angle of the rotation $g_\alpha \in \mathbb{Z}(\Gamma_N, O^\circ(1,m))$.

In the second case of HNN-extension, $\Gamma = \Gamma_0 *_{\Gamma_N} = \langle \Gamma_0, \gamma \mid \gamma \Gamma_N \gamma^{-1} = \Gamma_{N'} \rangle$, $\Gamma_{N'} \subset \Gamma_0$, we have constructed a non-trivial smooth curve through $[\rho_0] \in \mathcal{T}_{m-1}(\Gamma)$ so that its points correspond to conjugacy classes of the following representations:

$$\rho_\alpha \Gamma = \langle \Gamma_0, \gamma_\alpha \mid \gamma_\alpha \Gamma_N \gamma_\alpha^{-1} = \Gamma_{N'} \rangle, \tag{3.2}$$

where the element $\gamma_\alpha = g_\alpha \gamma (g'_\alpha)^{-1} \in O^\circ(1,m)$ and the small enough parameter α, $-\epsilon < \alpha < \epsilon$, is the angle of the rotation $g_\alpha \in \mathbb{Z}(\Gamma_N, O^\circ(1,m))$ and $g'_\alpha \in \mathbb{Z}(\Gamma_{N'}, O^\circ(1,m))$.

However, despite such nontriviality of the representation variety and in contrast to the classical Teichmüller theory for surface group representation, we have the next surprising result, called the weak rigidity theorem of J. Morgan, P. Shalen, and W. Thurston [16–18]. Due to this theorem, the space $\mathcal{H}_m(\Gamma)$ of conjugacy classes of faithful discrete representations of an uniform lattice $\Gamma \subset O^\circ(1,3)$ into $O^\circ(1,m)$, $m \geq 3$, is compact, and we have (cf. [8]):

Theorem 3.1. *Given uniform lattice $\Gamma \subset O^\circ(1,3)$ and $m \geq 3$, the variety $\mathcal{T}_{m-1}(\Gamma)$ has a natural compactification $\overline{\mathcal{T}_{m-1}(\Gamma)}$ whose boundary representations are faithful and have discrete images.*

A special interest for us is a question about the action of boundary representations of Γ (distinct from cusps) on the sphere at infinity of $\mathbb{H}^m$, $m \geq 4$.

Is it always discontinuous? (For boundary cusp representations, see [7, 13].)

For study of such kind of questions as well as question [5] about singularities of the representation varieties $\mathcal{T}_{m-1}(\Gamma) \subset \mathcal{H}_m(\Gamma)$ for a hyperbolic 3-lattice Γ, it is important to find new kinds of deformation of hyperbolic lattices which are different from "bendings" defined in (3.1) and (3.2) (see [7, 19–21]). Actually, using of the so-called stamping deformations defined by the author [19] in connection with cone hyperbolic structures on the manifold $M = \mathbb{H}^3/\Gamma$ together with a new Apanasov–Tetenov's construction [20] of "bendings" along geodesic hypersurfaces with boundary (or along pleated hypersurfaces) in M shows the main obstacle for smoothness of the varieties which has been discovered [5] for some lattices $\Gamma \subset O^{\circ}(1,n)$, $n \geq 4$, might be overcame in dimension 3.

4. Super-rigid groups and Platonov conjecture on arithmeticity

For a given finitely generated group Γ, we may consider another affine algebraic variety over the complex field $\mathbb{C}$, $\mathcal{R}_n(\Gamma) = \mathrm{Hom}(\Gamma, GL_n(\mathbb{C}))$ as well as its open subvariety $\mathcal{R}_n(\Gamma)^s$ of irreducible representations and the quotient $\mathcal{S}_n(\Gamma) = \mathcal{R}_n(\Gamma)^s/PGL_n(\mathbb{C})$ by the conjugation action. The points of $\mathcal{S}_n(\Gamma)$ correspond to isomorphism classes of simple n-dimensional representations of Γ. As before, a curve in $\mathcal{S}_n(\Gamma)$ through the (isomorphism class $[\rho]$ of the) simple representation ρ can be thought of as a deformation of ρ.

Groups Γ for which $\mathcal{S}_n(\Gamma)$ is zero-dimensional for all n are of special interest. They are called *super-rigid* because of an analogy with Margulis super-rigidity theorem [11]. This class includes $SL(n,\mathbb{Z})$, the reductive groups (groups for which every representation is semisimple, due to [3, Corollary 2.7] and a result of H. Bass [22]), the groups of *integral representation type* (that is, finitely generated groups such that all the values of all the characters of all their representations are integral over $\mathbb{C}$). Additionally, Bass [22] has shown that super-rigid groups also have the property (F-Ab), namely that every finite index subgroup has finite abelianization. Recently, A. Rapinchuk [23, Theorem 3] has shown that (F-Ab) groups of bounded generation are super-rigid. Here a group Γ has *bounded generation* if there exist elements $\gamma_1, \ldots, \gamma_t \in \Gamma$ with $\Gamma = \langle\gamma_1\rangle \ldots \langle\gamma_t\rangle$ where $\langle\gamma_i\rangle$ is the cyclic subgroup generated by γ_i.

Analyzing such known examples of finite representation-type groups, V. P. Platonov [24, 25] formulated the following

Platonov Conjecture. *Finitely generated linear super-rigid groups are of arithmetic type.*

Here, by an arithmetic-type group we mean a group commensurable with a direct product of a finite number of S-arithmetic groups (perhaps, for different number sets S). A survey of the background on the Platonov conjecture and its connection with the Grothendieck's problem on profinite completions can be found in [26]. It is also known [27, Theorem 1] that linear groups of arithmetic type are of bounded generation.

Consider a hyperbolic lattice as the finitely generated linear group Γ above. It is naturally to ask whether nonarithmetic lattices $\Gamma \subset O^{\circ}(1,n)$ are super-rigid. Concerning the well known M. Gromov–I. Piatetski-Shapiro interbreeding construction [28] of non-arithmetic hyperbolic lattices in any dimension $n \geq 3$, we see non-rigidity of such n-lattices in $GL_{n+2}(\mathbb{R})$ because of the bending deformations (3.1) and (3.2). Due to an observation by V. P. Platonov [29], it implies non-rigidity of such lattices in $GL_{n+2}(\mathbb{C})$, too.

Nevertheless, there are nonarithmetic hyperbolic lattices Γ which are "hyperbolically super-rigid" (in the sense of rigidity of their representations in any $O^\circ(1,m)$, $m \geq n$).

Theorem 4.1. *Let $\Gamma \subset O^\circ(1,3)$ be the uniform lattice (with the basic field $\mathbb{Q}(\sqrt{2},\sqrt{5})$) generated by four reflections in sides of a hyperbolic simplex $\Delta \subset \mathbb{H}^3$ with dihedral angles $\pi/2, \pi/3, \pi/5, \pi/2, \pi/3$ and $\pi/4$, i.e.,*

$$\Gamma = \langle \gamma_1, \ldots, \gamma_4 \mid \gamma_i^2 = (\gamma_i\gamma_j)^{m_{ij}} = 1, \quad m_{ij} = 2,3,4 \text{ or } 5; \quad i,j = 1,2,3,4\rangle, \quad (4.1)$$

which corresponds to the following Coxeter diagram:

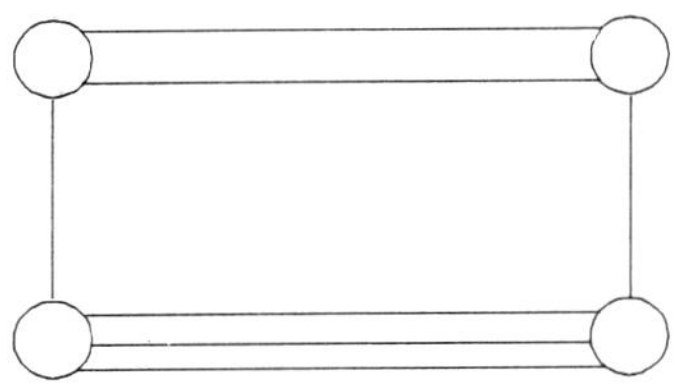

Then Γ is nonarithmetic and hyperbolically super-rigid.

Actually, due to a result of E. B. Vinberg [30], our group Γ in (4.1) is the unique nonarithmetic uniform hyperbolic lattice generated by reflections in sides of a hyperbolic (bounded) n-simplex $\Delta \subset \mathbb{H}^n$. There are few additional hyperbolic (co-finite) lattices which are nonarithmetic: six of them are in $O^\circ(1,3)$ and one in $O^\circ(1,5)$ [30]. All of these lattices are hyperbolically super-rigid. Namely, due to the author observation [31], they are rigid in $O^\circ(1,n+1)$: all their faithful representations in $O^\circ(1,n+1)$ are conjugated to the inclusion. General rigidity of Γ in $O^\circ(1,m)$, $m \geq n$, follows from the fact that a subgroup of $O^\circ(1,m)$ generated by $(n+1)$ reflections lies (up to conjugation in $O^\circ(1,m)$) either in $O^\circ(1,n)$ or in $O(n+1)$ or in $\mathrm{Isom}\,\mathbb{R}^{n+1}$.

These hyperbolically super-rigid but nonarithmetic lattices are likely providing counter examples to the above conjecture because they cannot be S-arithmetic for any finite number set S (for Γ in (4.1), $S = \{\sqrt{2}, \sqrt{5}\}$) while they are rigid in the most probable non-rigid cases of the hyperbolic linear groups $O^\circ(1,m)$ (cf. Theorem 1.1).

5. Global properties of representation varieties

Here we show how may geometric structures on manifolds can be used to shed light on topology of varieties of group representations. Namely, we are studying a natural question whether or not the varieties $\mathcal{T}_{m-1}(\Gamma) \subset \mathcal{H}_m(\Gamma)$ are connected (like Teichmüller spaces for surface groups, which are cells). Here we consider uniform hyperbolic lattices $\Gamma \subset O^\circ(1,3)$ (as before, cf. Theorem 1.1).

In contrast to Teichmüller spaces of hyperbolic surfaces, Apanasov's construction [32] of strange conformal structures on closed hyperbolic 3-manifolds M, $\pi_1(M) = \Gamma$ (with large enough number of disjoint totally geodesic surfaces), and using of the Sullivan's stability theorem for geometrically finite hyperbolic groups [33] show that the space of uniformized conformal structures on M with faithful holonomy representations $d_* : \pi_1(M) \to O^\circ(1,4)$ is not connected. This implies disconnectedness of our varieties:

Theorem 5.1. *There are (uniform) hyperbolic lattices $\Gamma \subset O^\circ(1,3)$ whose faithful discrete representation varieties $\mathcal{T}_3(\Gamma) \subset \mathcal{H}_4(\Gamma)$ are disconnected.*

The nature of that disconnectedness of the varieties is a "wild" geometric realization of some representations $\rho^* : \Gamma \to O^\circ(1,4)$. In fact, both discrete groups Γ and $\rho^*(\Gamma)$ act discontinuously in some 3-dimensional balls $B = B^3(0,1)$ and B^* in the 3-sphere S^3. Moreover, there is a (*quasi-symmetric*, see [34]) homeomorphism $f : \overline{B} \to \overline{B^*}$ between their closures, which is compatible with the group action:

$$\rho^*(\gamma) = f^*(\gamma) = f\gamma f^{-1}, \quad \gamma \in \Gamma. \tag{5.1}$$

However, this embedding f of the closed round ball $\overline{B}$ is wild everywhere on the boundary sphere ∂B so that the complement $S^3\overline{B^*}$ is not simply connected (see Fig. below).

Algebraically, this "wild" representation ρ^* is obtained from the inclusion ρ_0 by means of (large enough number of) consecutive bending deformations (3.1) (geometrically, only in the ball B and with some special values of parameters α_i) and a final bending deformation of the type (3.2).

It is still unknown, how many connected components are there in our varieties, and what invariants may distinguish them? One of possible approaches is using such geometric invariants (for conformal structures on corresponding 3-manifolds) as the η- and Chern–Simons invariants which are locally constant in the space of conformally flat structures [35, 36] (cf. [37]).

Now, passing to the dimension $m = 5$, we firstly remark that the composition of the constructed representation (5.1) and the inclusion $O^\circ(1,4) \subset O^\circ(1,5)$ provides only a representation lying in the same component of $\mathcal{T}_4(\Gamma) \subset \mathcal{H}_5(\Gamma)$ that contains the inclusion representation ρ_0 because the nerve of the knotting of the wild ball B^* in S^3 is one-dimensional (it is unknotted in S^4). That is why the problem of connectedness of the varieties $\mathcal{T}_4(\Gamma) \subset \mathcal{H}_5(\Gamma)$ cannot be reduced to Theorem 5.1.

However, the disconnectedness effect takes still place for these varieties $\mathcal{T}_4(\Gamma) \subset \mathcal{H}_5(\Gamma)$. It has been recently discovered by Apanasov's construction [38] of hyperbolic 3-lattice representations in $O^\circ(1,5)$. This construction also establishes a connection between variety components and some 2-dimensional knots in S^4.

We shall describe the main idea of our construction in general situation of faithful discrete representations of hyperbolic n-dimensional lattices Γ, $n \geq 3$, into

$O^{\circ}(1, n+2)$. This construction is based on the author's "block-building method" (cf. [32, 34]) and PL-approximation of smooth $(n-1)$-knots in the $(n+1)$-sphere S^{n+1} with a geometric control. It is roughly as follows.

Suppose that a tame $(n-1)$-knot $K \subset S^{n+1}$ belongs to the union of boundaries of a finite set of metric n-balls $\mathcal{C} = \{B_1, \ldots, B_k\}$ such that $\partial B_i \cap \partial B_j \neq \emptyset$ if the balls B_i, B_j are adjacent in the sense $j = i + 1 (\mathrm{mod}\, k)$. Additionally sup-

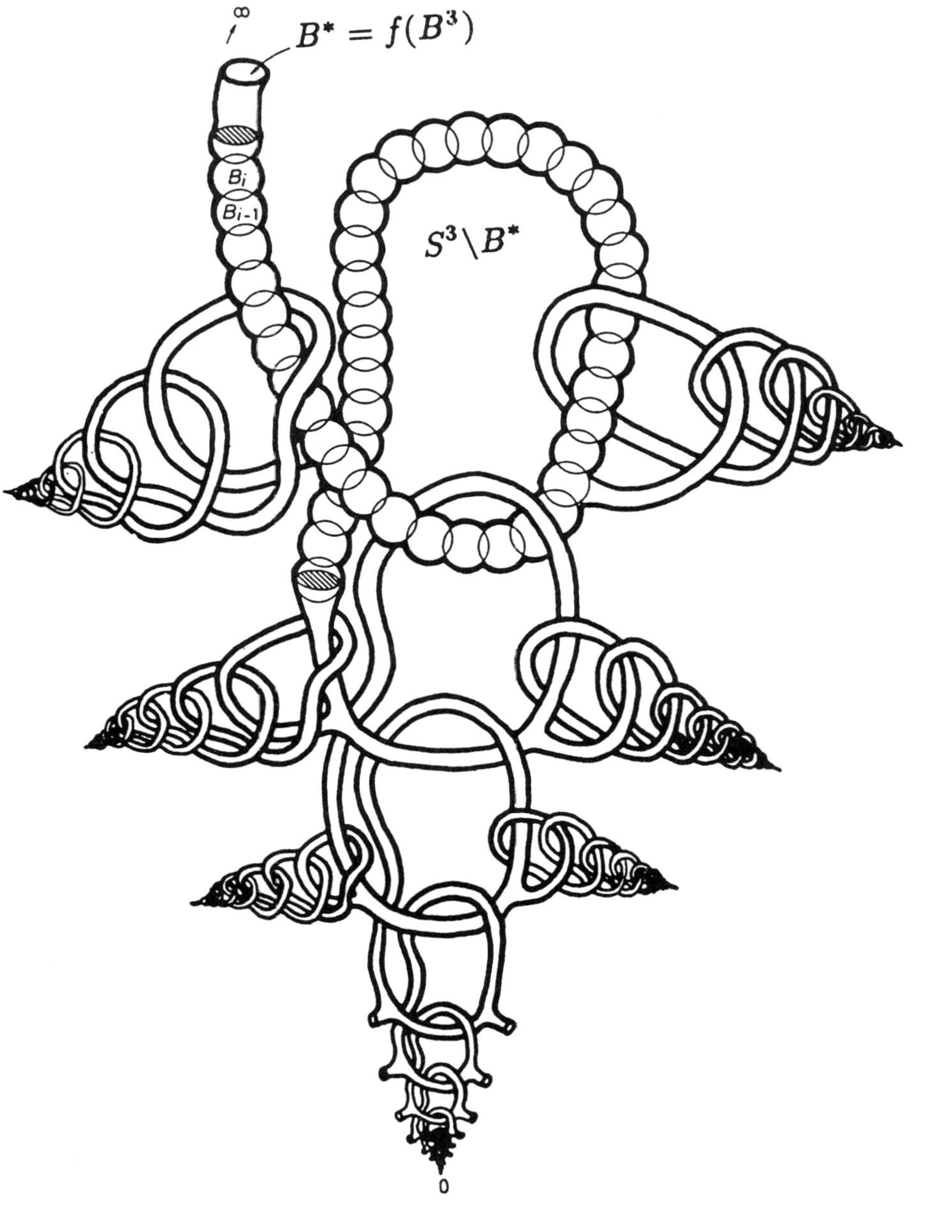

pose that, in each ball B_i, there is a discrete hyperbolic isometry group action $G_i \subset O^\circ(1,n) \subset O^\circ(1,n+2)$ such that, upon a geometric control on the isotopy class of K and the family $\mathcal{C}$, the (amalgamated) product $G \subset O^\circ(1,n+1)$ of groups $j_i^*(G_i)$ ($j_i : B_i \to S^{n+1}$ are inclusions) is a discrete group isomorphic to our given lattice $\Gamma \subset O^\circ(1,n)$. Then the limit set $\Lambda(G) \subset S^{n+1}$ of the group G is a certain infinite compounding K_∞ of the original knot K but still homeomorphic to $(n-1)$-sphere because of Tukia isomorphism theorem [39]. Due to the ergodicity of G-action on K_∞, this knot should be wild everywhere if K is not trivial. To prove this fact, we have to use additivity of Alexander knot invariants with respect to connected sum of knots [40]. Then the Sullivan stability theorem for geometrically finite hyperbolic groups [33] completes the proof on disconnectedness of the varieties $\mathcal{T}_{n+1}(\Gamma) \subset \mathcal{H}_{n+2}(\Gamma)$ because the limit set $\Lambda(\rho(\Gamma))$ of each quasi-Fuchsian representation is an unknotted 2-sphere in S^4.

Particularly, due to Andreev–Rivin characterization of hyperbolic 3-polyhedra [41, 42], this approach is very successful for non-trivial 2-knots, especially for ribbon knots [40]. Namely, we have the following result [38]:

Theorem 5.2. *For a given non-trivial ribbon 2-knot $K \subset S^4$ there exists a discrete faithful representation $\rho : \Gamma \to G = \rho(\Gamma) \subset O^\circ(1,5)$ of an uniform lattice $\Gamma \subset O^\circ(1,3)$ with ergodic action of G on the infinitely compounded, everywhere wild 2-knot $K_\infty = \Lambda(G) \subset S^4$. Furthermore, this action is compatible with a quasi-symmetric everywhere wild embedding $f : S^2 \hookrightarrow K_\infty \subset S^4$.*

Corollary 5.3. *Let $\Gamma \subset O^\circ(1,3)$ be any uniform lattice from Theorem 5.2. Then the varieties $\mathcal{T}^4(\Gamma) \subset \mathcal{H}^5(\Gamma)$ of discrete faithful representations are disconnected.*

Another kind of discrete faithful representations of a hyperbolic 3-lattice Γ into $O^\circ(1,5)$ has a close relation to first author's constructions [38] of non-trivial fiber bundles over closed hyperbolic 3-manifolds (or orbifolds) with geometric structures. Namely, given a closed hyperbolic 3-manifold (orbifold) M with $\pi_1(M) = \Gamma \subset O^\circ(1,3)$, each conjugacy class $[\rho]$ in the variety $\mathcal{T}^4(\Gamma) \subset \mathcal{H}^5(\Gamma)$ defines a 5-dimensional hyperbolic manifold $E = \mathbb{H}^5/\rho(\Gamma)$ and a closed conformal 4-manifold N at infinity of E, $N = \Omega(\rho(\Gamma))/\rho(\Gamma)$ ($\Omega(\rho(\Gamma))$ is the discontinuity set of $\rho(\Gamma)$ in the sphere at infinity $S^4 = \partial\mathbb{H}^5$), whose fundamental group satisfies to the exact sequence (2.2). Moreover, each such manifold E coming from our general construction of the group $G = \rho(\Gamma) \subset O^\circ(1,5)$ in Theorem 5.2 (see [38]) has a PL-embedding $M \hookrightarrow E$ (in general, non-locally-flat) which is a homotopy equivalence. The manifolds E and N themselves are 2-plane, and S^1 bundles over M if the exterior of the knot is homeomorphic to $B^3 \times S^1$ (i.e. the knot is trivial [43]). Moreover, it turns out that (besides trivial bundles corresponding to quasi-Fuchsian representation of $\Gamma = \pi_1(M)$) there are non-trivial fiber bundles over M associated with compact convex hyperbolic 3-polyhedra P whose sides have compatible involutions. We construct them (for hyperbolic uniform lattices Γ_0 generated by that involutions in sides of P) by using non-trivial automorphisms $\eta(x,\theta) = (\tau(\theta)(x), \theta)$ of $S^2 \times S^1$ compatible with $\Gamma_0 \supset \Gamma$ (cf. [7, p. 419–423; 44]).

Here $\tau(\theta)$ is a rotation of S^2 about its polar S^0 so that $G = \rho(\Gamma)$ is the free acting subgroup of the group $\rho(\Gamma_0)$ generated by compositions of Γ-generating involutions and rotations τ_i associated with ith sides of P (see [38]).

Due to [43], our η-Dehn surgery defines unknotted $S^2 \subset S^4$, so the limit set $\Lambda(G)$ is unknotted, hence $\Omega(G) = S^4\backslash\Lambda(G) \cong B^2 \times S^2$ is homotopy equivalent to S^1, and the exact sequence (2.2) is just the one coming from the fibration $S^1 \to N \to M$.

Nevertheless, fiber bundles over 3-manifolds M are more rigid than their analogues for surfaces (see [45]). Namely, the group of pseudo-isotopy classes of homeomorphisms of $S^2 \times S^1$ (the boundary of a regular neighborhood of $S^2 \subset S^4$) is $\mathbb{Z}_2 \times \mathbb{Z}_2 \times \mathbb{Z}_2$, i.e. there are at most two non-equivalent locally flat embeddings of S^2 in S^4 with homeomorphic exteriors; see [43, 46]. It implies:

Theorem 5.4. *For a given closed hyperbolic 3-manifold M there are at most two non-equivalent circle (or 2-plane) bundles over M allowing an uniformizable conformal structure developed onto a 2-knot complement (a complete hyperbolic metric, correspondingly). Moreover, there are at most finite number of equivalence classes of conformal 4-manifolds N (hyperbolic 5-manifolds E, $\partial E = N$) homotopy equivalent to M and whose development is onto a 2-knot complement.*

We finally remark that the number of connected components of the varieties $\mathcal{T}^4(\Gamma) \subset \mathcal{H}^5(\Gamma)$ in Corollary 1 (as well as the number of manifold equivalence classes in Theorem 5.4) depends of geometry of $M = H^3/\Gamma$, especially, of the number $b(M)$ of disjoint totally geodesic surfaces in M. In our examples of disconnected varieties in [38], this number $b(M)$ is more than 100 and, we think, it cannot be less than $C_n \cdot e$ where $C_n > 1$ is an universal constant and $e = e(K)$ is the potential energy of the n-knot K [47] (actually, for the used trefoil knot $k \subset S^3$, $e(k) \approx 74$).

References

[1] Mostow, G. D., Quasiconformal mappings in n-space and the rigidity of hyperbolic space forms. Publ. Math. IHES 34 (1968), 53–104.

[2] Margulis, G. A., The isometry of closed manifolds of constant negative curvature with the same fundamental group. Dokl. Akad. Nauk SSSR 192 (1970), 736–737. English translation: Sov. Math. Dokl. 11 (1970), 722–723.

[3] Lubotzky, A., Magid, A., Varieties of representations of finitely generated groups. Mem. Am. Math. Soc. 336 (1985).

[4] Geometry of Group Representations (ed. by W. Goldman, A. Magid). Contemp. Math. 74. Am. Math. Soc., Providence, RI, 1988.

[5] Johnson, D., Millson, J., Deformation Spaces Associated to Compact Hyperbolic Manifolds. In: Discrete Groups in Geometry and Analysis (ed. by R. Howe),

48–106. Papers in Honor of G. D. Mostow on His Sixtieth Birthday. Birkhäuser, Boston 1987.

[6] Goldman, W., Millson, J., Local rigidity of discrete groups acting on complex hyperbolic space. Invent. Math. 88 (1987), 495–520.

[7] Apanasov, B., Discrete Groups in Space and Uniformization Problems. Math. Appl. 40. Kluwer Academic Publishers, Dordrecht 1991.

[8] — Hyperbolic cobordisms and conformal structures. In: Topology'90, 27–36. Walter de Gruyter, Berlin–New York 1992.

[9] Kamishima, Y., Conformally flat manifolds whose development maps are not surjective. Trans. Am. Math. Soc. 294 (1986), 607–624.

[10] Kulkarni, R., Pinkall, U., Uniformization of geometric structures with applications to conformal geometry. Lect. Notes Math. 1209 (1986), 190–210.

[11] Margulis, G. A., Discrete groups of motions of manifolds of nonpositive curvature. Trans. Am. Math. Soc. 109 (1977), 33–45.

[12] Borel, A., Wallach, N., Continuous Cohomology, Discrete Subgroups, and Representations of Reductive Groups. Princeton University Press, Princeton 1980.

[13] Apanasov, B., Nontriviality of Teichmüller space for Kleinian group in space. In: Riemann Surfaces and Related Topics: Proc. of the 1978 Stony Brook Conf. (ed. by I. Kra, B. Maskit), 21–31. (Ann. Math. Stud., 97). Princeton University Press, Princeton 1981.

[14] — Thurston's bends and geometric deformations of conformal structures. In: Complex Analysis and Applications'85, 14–28. Sofia 1986.

[15] Kourouniotis, C., Deformations of hyperbolic structures. Math. Proc. Camb. Philos. Soc. 98 (1985), 247–261.

[16] Morgan, J., Shalen, P., Valuations, trees and degenerations of hyperbolic structures. Ann. Math. 120 (1984), 401–476.

[17] Morgan, J., Trees and hyperbolic geometry. In: Proc. Int. Congr. of Math. 1, 590–597. Berkeley 1987.

[18] Thurston, W., Hyperbolic structures on 3-manifolds I: deformations of acylindrical manifolds. Ann. Math. 124 (1986), 203–246.

[19] Apanasov, B., Deformations of conformal structures on hyperbolic manifolds. J. Diff. Geom. 35 (1992), 1–20.

[20] Apanasov, B., Tetenov, A., Deformations of hyperbolic structures along pleated surfaces. Prepr. In: Proc. Low-Dim. Topology Conf. (Knoxville, May, 1992). University of Oklahoma at Norman, 1992.

[21] Tan, S. P., Deformations of conformal structures on hyperbolic 3-manifolds. J. Diff. Geom. 37 (1993), 161–176.

[22] Bass, H., Groups of integral representation type. Pacific J. Math. 86 (1980), 15 – 51.

[23] Rapinchuk, A., Combinatorial theory of arithmetic groups. Prepr. Inst. Math. of Byelorussian Acad. Sci., Minsk 1990.

[24] Platonov, V. P., Adele groups and class numbers. Trudy Math. Inst. Steklov 163 (1984), 205–214.

[25] — Rings and varieties of representations of finitely generated groups. Voprosy

Algebry 4 (1988), 36–40.

[26] Platonov, V., Tavgen', O., Grothendieck's problem on profinite completions and representations of groups. K-Theory 4 (1990), 89–101.

[27] Tavgen', O. I., Finite width of arithmetic subgroups of Chevalley groups of rank ≥ 2. Dokl. Akad. Nauk SSSR 310:4 (1990), 802–806. English translation: Sov. Math. Dokl. 41:1 (1990), 136–140.

[28] Gromov, M., Piatetski-Shapiro, I., Nonarithmetic groups in Lobachevsky spaces. Publ. Math. IHES 66 (1988), 93–103.

[29] Platonov, V. P., Representation theory of finitely generated groups. Address on the workshop on Platonov Conjecture. University of Oklahoma at Norman, 1992.

[30] Vinberg, E. B., Discrete groups generated by reflections in Lobačevskiǐ spaces. Mat. Sbornik 72 (1967), 471–488.

[31] Apanasov, B., Kleinian groups, Teichmüller space, and the rigidity theorem of Mostow. Sib. Math. Zh. 21:4 (1980), 3–15. English translation: Sib. Math. J. 21:4 (1980), 483–491.

[32] — Non-standard uniformized conformal structures on hyperbolic manifolds. Invent. Math. 105:1 (1991), 137–152.

[33] Sullivan, D., Quasiconformal homeomorphisms and dynamics, Pt 2: Structural stability implies hyperbolicity for Kleinian groups. Acta Math. 155 (1985), 243–260.

[34] Apanasov, B., Quasi-symmetric embeddings of a closed ball inextensible in neighbourhoods of any boundary points. Ann. Acad. Sci. Fenn., Ser. A1 Math. 14 (1989), 243–255.

[35] Atiyah, M. F., Patodi, K. V., Singer, I. M., Spectral asymmetry and Riemannian geometry. I,II. Math. Proc. Camb. Philos. Soc. 77 (1975), 43–69; 78 (1975), 405–432.

[36] Chern, S. S., Simons, J., Characteristic forms and geometric invariants. Ann. Math. 99 (1974), 48–69.

[37] Apanasov, B., Kobayashi conformal metric on manifolds, Chern-Simons and η-invariants. Int. J. Math. 2:4 (1991), 361–382.

[38] — Representations of Hyperbolic Groups and Two-Dimensional Knots. Prepr. University of Oklahoma at Norman 1993.

[39] Tukia, P., On isomorphisms of geometrically finite Kleinian groups. Publ. Math. IHES 61 (1985), 171–214.

[40] Rolfson, D., Knots and Links. Perish Press, Berkeley 1976.

[41] Andreev, E. M., Convex polyhedra in Lobačevskiǐ spaces. Mat. Sbornik 81 (1970), 445–478. English translation: Math. USSR Sbornik 10 (1970), 413–440.

[42] Rivin, I., On geometry of convex polyhedra in hyperbolic 3-space. Ph. D. Thesis. Princeton 1986.

[43] Gordon, C. McA., Knots in the 4-sphere. Comment. Math. Helv. 51, 585 –596.

[44] Kuiper, N., Hyperbolic 4-manifolds and tesselations. Publ. Math. IHES 68 (1988), 47–76.

[45] Gromov, M., Lawson, B., Thurston, W., Hyperbolic 4-manifolds and conformally flat 3-manifolds. Publ. Math. IHES 68 (1988), 27–45.

[46] Gluck, H., The embedding of two-spheres in the four-sphere. Trans. Am. Math. Soc. 104 (1962), 308–333.

[47] Bryson, S., Freedman, M. H., Zheng–Xu He, Zhenghan Wang, Möbius invariance of knot energy. Bull. Am. Math. Soc. 28 (1993), 99–103.

On computable minimal enumerations

Serikzhan A. Badaev and Sergeĭ S. Goncharov *

Abstract. We consider some internal problems of the theory of numberings connected with existence and the number of computable minimal enumerations of classes of recursively enumerable sets.

1991 Mathematics Subject Classification: 03D45, 03D25.

1. Introduction. Preliminaries

Computable minimal enumerations, in particular Friedberg enumerations, have successful applications in the recursive model theory, theoretical computer science, and classical recursion theory (see, e.g. [1–4]). This lecture is devoted mainly to inner problems of the theory of numberings connected with existence and the number of computable minimal enumerations of classes of recursively enumerable (r. e.) sets.

For the notions of the theory of recursive functions we refer to the textbooks [5, 6]. For the convenience of the reader we recall some basic notions of the theory of numberings [7].

An arbitrary surjective mapping of the set N of natural numbers onto a finite or countable set $\mathcal{A}$ is called a *numbering* of $\mathcal{A}$.

A numbering of $\mathcal{A}$ is called a *numbering without repetition* if it is a bijective mapping of N onto $\mathcal{A}$.

We prefer to use the term "enumeration" instead of the term "numbering," if $\mathcal{A}$ is a class of r. e. subsets of N . In particular, $\mathcal{A}$ may consist of unary recursive functions. In this case we identify a function with its graph and pairs of natural numbers with their codes under the standard pairing function.

An enumeration $\alpha : N \to \mathcal{A}$ of $\mathcal{A}$ is called *computable* or *recursive* if $\{\langle n, x\rangle : x \in \alpha n\}$ is recursively enumerable, i.e., there is a uniformly effective procedure for enumerating the sequence $\alpha 0, \alpha 1, \alpha 2, \ldots$ of sets of $\mathcal{A}$.

We call a class of r. e. sets with computable enumeration a *computable class.* In the sequel we consider only computable classes.

A numbering α is said to be *reducible to a numbering* β (written $\alpha \leq \beta$) if there is a recursive function f such that $\alpha x = \beta f(x)$ for all $x \in N$.

* Supported by the Russian Fund for Fundamental Research (grant 93-01-01506).

Numberings α, β are said to be *equivalent* if $\alpha \leq \beta$ and $\beta \leq \alpha$.

In the sequel, $\mathcal{F}$ stands for the class of all partial recursive functions and $\mathcal{R}$ denotes the class of all recursively enumerable sets.

The notions of equivalent enumerations and their reducibility were introduced by H. Rogers (jr.) [8] for Gödel enumerations of $\mathcal{F}$. He proved that Gödel enumerations of $\mathcal{F}$ are pairwise equivalent [8]. It is obvious that equivalent enumerations of $\mathcal{F}$ are identical from the algorithmic point of view.

In opposite to Gödel enumerations of $\mathcal{F}$ in which every partial recursive function has infinitely many indices, R. F. Friedberg [9] constructed computable enumeration of $\mathcal{F}$ without repetition. So the class $\mathcal{F}$ admits nonequivalent computable enumerations.

A numbering α of a set A is said to be *minimal* if $\beta \leq \alpha$ implies $\alpha \leq \beta$ for every numbering β of A.

The main subject of the theory of numberings is to study partially ordered set $\langle \mathcal{L}(\mathcal{A}), \leq \rangle$, where $\mathcal{L}(\mathcal{A})$ consists of equivalent classes of computable enumerations of $\mathcal{A}$ and $\leq$ is induced by the reducibility of enumerations. Minimal enumerations are just those enumerations which generate minimal elements in $\mathcal{L}(\mathcal{A})$. The partially ordered set $\mathcal{L}(\mathcal{F})$ has a maximal element generated by Gödel enumerations. M. B. Pour-El [10] noted that computable enumerations without repetition generate minimal elements in $\mathcal{L}(\mathcal{F})$ and called them *Friedberg enumerations.*

The partially ordered set $\mathcal{L}(\mathcal{A})$ contains either one or countably many elements [11]. In the second case it is an upper but not lower semilattice [12].

The algebraic structure $\mathcal{L}(\mathcal{A})$ is called the *Rogers semilattice* of a class $\mathcal{A}$ of r. e. sets.

The Rogers semilattice can be viewed as an algorithmic complexity of a class $\mathcal{A}$ in whole while minimal enumerations can be viewed as its simplest enumerations.

The aim of this paper is to describe infinite classes of r. e. sets admitting minimal computable enumerations and to find the possible number of minimal elements in Rogers semilattices of various infinite computable classes. Note that, for a finite class $\mathcal{A}$ of r. e. sets, the semilattice $\mathcal{L}(\mathcal{A})$ always contains the unique minimal element [7].

2. Existence of minimal enumerations

The most important classes of r. e. sets ($\mathcal{F}$ and $\mathcal{R}$) possess minimal computable enumerations [9]. On the other hand, not every computable class of r. e. sets has computable Friedberg enumerations. Such a class was built by M. B. Pour-El and H. Putnam [13]. Let K be a creative set. Then the computable class

$$\{\{2x, 2x+1\} : x \in K\} \cup \{\{2x\}, \{2x+1\} : x \notin K\}$$

cannot be enumerated without repetition.

The problem of characterization of classes formed by r. e. sets with computable Friedberg enumerations is one of the classical topics of the recursion theory. We will list only a few articles on this subject and restrict our attention to other minimal enumerations. Sufficient conditions of existence of Friedberg enumerations were obtained by S C. Liu [14], M. B. Pour-El and W. A. Howard [15], M. B. Pour-El and H. Putnam [13], A. H. Lachlan [16, 17], C. E. Wolf [18], A. I. Mal'cev [19], Yu. L. Ershov [20], A. B. Khutoretskiĭ [21], S. S. Marchenkov [22], M. Kummer [23–25].

A. I. Mal'cev [26] introduced *decidable* and *positive* numberings as numberings α such that the equivalence

$$\Theta_\alpha \rightleftharpoons \{\langle x, y\rangle \; \alpha x = \alpha y\}$$

is recursive or recursively enumerable correspondingly. He has shown that decidable and positive numberings are minimal and that any decidable numbering of infinite set is equivalent to numbering without repetition. The classes $\mathcal{R}$ and $\mathcal{F}$ have computable positive enumerations not equivalent to Friedberg ones [21, 27]. Let

$$\mathcal{A} = \{A\} \cup \{\{x\} : x \in N \setminus A\}$$

where A is recursively enumerable but not recursive set. Then $\mathcal{A}$ has a computable positive enumeration but does not admit computable Friedberg enumeration. The description of classes of r. e. sets admitting computable positive enumerations is advanced due to A. I. Mal'cev [26], Yu. L. Ershov [27], A. B. Khutoretskiĭ [21] S. S. Marchenkov [22], V. V. V'yugin [28], S. A. Badaev [29], S. S. Goncharov [30].

Now we consider minimal but not positive enumerations. The classes $\mathcal{F}$, $\mathcal{R}$ [22, 31] and the class of all finite sets [32] admit such computable enumerations. There is a class $\mathcal{A}$ without computable positive enumeration although $\mathcal{L}(\mathcal{A})$ has minimal element [33] and there is a computable class $\mathcal{A}$ without minimal computable enumeration [34, 35].

In distinction to the Friedberg and positive enumerations, the notion of minimal enumeration is not inner, since, to distinguish a minimal enumeration, one must compare it with all enumerations of the class. This is the principal reason for the noticeable distinction between the study of Friedberg and positive enumerations from the one side and the study of minimal nonpositive enumerations from another side.

The notions of decidable (Friedberg) and positive numberings are expressed in terms of the equivalence Θ_α closely connected with a numbering α.

Minimality criterion for numberings can be also stated in terms of Θ_α in the following way. Let $[A]_\eta$ denote the closure of a set A under an equivalence η.

Theorem 2.1 [36, 37]. *A numbering α is minimal iff, for every recursively enumerable set W with $[W]_{\Theta_\alpha} = N$, there exists a recursively enumerable equivalence $\eta \subseteq \Theta_\alpha$ such that $[W]_\eta = N$.*

The following two theorems demonstrate the importance of this criterion.

Theorem 2.2 [37, 38]. *There exist computable discrete classes of recursively enumerable sets such that their Rogers semilattices have no minimal elements.*

A class $\mathcal{A}$ is called *discrete* if there is a *separating family* $\mathcal{S}$ for $\mathcal{A}$, i.e., a family of finite sets such that every set in $\mathcal{A}$ has a subset in $\mathcal{S}$ and every set in $\mathcal{S}$ has at most one superset in $\mathcal{A}$ [26].

A class $\mathcal{A}$ is called *effectively discrete* [27] if some strong array is a separating family for $\mathcal{A}$, and *weakly effective discrete* [28] if there exists a computable separating family for $\mathcal{A}$.

Theorem 2.2 and V'yugin's theorem on existence of computable classes without minimal computable enumerations [34, 35] are analogous to the well-known speed-up theorems of the recursion theory.

In contrast to the V'yugin construction, our proof is based on the priority-free simple diagonal method. On the other hand, the classes constructed by V'yugin are not discrete.

Theorem 2.3 [36]. *There exist discrete classes of finite sets such that they have computable enumeration without repetition and their Rogers semilattices contain lower cones without minimal elements.*

Theorem 2.3 gives a negative answer to Problem 58 posed by S. S. Goncharov [39]. Some sufficient conditions for Rogers semilattices to have lower cones without minimal elements were obtained in [40].

Definition [37, 38]. A numbering α is said to be *strictly minimal* if there exists a recursively enumerable equivalence $\eta \subseteq \Theta_\alpha$ such that $[W]_{\Theta_\alpha} = N$ implies $[W]_\eta = N$ for all recursively enumerable sets W.

Definition [37, 38]. A numbering α is said to be *effectively minimal* if there exists a computable sequence $\{\eta_n\}_{n\in N}$ of recursively enumerable equivalences such that, for all n, the relation $[W_n]_{\Theta_\alpha} = N$ implies $\eta_n \subseteq \Theta_\alpha$ and $[W_n]_{\eta_n} = N$.

Denote the classes of decidable, positive, strictly minimal, effectively minimal, and minimal computable enumerations by $\mathcal{D}$, $\mathcal{P}$, $\mathcal{SM}$, $\mathcal{EM}$, and $\mathcal{M}$ respectively. It is clear that $\mathcal{D} \subseteq \mathcal{P} \subseteq \mathcal{SM} \subseteq \mathcal{EM} \subseteq \mathcal{M}$. Recall that $\mathcal{D} \subset \mathcal{P}$ and $\mathcal{P} \subset \mathcal{M}$.

Theorem 2.4 [37, 38]. *There exist weakly effective discrete classes of recursively enumerable sets possessing enumerations from the class $\mathcal{SM} \setminus \mathcal{P}$.*

Theorem 2.5 [37, 38]. *There exist discrete classes of finite sets possessing enumerations from the class $\mathcal{M} \setminus \mathcal{EM}$.*

Corollary 1. *The following inclusions hold: $\mathcal{D} \subset \mathcal{P} \subset \mathcal{SM} \subseteq \mathcal{EM} \subset \mathcal{M}$.*

Corollary 2. *The classes $\mathcal{F}$ and $\mathcal{R}$ have ω pairwise nonequivalent computable enumerations in each of the classes $\mathcal{SM} \setminus \mathcal{P}$ and $\mathcal{M} \setminus \mathcal{EM}$.*

The classes in Theorems 2.4 and 2.5 possess computable positive enumerations. This assertion follows from [22, 28].

Now we formulate several open problems concerning the above considerations.

Problem 1 (Mal'cev [19]). Find necessary and sufficient structural conditions for a class of recursively enumerable sets to have computable Friedberg enumerations.

Problem 2 (Ershov; see [49]). Find necessary and sufficient structural conditions for a class of recursively enumerable sets to have computable positive enumerations.

Problem 3 (Ershov; see [41]). Find necessary and sufficient structural conditions for a class of recursively enumerable sets to have computable minimal enumerations.

Problem 4 (Badaev [37, 38]). Is there a discrete class of finite sets possessing computable enumerations in $\mathcal{SM} \setminus \mathcal{P}$?

Problem 5 (Badaev [37, 38]). Is there a (discrete) class of recursively enumerable (finite) sets possessing computable enumerations in $\mathcal{EM} \setminus \mathcal{SM}$?

3. The number of minimal enumerations

Investigations on the problem of a possible number of minimal elements in Rogers semilattices of r. e. sets resembles the solving of the problem of their existence. Both are well advanced for Friedberg and positive enumerations, but there are difficulties in the case of minimal nonpositive enumerations.

From the methodological point of view the study of the problem may be split into two periods with the border line 1980.

M. B. Pour-El [10] noted that the class $\mathcal{F}$ admits at least two nonequivalent computable Friedberg enumerations and posed the problem about the number of such enumerations.

Yu. L. Ershov [20] proved that the Rogers semilattice of the class of all primitive recursive functions contains ω minimal elements. He also stated the problem about the description of spectrum of number of minimal elements in Rogers semilattices. For the classes $\mathcal{F}$ and $\mathcal{R}$, the number of nonequivalent computable enumerations of the classes $\mathcal{D}$, $\mathcal{P} \setminus \mathcal{D}$, and $\mathcal{M} \setminus \mathcal{P}$ is equal to ω [21, 22, 31].

In our opinion, the most important result on the number of minimal enumerations obtained in the first period is due to S. S. Marchenkov.

Theorem 3.1 [42]. *The Rogers semilattice of a computable class of total recursive functions has either the least element or ω minimal elements.*

Both alternatives are realized and the following algorithmic criterion distinguishes them.

Theorem 3.2 [43]. *The Rogers semilattice of a class of total recursive functions has ω minimal elements iff there is a computable undecidable enumeration of this class in which every function has recursively enumerable set of indices.*

By the same method as in Theorem 3.1 we prove that weakly effective discrete classes of r. e. sets have up to equivalence either 1 or ω computable positive enumerations [29]. The results similar to the Marchenkov theorem were obtained by V. L. Selivanov [32] for various canonically enumerable classes of finite sets.

There is a fundamental methodological reason for the Rogers semilattices considered up to beginning of eighties to have exactly 0, 1 or ω minimal elements. In this period, the main tool for constructing minimal computable enumerations was the finite injury priority method. S. S. Goncharov has shown an inadequacy of this method to solve the problem of a possible number of computable Friedberg (positive) enumerations in general.

Definition [44]. A numbering α is said to be *limitly reducible to a numbering* β if there exists a $0'$-recursive function f such that $\alpha x = \beta f(x)$ for all $x \in N$. Numberings α and β are said to be *limitly equivalent* if they are limitly reducible to each other.

Theorem 3.3 [44]. *If a class of recursively enumerable sets admits two nonequivalent but limitly equivalent computable Friedberg (positive) enumerations then it has ω nonequivalent enumerations of this type.*

Computable enumerations of classes of finite sets, total recursive functions, and weakly effective discrete classes of r. e. sets are pairwise limitly equivalent [44]. Hence, the above-mentioned facts on the number of nonequivalent computable Friedberg (positive) enumerations represent special cases of Theorem 3.3.

On the other hand, Theorem 3.3 implies that, for construction of classes admitting two nonequivalent computable Friedberg (positive) enumerations, it is necessary to employ infinite injury priority arguments.

Theorem 3.4 [45]. *For a natural number n, there exists a class of recursively enumerable sets admitting up to equivalence exactly n computable Friedberg enumerations.*

Now, we indicate two fundamental applications of Theorem 3.4. The problem on the number of nonequivalent computable Friedberg enumerations is close to the well-known problem on the spectrum of nonautoequivalent constructivizations of recursive models [2, 3]. Applying Theorem 3.4, S. S. Goncharov [2, 3] proved that the spectrum is equal to $\{\omega, 0, 1, 2, \dots\}$. The second application of Theorem 3.4 was indicated by M. Kummer [4]. For a natural number $n > 1$, with the help of Theorem 3.4 he constructs recursive functions having exactly n types of recursive isomorphism. For details we refer to [5, Chapters 4 and 11].

Theorem 3.5 [30]. *For each natural number n there exist classes of recursively enumerable sets admitting up to equivalence exactly n computable positive enumerations.*

Now we consider in detail classes $\mathcal{A}$ of r. e. sets admitting unique up to equivalence computable Friedberg (positive) enumeration. It is not difficult to find classes with computable Friedberg enumeration generating the least element in the Rogers semilattice. For example every computable Friedberg enumeration of the class $\mathcal{A} \rightleftharpoons \{\{x\} : x \notin A\} \cup \{\{x, a\} : x \in A\}$, where A is a r. e. but nonrecursive set and a is a fixed element from A, is obviously reducible to all computable enumerations of $\mathcal{A}$. Using Theorem 4.4 below it is easy to see that the Rogers semilattice $\mathcal{L}(\mathcal{A})$ is not trivial. Thus, the semilattice $\mathcal{L}(\mathcal{A})$ has the least element and is not trivial.

Theorem 3.6 [46]. *Suppose that a class of recursively enumerable sets admits computable Friedberg enumerations. Then the Rogers semilattice has either the least element or ω minimal elements which are generated by positive enumerations.*

Theorem 3.7 [47]. *There exists a class of recursively enumerable sets admitting exactly one up to equivalence computable Friedberg enumeration which does not generate the least element in Rogers semilattice.*

Recently a simple example of a similar class was found by M. Kummer.

Theorem 3.8 [41]. *There exist classes of recursively enumerable sets admitting up to equivalence exactly one computable positive enumeration which does not generate the least element in Rogers semilattice.*

It is interesting to find a simple example of a class similar to the one from the previous theorem.

An analog of Theorem 3.6 is valid for positive enumerations but in some local setting. Let α be a computable positive enumeration of a class $\mathcal{A}$ of r. e. sets and $\mathcal{K}(\alpha)$ be the class of all computable enumerations which are limitly equivalent to α. Then $\langle \mathcal{K}(\alpha), \leq \rangle$ is a subsemilattice of the Rogers semilattice $\langle \mathcal{L}(\mathcal{A}), \leq \rangle$.

Theorem 3.9 [48]. *For arbitrarily chosen computable positive enumeration α, the semilattice $\mathcal{K}(\alpha)$ has either the least element or ω minimal elements.*

Corollary. *Let $\mathcal{A}$ be one of the following computable classes:*

a class of finite sets,

a class of total recursive functions,

a weakly effective discrete class.

Then the Rogers semilattice $\mathcal{L}(\mathcal{A})$ has either the least element or ω minimal elements.

This statement immediately follows from Theorem 3.9 since $\mathcal{K}(\mathcal{A}) = \mathcal{L}(\mathcal{A})$ for each of the desired classes $\mathcal{A}$.

Finally, we recall some open problems.

Problem 6 (Ershov, Marchenkov; see [44, 45]). Find a possible number of minimal elements in Rogers semilattices of classes of recursively enumerable sets.

Problem 7 (Goncharov [39, 46]). Is the following hypothesis valid: if a class of recursively enumerable sets admits two nonequivalent but limitly equivalent computable minimal enumerations, then its Rogers semilattice has countable number of minimal elements ?

Problem 8 (Goncharov [39, 46]). Describe classes of recursively enumerable sets whose Rogers semilattice has the least element generated by computable Friedberg (positive) enumeration and characterize these semilattices.

Problem 9 (Goncharov [39, 46]). Describe classes of recursively enumerable sets whose Rogers semilattice has the least element and characterize these semilattices.

Problem 10 (Goncharov [41]). How many minimal elements are in the Rogers semilattice of a class of recursively enumerable sets which admits exactly one up to equivalence computable positive enumeration ?

4. Classes with trivial Rogers semilattices

First of all, recall that Rogers semilattices of computable classes of r. e. sets contain either 1 (in this case we call it *trivial*) or ω elements [11]. The problem of describing classes with trivial Rogers semilattice was formulated by Yu. L. Ershov [20] and until now it is one of the most important problems in the theory of numberings (see [7, 49] for motivations and details).

In most known cases classes of r. e. sets and classes of total recursive functions with trivial Rogers semilattice are discrete [20, 26, 35, 43, 50]. For classes of total recursive functions the discreteness of a class is necessary for Rogers semilattice to be trivial [20]. At the same time there exist nondiscrete classes of r. e. sets with trivial Rogers semilattice [12]. It is clear that if all computable enumerations of a class are minimal then its Rogers semilattice is trivial. In fact the single element of Rogers semilattices considered in papers cited above is generated by positive enumerations.

Theorem 4.1 [37, 38]. *There exist classes of recursively enumerable sets such that all their computable enumerations belong to* $\mathcal{SM} \setminus \mathcal{P}$.

Theorem 4.2 [37, 38]. *There exist classes of recursively enumerable sets such that all their computable enumerations belong to* $\mathcal{M} \setminus \mathcal{EM}$.

Remark. In Theorems 4.1 and 4.2 the desired classes may be built both discrete and nondiscrete.

The singularity of classes of total recursive functions is that for any computable enumeration one can find an effective procedure for separating any two different functions. Discrete classes of r. e. sets have a similar property. This allows one to succeed separation property in a study of classes of total recursive functions and discrete classes of r. e. sets.

Definition [30]. An enumeration α of a class of total recursive functions is said to be *bounded* if there is a natural number a and monotone nondecreasing recursive function f such that $\lim_{n\to\infty} f(n) = \infty$ and, for all n, x, y, if $a \leq x, y \leq f(n)$ and $[\alpha(x)](i) = [\alpha(y)](i)$ for each $i \leq n$ then $\alpha x = \alpha y$

Theorem 4.3 [30]. *The Rogers semilattice of a computable class of total recursive functions is trivial iff all computable enumerations of this class are bounded enumerations.*

Definition [37, 38]. The class of recursively enumerable sets is called *essentially nondiscrete* if it contains a pair of embedded sets.

Note that the classes with trivial Rogers semilattice in Theorems 4.1 and 4.2 as well as the similar classes of [12] are not essentially nondiscrete. The main difficulty in solving of the Ershov problem for classes of r. e. sets apparently arises from essentially nondiscrete classes. As is known, the Rogers semilattices of the finite essentially nondiscrete classes are infinite [7]. For infinite essentially nondiscrete classes, some sufficient structural conditions for Rogers semilattices to be nontrivial one can obtain from the results of [21, 22, 29, 31, 51, 52]. For example, if a class is not closed under unions of increasing chains of sets [49] or contains the union of all its elements [29] then its Rogers semilattice is infinite.

Theorem 4.4 [53]. *Suppose that a computable class $\mathcal{A}$ of recursively enumerable sets contains a pair of embedded sets $A \subset B$. If there exists a recursive set C such that $A \subseteq C \subseteq B$ then the Rogers semilattice $L(A)$ is not trivial.*

Theorem 4.5 [54]. *There exist essentially nondiscrete classes of recursively enumerable sets with trivial Rogers semilattice.*

Problem 11 (Ershov [20]). Describe computable classes of recursively enumerable sets with trivial Rogers semilattice.

Problem 12 (Badaev [37, 38]). Does there exist classes of recursively enumerable sets such that all their computable enumerations belong to $\mathcal{EM} \setminus \mathcal{SM}$?

References

[1] Ershov, Yu. L., Decision Problems and Constructivizable Models. Nauka, Moscow 1980.

[2] Goncharov, S. S., The problem of the number of non-autoequivalent constructivizations. Dokl. Akad. Nauk SSSR 251:2 (1980), 271–274. English translation: Sov. Math., Dokl. 21 (1980), 411–414.

[3] — Problem of the number of non-self-equivalent constructivizations. Algebra Logika 19:6 (1980), 621–639. English translation: Algebra Logic 19:6 (1980), 401–414.

[4] Kummer, M., Some applications of computable one-one numberings. Arch. Math. Logic 30:4 (1990), 219–230.

[5] Rogers, H. (jr.), Theory of recursive function and effective computability. McGraw-Hill Book Co., New York 1967.

[6] Soare, R. I., Recursively Enumerable Sets and Degrees. Springer-Verlag, Berlin 1987.

[7] Ershov, Yu. L., Theory of Numerations. Nauka, Moscow 1977.

[8] Rogers, H. (jr.), Gödel numberings of partial recursive functions. J. Symb. Log. 23:3 (1958), 331–341.

[9] Friedberg, R. F., Three theorems on recursive functions. J. Symb. Log. 23:3 (1958), 331–341.

[10] Pour-El, M. B., Gödel numberings versus Friedberg numberings. Proc. Am. Math. Soc. 15:2 (1964), 252–255.

[11] Khutoretskiĭ, A. B., On the cardinality of the upper semilattice of computable enumerations. Algebra Logika 10:5 (1971), 561–569. English translation: Algebra Logic 10:5 (1971), 348–352.

[12] Selivanov, V. L., Two theorems on computable numerations. Algebra Logika 15:4 (1976), 470–484. English translation: Algebra Logic 15:4 (1976), 297–306.

[13] Pour-El, M. B., Putnam, H., Recursively enumerable classes and their applications to sequences of formal theories. Arch. Math. Logik Grundlagenforsch 8 (1965), 104–121.

[14] Liu, S. C., An enumeration of the primitive recursive functions without repetition. Tohôku Math. J., II. Ser. 12:3 (1960), 400–403.

[15] Pour-El, M. B., Howard, W. A., A structural criterion for recursive enumeration without repetition. Z. Math. Logik Grundlagen Math. 10:2 (1964), 105–114.

[16] Lachlan, A. H., On recursive enumeration without repetition. Z. Math. Logik Grundlagen Math. 11:3 (1965), 209–220.

[17] — On recursive enumeration without repetition: A correction. Z. Math. Logik Grundlagen Math. 13:2 (1967), 99–100.

[18] Wolf, C. E., Standard recursively enumerable classes. Notices Am. Math. Soc. 11 (1964), 386.

[19] Mal'cev, A. I., Algorithms and Recursive Functions. Nauka, Moscow 1965. English translation: Wolters-Noordhoff Publishing, Groningen 1970.

[20] Ershov, Yu. L., Numerations of families of general recursive functions. Sib. Mat. Zh. 8:5 (1967), 1015–1025. English translation: Sib. Math. J. 8:5 (1967).

[21] Khutoretskiĭ, A. B., On the reducibility of computable numerations. Algebra Logika 8:2 (1969), 251–264. English translation: Algebra Logic 8:2 (1969), 145–151.

[22] Marchenkov, S. S., The minimal numerations of systems of recursively enumerable sets. Dokl. Akad. Nauk SSSR 198:3 (1971), 530–532. English translation: Sov. Math., Dokl. 12 (1971), 843–846.

[23] Kummer, M., Numberings of $R_1 \cup F$. In: Lect. Notes Comput. Sci. 385, 166–186. Springer-Verlag, Berlin 1989.

[24] — An easy priority-free proof of a theorem of Friedberg. Theor. Comput. Sci. 74:2 (1990), 249–251.

[25] — Recursive enumeration without repetition revisited. In: Lect. Notes Math. 1432, 255–275, Springer-Verlag, Berlin 1990.

[26] Mal'cev, A. I., Positive and negative enumerations. Dokl. Akad. Nauk SSSR 160:2 (1965), 278–280. English translation: Sov. Math., Dokl. 6 (1965), 75–77.

[27] Ershov, Yu. L., Computable enumerations. Algebra Logika 7:5 (1968), 71–99. English translation: Algebra Logic 7:5 (1968), 34–38.

[28] V'yugin, V. V., On discrete families of recursively enumerable sets. Algebra Logika 11:3 (1972), 243–256. English translation: Algebra Logic 11:3 (1972), 137–144.

[29] Badaev, S. A., Positive enumerations. Sib. Mat. Zh. 18:3 (1977), 483–496. English translation: Sib. Math. J. 18:3 (1977), 343–352.

[30] Goncharov, S. S., Positive computable enumerations. Dokl. Ross. Akad. Nauk 332:2 (1993). English translation: Sov. Math., Dokl. (1993).

[31] Khutoretskiĭ, A. B., Two existence theorems for computable numerations. Algebra Logika 8:4 (1969), 483–492. English translation: Algebra Logic 8:4 (1969), 277–282.

[32] Selivanov, V. L., Enumerations of canonically calculable families of finite sets.

Sib. Mat. Zh. 18:6 (1977), 1373–1380. English translation: Sib. Math. J. 18:6 (1877), 973–979.

[33] Marchenkov, S. S., The existence of families without positive numerations. Mat. Zametki 13:4 (1973), 597–604. English translation: Math. Notes 13:4 (1973), 360–363.

[34] V'yugin, V. V., On minimal enumerations of computable classes of recursively enumerable sets. Dokl. Akad. Nauk SSSR 212:2 (1973), 273–275. English translation: Sov. Math., Dokl. 14 (1973), 1338–1340.

[35] — On some examples of upper semilattices of computable enumerations. Algebra Logika 12:5 (1973), 512–529. English translation: Algebra Logic 12:5 (1973), 277–286.

[36] Badaev, S. A., A problem of S. S. Goncharov. Sib. Mat. Zh. 32:3 (1991), 212–214. English translation: Sib. Math. J. 32:3 (1991), 532–534.

[37] — On minimal enumerations. Sib. Adv. Math. 2:2 (1992), 1–30.

[38] — Minimal enumerations. In: Tr. Inst. Mat. 25, 3–34. Institut Matematiki, Novosibirsk 1993.

[39] Logical Note-book. Unsolved Questions of Mathematical Logic. Institut Matematiki SO AN SSSR. Novosibirsk 1986.

[40] Badaev, S. A., On semilattices of computable enumerations having lower cones without minimal elements. In: Theory of Models, 3–6. Kazakh. Gos. Univ., Alma-Ata 1990.

[41] Goncharov, S. S., A unique positive enumeration. Sib. Adv. Math. 4:1 (1994), 65–94.

[42] Marchenkov, S. S., The computable enumerations of families of general recursive functions. Algebra Logika 11:5 (1972), 588–607. English translation: Algebra Logic 11:5 (1972), 326–336.

[43] Badaev, S. A., Computable enumerations of families of general recursive functions. Algebra Logika 16:2 (1977), 129–148. English translation: Algebra Logic 16:2 (1977), 83–98.

[44] Goncharov, S. S., Limiting equivalent constructivizations. In: Tr. Inst. Mat. 2, 4–12. Nauka, Novosibirsk 1982.

[45] — Computable single-valued numerations. Algebra Logika 19:5 (1980), 507–551. English translation: Algebra Logic 19:5 (1980), 325–356.

[46] — Positive numerations of families with one-valued numerations. Algebra Logika 22:5 (1983), 481–488. English translation: Algebra Logic 22:5 (1983), 345–350.

[47] — A family with a unique single-valued, but not the minimal numeration. In: Tr. Inst. Mat. 8, 42–48. Nauka, Novosibirsk 1988.

[48] Badaev, S. A., Minimal enumerations of positively enumerable families. Algebra Logika 33:3 (1994), 233–254. English translation: Algebra Logic 33:3 (1994).

[49] Lavrov, I. A., Computable numberings. In: Logic, Foundations of Mathematics and Computability Theory, Proc. 5th Int. Congr. Logic, Methodology and Philos. of Sci. (London-Ontario, 1975) (Univ. West. Ont. Ser. Philos. Sci. 9) pt 1, 195–206. D. Reidel Publishing Company, Dordrecht 1977.

[50] Selivanov, V. L., Enumerations of families of general recursive functions. Algebra

Logika 15:2 (1976), 205–226. English translation: Algebra Logic 15:2 (1976), 128–141.

[51] Lachlan, A. H., On the indexing of classes of recursively enumerable sets. J. Symb. Log. 31:1 (1966), 10–22.

[52] Marchenkov, S. S., The semilattices of computable numerations. Dokl. Akad. Nauk SSSR 198:4 (1971), 766–768. English translation: Sov. Math., Dokl. 12 (1971), 886–888.

[53] Badaev, S. A., On cardinality of semilattices of enumerations of nondiscrete families. Sib. Mat. Zh. 34:5 (1993), 3–10. English translation: Sib. Math. J. 34:5 (1993), 795–800.

[54] Goncharov, S. S., Badaev, S. A., Families whose Rogers semilattices consist of a single element (to appear).

Locally finite simple groups from the standpoint of the classification of finite simple groups

Vissarion V. Belyaev

Abstract. To partition the class of enormous groups in small ones we need notions taking into account global properties of groups and simultaneously reducible to local conditions by virtue of methods of studies in the class of locally finite groups. Such terminology and a characterization of the class of locally finite simple groups were invented by the author. There are two kinds of the account, one relies on the notion of a signalizer and the other on that of an inert subgroup. The first allows to formulate results in full forms without countability assumption, but it does not give a clear view of the situation and of perspectives of further studies. So we choose the second way here.

1991 Mathematics Subject Classification: 20F50, 20E32, 20D05.

There is a connection of a local nature between the class of locally finite simple groups and that of finite simple groups, which was called an approximation principle [1]. But the connection is not close enough to derive the classification of infinite locally finite simple groups starting from the classification of finite simple groups. Arbitrariness in constructing the infinite locally finite simple groups does not give possibility to carry out such a classification. However many problems of the theory of locally finite groups require to throw light on the structure of the class of locally finite simple groups.

The investigation was initiated at the beginning of eighties when using the classification of finite simple groups a locally finite simple group have been proved [2–5] to be either a finite rank group of Lie type defined over a locally finite field or an enormous group. The term *enormous group* was introduced in [1] for a locally finite simple group failing to satisfy some set of finiteness conditions. The partition of the class of all locally finite simple groups in two unlike parts answered questions on finiteness conditions on the one hand and on the other drew attention to the class of enormous groups which had been in a mess.

The class of Hall universal groups and alternating finitary permutation groups is a small part of the class of enormous groups. The local structure of a universal group is the same as that of alternating groups (each of them is a direct limit of finite alternating groups). But their global properties are quite different and even extremes in a sense. Although each countable locally finite group is embeddable in a Hall universal group, an infinite simple subgroup of an alternating group is isomorphic to an alternating group (perhaps, of another degree). Local closeness and unlikeliness of whole character is not a feature of the example but the usual thing in the class of enormous groups, which is a great difficulty for the study. That is why for partition of the class of enormous groups in small ones we need notions taking into account global properties of groups and simultaneously reducible to local conditions by virtue of methods of studies in the class of locally finite groups.

The terminology and the characterization of the class of locally finite simple groups were introduced by the author [6–10]. There are two ways of exposition of the obtained results: the first method is based on the notion of a signalizer while the notion of inert subgroup is mainly used in the second one. In contrast to the second method, the first allows us to formulate results in full form without countability assumption, but it does not clear up the situation and does not indicate perspectives of further study. Therefore we choose the second way.

Definition. A subgroup H of a group G is said to be *inert* if the intersection $H \cap H^g$ has finite index in H for every $g \in G$ (Note that this term was suggested by O. Kegel.)

Inert subgroups appear in different parts of group theory. For example, given a connected locally finite graph Γ, the stabilizer of any graph vertex in $\mathrm{Aut}\,\Gamma$ is an inert subgroup. A more complicated example is the following: the subgroup $\mathrm{SL}\,(n, Z)$ of $\mathrm{SL}\,(n, Q)$ and its conjugated subgroups intersect at finite index subgroups. The specific character of the embedding of $\mathrm{SL}\,(n, Z)$ in $\mathrm{SL}\,(n, Q)$ is used in arithmetic group theory to define a topological space structure on $\mathrm{SL}\,(n, Q)$ such that the relative completion of $\mathrm{SL}\,(n, Q)$ is a locally compact totally disconnected group. This fact is immediately connected with the notion of an inert subgroup. Indeed, an open subgroup of a locally compact group is inert.

The appearance of inert subgroups of a countable locally finite group G is due to the choice of a sequence $\{H_i\}$ of finite subgroups. Let $H_0 = 1$ and H_{n+1} be defined by induction so that $\langle H_0, \dots, H_n, g_0, \dots, g_n\rangle \le N_G(H_{n+1})$ and $\langle H_0, \dots, H_n\rangle \cap H_{n+1} = 1$.

For a group $H = \langle H_0, H_1, \dots\rangle$ and an element g_n the inclusion $H \cap H^{g_n} \ge \langle H_{n+1}, H_{n+2}, \dots\rangle$ is valid, hence H is an inert subgroup of G.

What is the structure of a proper inert subgroup of a simple group? In [7] it is proved that an inert subgroup is residually finite and possesses a number of specific properties which imply the following theorem.

Theorem 1. *Let H be an inert subgroup of a locally finite simple group G. Then one of the following conditions is true:*

1) $H = G$,

2) *H is a finite group,*

3) *H is an infinite locally normal group,*

4) *there is a prime p such that* $F(H) = O_p(H) \neq 1$, $H/O_p(H)$ *is a locally normal group, and* $FC(H) = FC(O_p(H)) = 1$,

5) $F(H) = FC(H) = 1$, $\mathrm{Sol}\,(H) \neq 1$, $H/\mathrm{Sol}\,(H)$ *is a locally normal group,*

6) $FC(H) = F(H) = \mathrm{Sol}\,(H) = 1, \quad 1 \neq H \neq G$,

where $F(H)$ *is the locally nilpotent radical of* H, $\mathrm{Sol}\,(H)$ *the locally solvable radical of* H, $FC(H) = \{g \in H \mid |H : C_H(g)| < \infty\}$.

Certainly, a locally finite simple group may contain inert subgroups of different types. It is easy to prove that a Hall universal group contains inert subgroups of all types 1–6. From the suggested standpoint of characterization of inert subgroups it is interesting to describe what combinations of types are realized in locally finite simple groups. The first step in this direction was done in [6] where countable locally finite simple groups with nontrivial inert subgroups only were studied, i.e., with inert subgroups of types 1, 2. In this case the following assertion holds.

Theorem 2. *For a countable locally finite simple group G the following conditions are equivalent:*

1) *G is a group of Lie type,*

2) *G contains inert subgroups of types* 1, 2 *only.*

Hence each countable enormous group contains a nontrivial inert subgroup. The following theorem [6] illustrates this fact.

Theorem 3. *A countable locally finite simple group G is tightly embeddable in a nondiscrete locally compact group iff G is enormous. In particular, a countable enormous group is nontrivially topologized.*

The second step in the study of types of inert subgroups in countable locally finite simple groups was done in [9] where the following theorem has been proved.

Theorem 4. *For a countable locally finite simple group G the following conditions are true:*

1) *G is an alternating group,*

2) *G contains inert subgroups of types* 1–3 *only.*

It can be shown that the countable locally finite simple groups with inert subgroups of types 1–4 only have faithful linear representations of a special type. Moreover, their characterization in terms of inert subgroups requires a careful study of the infinite-dimensional linear groups called *groups of finitary linear transformations.* Note that the author used the term "a group of finite linear transformations" in the earlier paper [8]. The reason for introduction of the above term in [13] is to correlate the used terminology with that of [11, 12].

Definition. A linear transformation g of a vector space V is said to be *finitary* if g acts transitively on a subspace of finite condimension in V. The subgroup of $\mathrm{GL}(V)$ consisting of finitary linear transformations is called the *group of finitary linear transformations of* V. A group is said to be *finitary* if its faithful linear representation is a group of finitary linear transformations.

Till the late eighties the class of finitary groups was insufficiently explored. A full-scale investigation of periodic finitary groups was begun in [11, 12], where the first important structure results were obtained with the help of the classification of finite simple groups. However they were not enough to give a characterization of periodic simple finitary groups in terms of inert subgroups. For this purpose, the author [14–16] constructs the general structure theory of periodic finitary groups. Note that the periodic finitary groups were also discussed in [12, 13, 17–24]. We emphasize that the different method of characterization of periodic simple groups was suggested in [2].

In particular, the structure theory described in [14–16] implies the following theorem.

Theorem 5. *For a countable locally finite simple group G the following conditions are equivalent:*

1) *G is finitary,*

2) *G does not contain inert subgroups of type* 6,

3) *G does not contain inert subgroups of types* 5, 6.

Theorems 2, 4, 5 were used to obtain a number of interesting results [8–10]. We formulate two of them.

Theorem 6. *Let G be a locally finite simple group. Then the following conditions are true:*

1) *G is a group of Lie type iff every residually finite subgroup of G is nilpotent-by-finite,*

2) *G is alternating iff every residually finite subgroup of G is locally normal,*

3) *G is finitary iff every residually finite subgroup with the trivial locally solvable radical is locally normal.*

Theorem 7. *Let G be a countable locally finite simple nonfinitary group. Then $G = AB$ for proper inert subgroups A and B.*

References

[1] Kegel, O. H., Wehrfritz, B. A. F., Locally Finite Groups (North-Holland Math. Libr. 3). North-Holland, Amsterdam–London 1973.

[2] Belyaev, V. V., Locally finite Chevalley groups. In: Investigations in Group Theory, Collect. Artic., 39–50. Ural'skiĭ Naushnyĭ Tsentr AN SSSR, Sverdlovsk 1984.

[3] Borovik, A. V., Embeddings of finite Chevalley groups and periodic linear groups. Sib. Mat. Zh. 24:6 (1983), 26–35. English translation: Sib. Math. J. 24:6 (1983), 843–851.

[4] Hartley, B., Shute, G., Monomorphisms and direct limits of finite groups of Lie type. Quart. J. Math., Oxf. II. Ser. 35:137 (1984), 49–71.

[5] Thomas, S., The classification of the simple periodic linear groups. Arch. Math. 41:2 (1983), 103–116.

[6] Belyaev, V. V., Locally finite groups containing a finite inseparable subgroup. Sib. Mat. Zh. 34:2 (1993), 23–41. English translation: Sib. Math. J. 34:2 (1993), 218–232.

[7] — Inert subgroups of infinite simple groups. Sib. Mat. Zh. 34:4 (1993), 17–23. English translation: Sib. Math. J. 34:4 (1993), 606–611.

[8] — Locally finite simple groups as a product of two inert subgroups. Algebra Logika 31:4 (1992), 360–368. English translation: Algebra Logic 31:4 (1992), 216–220.

[9] — Local characterization of infinite alternating and Lie type groups. Algebra Logika 31:4 (1992), 369–390. English translation: Algebra Logic 31:4 (1992), 221–234.

[10] — Local characterizations of periodic simple groups of finitary transformations. Algebra Logika 32:3 (1993), 231–250. English translation: Algebra Logic 32:3 (1993), 123–132.

[11] Hall, J. I., Infinite alternating groups as finitary linear transformation groups. J. Algebra 119:2 (1988), 337–359.

[12] Phillips, R. E., The structure of groups of finitary transformations. J. Algebra 119:2 (1988), 400–448.

[13] Belyaev, V. V., Irreducible periodic groups of finitary transformations. Algebra Logika 33:2 (1994), 109–134. English translation: Algebra Logic 33:2 (1994).

[14] — The structure of p–groups of finitary transformations. Algebra Logika 31:5 (1992), 453–478. English translation: Algebra Logic 31:5 (1992), 271–283.

[15] — Locally solvable periodic groups of finitary transformations. Algebra Logika 31:6 (1992), 569–591. English translation: Algebra Logic 31:6 (1992), 331–341.

[16] — Semisimple periodic groups of finitary transformations. Algebra Logika 32:1 (1993), 17–33. English translation: Algebra Logic 32:1 (1993), 8–16.

[17] — Finitary linear representations of infinite symmetric and alternating groups. Algebra Logika 32:6 (1993), 591–606. English translation: Algebra Logic 32:6 (1993), 319–327.

[18] — The structure of periodic finitary transformation groups. Algebra Logika 33:4 (1994), 347–366. English translation: Algebra Logic 33:4 (1994).

[19] Cameron, P. J., Hall, J. I., Some groups generated by transvection subgroups. J. Algebra 140:1 (1991), 184–209.

[20] Hall, J. I., Finitary linear transformation groups and elements of finite local degree. Arch. Math. 50:4 (1988), 315–318.

[21] Hall, J. I., Hartley, B., A group theoretical characterization of simple locally finite, finitary linear groups. Arch. Math. 60:2 (1993), 108–114.

[22] Hartley, B., Zalesskiǐ, A. E., The ideal lattices of the complex group rings of finitary special and general linear groups over finite fields. Preprint.

[23] Kegel, O. H., Schmidt, D., Existentially closed finitary linear groups. In: Groups 2, Proc. Int. Conf. (St. Andrews/UK 1989) (Lond. Math. Soc. Lect. Note Ser. 160), 353-362. Cambridge University Press, Cambridge 1991.

[24] Meierfrankenfeld, U., Phillips, R. E., Puglisi, O., Locally solvable finitary linear groups (to appear).

Some complex sublattices of clone lattices

Andreĭ A. Bulatov

Abstract. Lattices of clones of finite-valued functions are considered in the paper. All clones of k-valued functions form a complete lattice which is usually denoted by $\mathcal{L}_k$. Any lattice being at most countable direct product of finite lattices is proved to be isomorphically embeddable into $\mathcal{L}_4$. The proof is rather long and sophisticated, so a rough sketch of the proof only is given.

1991 Mathematics Subject Classification: 06A23, 06B05.

As it is seen from the title, the main object of the article is lattices of clones of finite-valued functions. All the functions under consideration are defined on the set $E_k = \{0, \ldots, k-1\}$. Recall that a *clone* is a set of k-valued functions which contains all projections (i.e. functions of the form $e_i^n(x_1, \ldots, x_n) = x_i$) and is closed under all compositions. All clones of k-valued functions form a complete lattice denoted by $\mathcal{L}_k$. Along with the term "clone," the similar term "iterative algebra" is often used [1]. By an *iterative algebra* we mean a set of k-valued functions which is closed under composition, permutation, and identification of variables as well as adding fictitious arguments. If an iterative algebra contains all projections then it is exactly a clone. All results of the article are formulated for clone lattices. However, it is convenient to prove them in terms of iterative algebras. Denote by $\mathcal{L}'_k$ the lattice of all iterative algebras of k-valued functions. The lattice $\mathcal{L}_k$ is isomorphic to the principal filter of $\mathcal{L}'_k$ generated by the algebra of all projections. The iterative algebra of all k-valued functions is denoted by P_k.

In investigation of such complex lattice as $\mathcal{L}_k$, many natural problems arise. One of them is to find families of sublattices of some special kind. We can regard this problem being solved for $\mathcal{L}_2$ because the full description of $\mathcal{L}_2$ is known [2]. There are only a few advancements in solution of this problem for $\mathcal{L}_k$ if $k \geq 3$. In particular, it was proved [3] that the power set of a countable set is embeddable as an interval into $\mathcal{L}_3$. Hence it is embeddable into $\mathcal{L}_k$ for $k \geq 3$ since $\mathcal{L}_3$ is isomorphic to some interval in $\mathcal{L}_k$. Thus, any finite or countable distributive lattice is embeddable into $\mathcal{L}_3$. It was noted in [4] that, for any finite lattice L, there is k (rather large) such that L is a sublattice of $\mathcal{L}_k$. By [5], the subsemigroup lattice of the free semigroup of countable rank is isomorphically embeddable into $\mathcal{L}_4$. In view of this fact and the results of [6, 7], we can conclude that a wide class of finite lattices is embeddable into $\mathcal{L}_k$ for $k \geq 4$.

The starting point of our paper is the question: is an arbitrary finite lattice embeddable into $\mathcal{L}_k$ for some fixed k? A stronger assertion may be proved. It implies the solution for the problem to find all finite sublattices in $\mathcal{L}_k$. Now we formulate the main result of the article.

Theorem. *If a lattice is at most countable direct product of finite lattices then it is isomorphically embeddable into* $\mathcal{L}_4$.

The theorem gives us examples of continual lattices embeddable into $\mathcal{L}_4$. If we consider countable lattices of the kind only, then we get the well known class of countable residually finite lattices.

Corollary 1. *A countable residually finite lattice is isomorphically embeddable into* $\mathcal{L}_4$.

Let $\mathcal{X}$ be a lattice variety generated by its finite members. As is known, the $\mathcal{X}$-free lattice of the countable rank is residually finite. This fact gives the following example of a sublattice of $\mathcal{L}_4$. (Note that the absolutely free lattice of countable rank is embeddable in view of [5–7].)

Corollary 2. *The* $\mathcal{X}$*-free lattice of countable rank is isomorphically embeddable into* $\mathcal{L}_4$.

Corollary 3. *A finite lattice is isomorphically embeddable into* $\mathcal{L}_4$.

Corollary 4. *The lattice* $\mathcal{L}_4$ *does not satisfy any nontrivial lattice quasi-identity.*

Similar assertion for nontrivial identities in $\mathcal{L}_3$ was proved in [5].

From the proof of the theorem we can obtain the assertion connected with the so-called monoidal intervals. Recall that the *monoidal interval* ($\operatorname{Int} M$) for a transformation monoid M on the set E_k is the family of clones such that its every member has M as the set of its unary operations (this family is an interval in $\mathcal{L}_k$). Question on the structure (in particular, cardinality) of monoidal intervals was discussed in the literature (see, for example [8]). All clones occurred in the embedding constructed in the proof of the theorem have the same set M of unary operations on E_4; here $M = \{e_1^1, f\}$, $f(2) = 2$, and $f(x) = 0$ for $x \neq 2$. Therefore the following statement holds.

Corollary 5. *If a lattice is at most countable direct product of finite lattices then it can be isomorphically embedded into* $\operatorname{Int} M$.

The author suggests that the embedding in the proof of the theorem can be modified so that the property from the corollary would be true not only for the mentioned monoid M but for some others, too.

The question on the validity of the theorem for $\mathcal{L}_3$ is of very interest. In other words, is the restriction $k \geq 4$ in the theorem of principle or it is caused by defects of authors technique only? No essential distinction between lattice properties of lattices $\mathcal{L}_3$ and $\mathcal{L}_k$ for $k \geq 4$ is known. The author has no hypothesis on this question. Two following problems are closely connected.

Problem 1. Can an arbitrary finite lattice be embedded into $\mathcal{L}_3$?

Problem 2. Does the lattice $\mathcal{L}_3$ satisfy any nontrivial quasi-identity?

It is of interest to get some information about mutual embeddability of lattices $\mathcal{L}_k$. As we know, $\mathcal{L}_k$ can be embedded (moreover, as an interval) into $\mathcal{L}_m$ if $m \geq k$ [9]. But, is the inverse assertion true? More precisely, are there two integers k and m such that $k < m$ and $\mathcal{L}_m$ is embeddable into $\mathcal{L}_k$? We formulate a more concrete question.

Problem 3. Is there an integer k such that $\mathcal{L}_m$ can be embedded into $\mathcal{L}_k$ for any m (it would be well if k equals 3 or 4)?

By Corollary 3, each finite lattice is isomorphic to a sublattice of $\mathcal{L}_4$. But it is doubtful whether the analogous fact is true for countable lattice. By the theorem, all lattices embedded into $\mathcal{L}_4$ have many congruences. So, we ought to find an unembeddable lattice among infinite lattices with few congruences, for example, among prime lattices. The most known lattices of such kind are the lattice of equivalence relations on a countable set and the prime lattice M_ω of height 2 with a countable set of incomparable elements.

Problem 4. Is the lattice M_ω or the lattice of equivalences on a countable set embeddable into $\mathcal{L}_k$ for some k (in particular, for $k = 3, 4$)?

The above-mentioned results of [3] give us an example of an embedding as an interval or a presentation by an interval in clone lattices. So, we can reformulate the above-posed problems for embeddings of this restricted kind. But, in Problem 1, we have to ask about all lattices $\mathcal{L}_k$ as well as about $\mathcal{L}_3$.

Now, we say a few words on the proof of the theorem. The proof is rather long and sophisticated, so we describe the concrete lattice embedded into $\mathcal{L}_4$ and give a sketch of the proof. As usual, the lattice of all sublattices (0, 1-sublattices) of L is denoted by $\operatorname{Sub} L$ ($\operatorname{Sub}_{0,1} L$). As was shown in [10], any finite lattice can be isomorphically embedded into $\operatorname{Sub} P(A)$ where $P(A)$ is the power set of a finite set A. It is easy to see that $\operatorname{Sub} P(A)$ is embeddable into $\operatorname{Sub}_{0,1} P(B)$ where $B = A \cup \{b, c\}$, $b, c \notin A$. Hence any finite lattice can be isomorphically embedded into the lattice of all 0, 1-sublattices of an appropriate finite Boolean algebra. Thus, it suffices to embed direct products of lattices of the form $\operatorname{Sub}_{0,1} P(A)$ where A is a finite set. Since $A \subseteq B$ implies that $\operatorname{Sub}_{0,1} P(A)$ is a sublattice of $\operatorname{Sub}_{0,1} P(B)$, it suffices to construct embedding into $\mathcal{L}_4$ of the lattice $\prod_{l \in \mathbb{N}} \operatorname{Sub}_{0,1} P(A_l)$ where $A_l = \{0, 1\}^l$.

Now, we must construct an embedding of the pointed lattice. For each $B \subseteq \{0,1\}^l$, $l \in \mathbb{N}$, we define a special function $f_B(x, y, z_1, \dots, z_l)$. These functions have many useful properties. For example, for any two sets $B_1, B_2 \subseteq \{0,1\}^l, l \in \mathbb{N}$, the following identities are true:

$$\begin{aligned} & f_{B_1}(f_{B_2}(x,y,z_1,\dots,z_l),y,z_1,\dots,z_l) \\ (1\cap) \qquad = \;& f_{B_2}(f_{B_1}(x,y,z_1,\dots,z_l),y,z_1,\dots,z_l) = f_{B_1\cap B_2}(x,y,z_1,\dots,z_l), \\ & f_{B_1}(x,f_{B_2}(x,y,z_1,\dots,z_l),z_1,\dots,z_l) \\ (1\cup) \qquad = \;& f_{B_2}(x,f_{B_1}(x,y,z_1,\dots,z_l),z_1,\dots,z_l) = f_{B_1\cup B_2}(x,y,z_1,\dots,z_l). \end{aligned}$$

So, compositions of the kind correspond to operations of the lattice $P(\{0,1\}^l)$. On the other hand, applying another "bad" compositions or identifications of variables, we get a "bad" function. Therefore, any 0, 1-sublattice Q of $P(\{0,1\}^l)$ is put into correspondence with an iterative algebra generated by functions which closely connected with functions of the set $\{f_B \mid B \in Q\}$. Finally, each element $(Q_1, Q_2, \dots)$ of $\prod_{l\in\mathbb{N}} \mathrm{Sub}_{0,1} P(\{0,1\}^l)$ corresponds to the lattice union of all images of 0, 1-sublattices $Q_1, Q_2, \dots$. The main part of the proof is devoted to the check on the constructed mapping to be stable and bijective.

References

[1] Mal'cev, A. I., Iterative Post Algebras (Library of the Department of Algebra and Mathematical Logic of Novosibirsk State Univ. 16). Novosibirsk. Gos. Univ., Novosibirsk 1976.

[2] Post, E., Two-valued Iterative Systems of Mathematical Logic. Princeton University Press, Princeton 1941.

[3] Haddad, L., Rosenberg, I., Un intervalle Boolean de clones sur un univers fini. Ann. Sci. Math. Qué. 12:2 (1988), 211–231.

[4] Negru, I. S., Algebraic properties of the lattice of Post classes and their many-valued generalizations. In: Studies in Nonclassical Logics and Formal Systems, 300–315. Nauka, Moscow 1983.

[5] Bulatov, A. A., Identities of closed class lattices. Diskretn. Mat. 4:4 (1992), 140–148.

[6] Repnitskiĭ, V. B., On sublattices of semigroup lattices. Order (submitted).

[7] Repnitskiĭ, V. B., On lattices which are embeddable in subsemigroups lattices. Semigroup Forum 46 (1993), 388–397.

[8] Szendrei, A., Clones in Universal Algebra. Les Presses de l'Université de Montrĕal, Montrĕal (Québec) 1986.

[9] Pöschel, R., Kaluznin, L. A., Funktionen- und Relationenalgebren. Berlin, VEB DVW 1979.

[10] Hong Dang, X., On sublattice varieties. Algebra Univers. 27:3 (1990), 411–412.

An example of an unconstructivizable semilattice of computable numberings

Vyacheslav P. Dobritsa

Abstract. The author raised a natural question [1] of whether it is possible to apply the effective methods of investigation of nonconstructive models to the study of classes of constructive models and further to the study of computable numberings of classes of constructive models. A sufficient condition for uncomputability of the set of all computable numberings of a class of constructive models is given in [1]. However, the following question is left open: is it possible to constructivize the semilattice of computable numberings of a class of constructive models provided that the carrier set of the semilattice, i.e. the set of computable numberings of this class, is computable? In this paper we prove that in general the answer is negative. This result was announced in [2].

1991 Mathematics Subject Classification: 03D45.

Throughout the paper we follow the notation of [3–5]. Recall some definitions. Denote by $\mathfrak{M}$ a model and by $|\mathfrak{M}|$ its carrier set. An enumeration $\nu : N \to |\mathfrak{M}|$ is called *constructive* if operations and predicates of $\mathfrak{M}$ (in particular, the predicate of equality), when assigned to numbers, are recursive. In this case the pair $(\mathfrak{M}, \nu)$ is called a *constructive model* and the model $\mathfrak{M}$ is called *constructivizable.* Two constructive enumerations ν and μ of $\mathfrak{M}$ are said to be *autoequivalent* if there exists an automorphism φ of $\mathfrak{M}$ and a recursive function f such that the following diagram is commutative:

$$\begin{array}{ccc} N & \xrightarrow{f} & N \\ \downarrow \nu & & \downarrow \mu \\ \mathfrak{M} & \xrightarrow{\varphi} & \mathfrak{M} \end{array}$$

Constructive enumerations of models are usually considered up to autoequivalence. A class $\mathcal{K}$ of constructive models is called *computable* if there exists an effective numbering γ of all models in $\mathcal{K}$. It is evident that the numbering $\gamma : N \to \mathcal{K}$ satisfies two conditions:

1) $\forall n\, \exists(\mathfrak{M}, \nu) \in \mathcal{K} \qquad (\gamma_n \underset{a}{\equiv} \nu),$

2) $\forall(\mathfrak{M}, \nu) \in \mathcal{K}\, \exists n \qquad (\gamma_n \underset{a}{\equiv} \nu).$

A computable numbering α of a class of constructive models $\mathcal{K}$ is said to be *reducible to a numbering* β of $\mathcal{K}$ if there exists a recursive function $f(x)$ such that $\alpha_n \underset{a}{\equiv} \beta_{f(n)}$. Numberings α and β are called *equivalent* (written $\alpha \equiv \beta$) if α is reducible to β ($\alpha \leqslant \beta$) and, conversely, β is reducible to α ($\beta \leqslant \alpha$).

Let $J(\mathcal{K})$ stand for the set of computable numberings of $\mathcal{K}$ which are considered up to equivalence. Reducibility of numberings in $J(\mathcal{K})$ defines the upper semilattice

$$\mathfrak{L}(\mathcal{K}) \leftrightharpoons \langle J(\mathcal{K}); \leqslant \rangle.$$

If there exists the greatest element $\overline{\gamma}$ of $\mathfrak{L}(\mathcal{K})$, the corresponding computable numbering γ of the class $\mathcal{K}$ is called *principal* and $\mathcal{K}$ is called *strictly computable.* In this case we have reducibility $\alpha \leqslant \gamma$ for any computable numbering α. A computable numbering β is called *least* if it is reducible to any computable numbering (of the same class). If a least computable numbering does exist, it defines the least element of $\mathfrak{L}(\mathcal{K})$.

A class of constructive models can be uncomputable (i.e., possessing no computable numbering). It also can be computable but having neither least nor principal numbering, or having exactly one of them, or having both of them simultaneously.

In most cases the set $J(\mathcal{K})$ is uncomputable [1]. However, a sufficient condition of computability of $J(\mathcal{K})$ is given by the following

Theorem 1 [6]. *If there exists a principal computable numbering and a least computable numbering of a class $\mathcal{K}$ of constructive models then the set $J(\mathcal{K})$ of all computable numberings of $\mathcal{K}$ is computable.*

If the set $J(\mathcal{K})$ is computable, a natural question arises whether it is possible to constructivize the corresponding semilattice $\mathfrak{L}(\mathcal{K})$. It turns out that in general it is impossible.

Theorem 2. *There exists a computable class $\mathcal{K}$ of constructive models with computable set $J(\mathcal{K})$ of computable numberings of $\mathcal{K}$ such that its upper semilattice of computable numberings is not constructivizable.*

Proof. Consider a class formed by two constructive models $\mathfrak{M}_1$ and $\mathfrak{M}_2$, each of them being a linearly ordered finite set. The model $\mathfrak{M}_1$ contains two elements and the model $\mathfrak{M}_2$ contains three elements:

$$\mathfrak{M}_1 \leftrightharpoons \{a_1 < a_2\}, \quad \mathfrak{M}_2 \leftrightharpoons \{a_1 < a_2 < a_3\}.$$

Each model is (strongly) constructivizable and autostable, i.e., it possesses the unique constructive enumeration up to autoequivalence. Denote by α_1 a constructive enumeration of $\mathfrak{M}_1$ and by α_2 that of $\mathfrak{M}_2$. The corresponding class will be denoted by $\mathcal{K} = \{(\mathfrak{M}_1, \alpha_1), (\mathfrak{M}_2, \alpha_2)\}$.

It is obvious that the class $\mathcal{K}$ is computable, and there exists a least computable numbering of $\mathcal{K}$ defined as follows:

$$\gamma_n \underset{a}{\equiv} \begin{cases} \alpha_1 & \text{for even } n, \\ \alpha_2 & \text{for odd } n. \end{cases}$$

For every computable numbering β of $\mathcal{K}$, the recursively enumerable set

$$I_\beta^{\alpha_2} = \{n \mid \beta_n \underset{a}{\equiv} \alpha_2\}$$

which is the index set of the constructive enumeration of the second model is defined in a natural way. The index set $I_\beta^{\alpha_1}$ of the first constructive model of $\mathcal{K}$ is the complement of $I_\beta^{\alpha_2}$, i.e.,

$$I_\beta^{\alpha_1} = N \setminus I_\beta^{\alpha_2}.$$

For two computable numberings γ and β of $\mathcal{K}$, the following equivalences are valid:

$$\gamma \leqslant \beta \iff I_\gamma^{\alpha_2} \leqslant_m I_\beta^{\alpha_2}, \tag{1}$$

$$\gamma \equiv \beta \iff I_\gamma^{\alpha_2} \equiv_m I_\beta^{\alpha_2}. \tag{2}$$

At the same time, it is possible to define a computable numbering ν of $\mathcal{K}$ such that $I_\nu^{\alpha_2} = A$ for a recursively enumerable set A. To this end, it suffices to calculate elements of A and simultaneously construct a numbering γ in accordance with the following algorithm which is uniform with respect to n:

1) the constructive enumeration ν_n with index n in numbering ν is constructed in accordance with the enumeration α_1 until it turns out that n belongs to the set A,

2) when we establish the membership of n in A, we insert the already constructed part of the model corresponding to ν_n into the model $(\mathfrak{M}_2, \alpha_2)$ and further construct the model corresponding to ν_n in accordance with the enumeration α_2.

In the above algorithm, we take a creative set K as the set A. Then we obtain a greatest numbering X of $\mathcal{K}$ in the sense of reducibility, i.e. it will be a principal computable numbering of $\mathcal{K}$. By Theorem 1 we conclude that the set $J(\mathcal{K})$ is computable.

Now we show that the corresponding upper semilattice $\mathcal{L}(\mathcal{K})$ of computable numberings is not constructivizable. By (1), (2) and the above construction of the computable numbering ν of $\mathcal{K}$ starting from arbitrary recursively enumerable set A, the upper semilattice $\mathcal{L}(\mathcal{K})$ of computable numberings of $\mathcal{K}$ is isomorphic to the upper semilattice of m-degrees of recursively enumerable sets, which is not constructivizable, as follows from the description [7] of initial segments of this semilattice.

Thus, we have constructed the class $\mathcal{K}$ of computable models which satisfies the required conditions.

References

[1] Dobritsa, V. P., The condition of effective infinity of sets of computable numberings. In: 11th Int. Conf. on Math. Logic, Abstracts. Kazan' 1992.

[2] — An example of non-constructivizable semilattice of computable numberings. In: 3rd Int. Conf. on Algebra Dedicated to Memory of M. I. Kargapolov (Krasnoyarsk, August 23–28, 1993), Abstracts. INOPROF, Krasnoyarsk 1993.

[3] Ershov, Yu. L., Decision Problems and Constructivizable Models. Nauka, Moscow 1980.

[4] Rogers, H. (jr), Theory of recursive functions and effective computability. McGraw–Hill, New York 1967.

[5] Dobritsa, V. P., Structural properties of computable classes of constructive models. Algebra Logika 26:1 (1987), 36–62. English translation: Algebra Logic 26:1 (1987), 24–41.

[6] — On computability of sets of computable numberings of a class of constructive models. Mat. Zametki 55:5 (1994), 43–50. English translation: Math. Notes 55:5 (1994).

[7] Lachlan, A. H., Recursively enumerable many-one degrees. Algebra Logika 11:3 (1972), 326–358. English translation: Algebra Logic 11:3 (1972), 186–202.

Enumeration of characteristic subgroups of unipotent Lie-type groups

Georgiĭ P. Egorychev and Vladimir M. Levchuk *

Abstract. The number of different characteristic subgroups of the classic-type group $UG(K)$ over a field K and that of normal D-invariant (D is the set of diagonal automorphisms) subgroups of the same group are given.

1991 Mathematics Subject Classification: 20D06, 20D25.

1. Introduction

Throughout the paper, K stands for a field of characteristic > 2. As in [1], $UG(K)$ denotes the maximal unipotent subgroup of a Lie-type group G over K. The main result of the paper cited is formulated as follows:

Theorem 1. *The number $c(G, K)$ of characteristic subgroups of the classic-type group $UG(K)$ over a field K of characteristic > 2 is equal to:*

$$\binom{n}{[n/2]n} \quad \textit{for} \quad G = A_{n-1} \textit{ or } {}^2A_{n-1},\ n > 1$$

$$\binom{2n}{n} \quad \textit{for} \quad G = B_n\ (n > 3,\ 2K = K), \textit{ or }$$
$$C_n\ (n \geqslant 2,\ 3K = 0,\ |K| > 3), \textit{ or }$$
$$D_{n+1}\ (n > 3), \textit{ or }$$
$${}^2D_{n+1}\ (n \geqslant 3,\ |K| > 4);$$

$$4^{n-1} + \binom{2n-1}{n} \quad \textit{for} \quad 2K = 0,\ G = B_n \textit{ or } C_n,\ n \geqslant 2;$$

$$\binom{2n}{n} - 1 \quad \textit{for} \quad 6K = K;$$

$$\binom{2n}{n} - 2 \quad \textit{for} \quad G = C_n,\ n > 2\ \ (|K| = 3).$$

*Supported by the Russian Fund for Fundamental Research (grants 93–011–1560 and 94–01–01084).

The problem of estimating the number of different subgroups of a group has already been studied. For example, an estimate for the number of fixed-order subgroups of a finite p-group is well known (G. Frobenius, Ph. Hall, et al.)

Note that, for $UG(K)$, the enumeration of characteristic subgroups is closely connected with that of normal subgroups invariant with respect to the subgroup D of all diagonal automorphisms. For $G = A_{n-1}$ ($|K| > 2$), the number $d(G, K)$ of such subgroups was determined out by G. P. Egorychev [2, Theorem 2.1.2] and B. A. Tolasov [3]. The results are based on the description [4, Section 4] of D-invariant normal subgroups of the unitriangular group $UT(n, K)$. The latter is isomorphic to the adjoint group of the ring $NT(n, K)$ of niltriangular matrices. The class of all D-invariant normal subgroups of the adjoint group coincides with the class of all D-invariant ideals in the ring $NT(n, K)$ and the associated Lie ring (see [5, Theorem 8]).

By analogy with the case of $G = A_{n-1}$ in Chevalley K-algebra of arbitrary type Φ (see, for example, [6, Section 4.4; 7]), we select a subalgebra $N\Phi(K)$, whose basis is formed by elements e_r of Chevalley basis relative to positive roots $r \in \Phi^+$. In [1, 5] the group $U\Phi(K)$ is treated as the adjoint group of the Lie ring $N\Phi(K)$.

The number $d(G, K)$ is determined in Theorem 2. The number $d(\Phi, K)$ is also found in the paper, including the most difficult case $p(\Phi)!K = 0$; here $p(\Phi)$ is defined as maximal ratio $(r, r)/(s, s)$, $(r, s \in \Phi)$.

This investigation is also interesting as an example of direct enumeration, where modern methods of calculation of combinatorial sums had to be involved for obtaining the answer in closed form.

Some results of this paper were announced in [8], where the list of the subgroups of groups $UG(K)$ of small Lie ranks was presented.

2. Construction and enumeration of subgroups

Theorem 2. *The number $d(G, K)$ of normal D-invariant subgroups of the classic-type group $U(G, K)$ over a field K of characteristic > 2 is equal to:*

$\frac{1}{n}\binom{2n}{n-1}$ *for* $G = A_{n-1}$, n> 1;

$\binom{n}{[n/2]}$ *for* $G = {}^2A_{n-1}$, $n > 1$;

$\binom{2n}{n}$ *for* $G = {}^2D_n$, $n \geqslant 2$, $|K| > 4$, *or*
$K = 2K$, $G = B_n$, *or*
C_n, $n \geqslant 1$;

$$\binom{2n-1}{n} + 4^{n-1} \qquad \text{for} \qquad 2K = 0,\ G = B_n \text{ or } C_n,\ n > 2;$$

$$\binom{2n-1}{n} + \binom{2n-2}{n} \qquad \text{for} \qquad G = D_n,\ n > 2.$$

We want to describe normal D-invariant subgroups of $UG(K)$ under the assumptions on G and K in Theorem 2.

In the sequel, we suppose that the roots $r, s \in \Phi$ satisfy the inequality $s \leqslant r$ if $r - s$ is a linear combination of fundamental roots with non-negative coefficients. The roots r and s are called *incident* if $r \leqslant s$ or $s \leqslant r$. The partial order $\leqslant$ is also defined on the system ${}^2\Phi$ composed of minimal nonempty subsets invariant under proper symmetry [9, Lemma 7].

For a set of pairwise nonincident roots $r_1, r_2, \ldots, r_m \in G$, we select a subgroup $T(r_1, \ldots, r_m)$. By definition, it is generated by all possible root subgroups corresponding to the roots s such that $s \leqslant r_i$, even in the case of a single i.

Lemma 1. *Subgroups $T(r_1, \ldots, r_m)$ of the classic type group $UG(K) \cong NG(K)$ are different for different sets of pairwise nonincident elements $r_1, r_2, \ldots, r_m \in G$. All these subgroups are normal D-invariant subgroups and exhaust them if $|K| > 2$, or $|K| > 4$ for $G = {}^2D_n$, $n \geqslant 3$, or $2K = K$ for $G = B_n$ and C_n.*

Proof. We only need to prove the last statement of the lemma. If $|K| > 2$ or $|K| > 4$ for $G = {}^2D_n$, $n \geqslant 3$, then the commutant of two root subgroups in $UG(K)$ is generated by root subgroups. Under the same conditions, any normal D-invariant subgroup is also generated by root subgroups (see [1, Section 5; 5, Theorem 9; 9, Theorem 1]). Let $2K = K$ for $G = B_n$ or C_n. By Lemmas 3–5 of [1], the normal closure of the r-root subgroup in $UG(K)$ contains all thes-root subgroups for $s \leqslant r$ and is generated by them. Lemma 1 is proved.

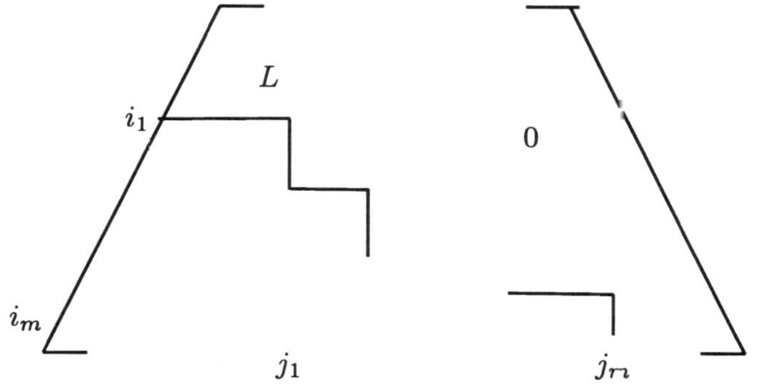

Fig. 1

Below we use a representation of $UG(K)$ in terms of B_n^+-, C_n^+-, and BC_n^+-matrices (see [1]) and the notation (see [10, Tabl. I–IV])

$$\varepsilon_i - v\varepsilon_j = p_{i,vj}, \quad 1 \leqslant j \leqslant i \leqslant n, \quad v \in \{0, -1, 1\}.$$

If $r_1 = p_{i_1,j_1}, \ldots, r_m = p_{i_m,j_m}$, then the task of enumeration of subgroups $T(r_1, \ldots, r_m)$ turns out to be equivalent (under the restrictions on G and K made in Lemma 1) to that of enumeration of the steplines L in corresponding Φ^+-matrix (see Fig. 1). Hence the numbers

$$d(^2A_{2n-1}, K), \quad d(B_n, K)\ (2K = K), \quad d(C_n, K)\ (2K = K), \quad d(^2D_{n+1}, K)$$

coincide. On the other hand, it turns out that $d(B_n, K)$ is equal to the number of integer-pair sequences $(i_1, j_1), (i_2, j_2), \ldots, (i_m, j_m)$ $(m \geqslant 0)$ if $-i_l < j_l < i_l$ for all l and $j_1 < j_2 < \cdots < j_m$, $i_1 < i_2 \cdots < i_m \leqslant n$.

The subgroup $T(r_1, \ldots, r_n)$ of the adjoint group $N\Phi(K) \cong U\Phi(K)$ of normal type is also a subalgebra of the Lie ring $N\Phi(K)$. Its basis is formed by elements e_s of Chevalley basis with $s \leqslant r_i$, even in the case of single i. By Lemma 2 of [1] the subalgebras $T(r_1, \ldots, r_m)$ are D-invariant ideals of $N\Phi(K)$.

Consider the auxiliary table

$$\begin{array}{c} a_{00} \\[2ex] a_{1\,-1}\ a_{10}\ a_{11} \\[2ex] \cdots\ \cdots\ \cdots \\[2ex] a_{n\,-n}\ \ldots\ a_{n\,-1}\ a_{n0}\ a_{n1}\ \ldots\ a_{nn} \end{array} \tag{1}$$

obtained from B_{n+1}^+-matrix [1] by replacing the roots p_{uv} with $a_{u-1,v}$. Denote by $N[(i, v); (k, m)]$ the number of all steplines from a_{iv} to a_{km} $(v \leqslant m \leqslant k \leqslant n)$, $|v| \leqslant i \leqslant k$. Lemma 1 implies the following assertion.

Lemma 2. *Let G and K be the same as in Lemma 1. Then*

$$d(A_{n-1}, K) = \sum_{i=1}^{n} N[(i, 1); (n, n)],$$

$$d(B_n, K) = \sum_{i=0}^{n} N[(i, -i); (n, n)],$$

$$d(^2A_{2n-2}, K) = \sum_{i=1}^{n} N[(i, -i + 1); (n, n)],$$

$$d(D_n, K) = d(^2A_{2n-2}; K) + d(A_{n-1}, K) - d(A_{n-2}, K)$$
$$+ \sum_{i=2}^{n-1} \sum_{t=i}^{n-1} \sum_{m=t+1}^{n} N[(i, -i + 1); (t, -1)] \cdot N[(m, 1); (n, n)].$$

Thus, the problem of evaluation of the numbers $d(G, K)$ under the assumptions of Lemma 1 is reduced to a combinatorial one. In Sections 3 and 4 we obtain its solution (as in the case $G = A_{n-1}$), using integral representation of sums (see [2]) and listing lattice paths (see [11]).

In Lemma 1, the case $2K = 0$, $G = B_n$, and C_n is essential since $[e_r, e_s] = e_r * e_s = 2e_{r+s} = 0$ for any short roots r and s such that $r + s$ is a long root.

Let T_0 be a subalgebra of the algebra $NC_n(K)$, $2K = 0$, whose basis is formed by e_s for all possible short roots s. It turns out that the minimal D-invariant ideal of the Lie ring $NC_n(K)$ containing an element e_r for arbitrary short root r coincides with the intersection $T(r) \cap T_0$. The latter coincides with the normal D-invariant closure of e_r if $r \leqslant p$ for a long fundamental root p (i.e., $p = p_{1,-1}$, $r = p_{i,-v}$, $0 < v < i$, in the notation of [1]). In the case $r \nleqslant p$, the normal closure coincides with $T(r)$ because

$$[Ke_{iv}, K_{v-v}] = Ke_{i\,-v} + Ke_{i\,-i} \quad \text{for } |K| > 2. \tag{2}$$

By Theorem 1 and Lemma 2 of [9], all D-invariant normal subgroups of the adjoint group of $NC_n(K)$ are K-linearly generated by elements of the form e_s, $s \in \Phi^+$ (see also [1, Section 5]), and can be written as

$$\sum_{k=1}^{m}(T_0 \cap T(r_k)) + \sum_{r \in M} Ke_r; \tag{3}$$

here $\{r_1, \ldots, r_m\}$ is an arbitrary (possibly empty) set of pairwise nonincident short roots from Φ^+ and M is a subset of long roots from Φ^+. Moreover, if $r \in M$ is not a maximal root, then, for some k, we have $r \leqslant r_k$ or $r + q \leqslant r_k$ for any simple root q.

Note that the above conditions on M and roots r_k are necessary and sufficient for a subalgebra (2) to be an ideal. In particular, if $r \in M$ is not a maximal root and $q, r + q \in \Phi^+$ then $r + q$ is a short root and the inclusion $H \ni e_{s+q} = e_q * e_s$ is equipotent to the inequality $s + q \leqslant r_k$ for some k.

Lemma 3. *Every normal D-invariant subgroup of an adjoint group is an ideal of the Lie ring $NC_n(K)$, $2K = 0$, $|K| > 2$, and can be written in the form (3). The subalgebra (3) is a normal subgroup of an adjoint group if and only if the following two conditions are satisfied:*

(a) *if $r \in M$ then all the long roots of height $> h(r)$ are in M,*

(b) *if $r_k \nleqslant p$, where $1 \leqslant k \leqslant m$, $p = p_{1,-1}$, is a long simple root, then every long root $r \leqslant r_k$ is in M.*

Proof. As was noted above, a D-invariant normal subgroup is K-linearly generated by elements of the form e_r, $r \in \Phi^+$. Moreover, a normal closure of Ke_r coincides with $T_0 \cap T(r)$ for short roots $r \leqslant p$ and with $T(r)$ for the rest of them by virtue of (2). Thus, the subgroup is written in the form (3). Conditions (a) and (b) are obtained with the help of (2). Lemma 3 is proved.

Lemma 4. *Let K be a field of characteristic 2 and of order > 2. Then*

$$d(C_n, K) = \sum_{i=1}^{n} \sum_{t=i}^{n} (t - i + 2) S(i, t, n) \tag{4}$$

where $S(i,t,n)$ is the number of sets of integer pairs (u_1, v_1), $(u_2, v_2), \ldots, (u_m, v_m)$ such that $i = u_1 < u_2 < \cdots < u_m = n$, $v_1 < v_2 < \cdots < v_m = n$, $-u_k < v_k \leqslant u_k$, and t is the least one among the numbers u_k such that the condition $v_k > 0$ is satisfied.

In Sections 3 and 4 it is proved that the right-hand side of (4) coincides with $4^{n-1} + \binom{2n-1}{n}$. For a perfect field of characteristic 2, the groups $UB_n(K)$ and $UC_n(K)$ are isomorphic. Consequently, $d(C_n, K) = d(B_n, K)$. Thus, Theorem 2 is proved.

Theorem 1 follows from Theorem 2 and the description of characteristic subgroups (see [1, Section 5]).

3. Combinatorial analysis of the number of paths and formulas for $d(D_n, K)$ and $d(C_n, K)$

We use the well-known result on the number of paths with diagonal steps on a lattice with integer (Cartesian) coordinates.

Lemma 5 [11, Theorem 1]. *The number of paths with diagonal steps laying above the real axis and leading from the origin to a fixed point (X, Y) is equal to:*

$$N_{X,Y} = \frac{Y}{X} \binom{X}{(X+Y)/2}. \tag{5}$$

By Lemma 5, the number of similar paths but laying above or on the real axis is equal to:

$$N^*_{X,Y} = \frac{Y+1}{X+1} \binom{X+1}{(X+Y+2)/2}. \tag{6}$$

3.1. The case $d(D_n, K)$. Due to the combinatorial description given in Section 2, we have

$$d(D_n, K) = d(^2A_{2n-2}, K) + \sum_{i=2}^{n-1}\sum_{t=i}^{n-1}\sum_{m=t+1}^{n} N_1(i,t)N_2(m,n)$$

$$+ d(A_{n-1}, K) - d(A_{n-2}, K), \quad n \geqslant 3, \qquad (7)$$

where (in a suitable coordinate system (x, y), see Fig. 2) $N_1(i,t)$ denotes the number of various paths with rectangular steps from points $A_i = (-i+1, i)$, $i \in \overline{2,n}$, to points $B_t = (-1, t)$, $t \geqslant i$, and $N_2(m,n)$ stands for the number of various paths from the point $(1, m)$ to the point $(n, 0)$ all of whose points are below or on the line L from $(n, 0)$ to $(0, n)$.

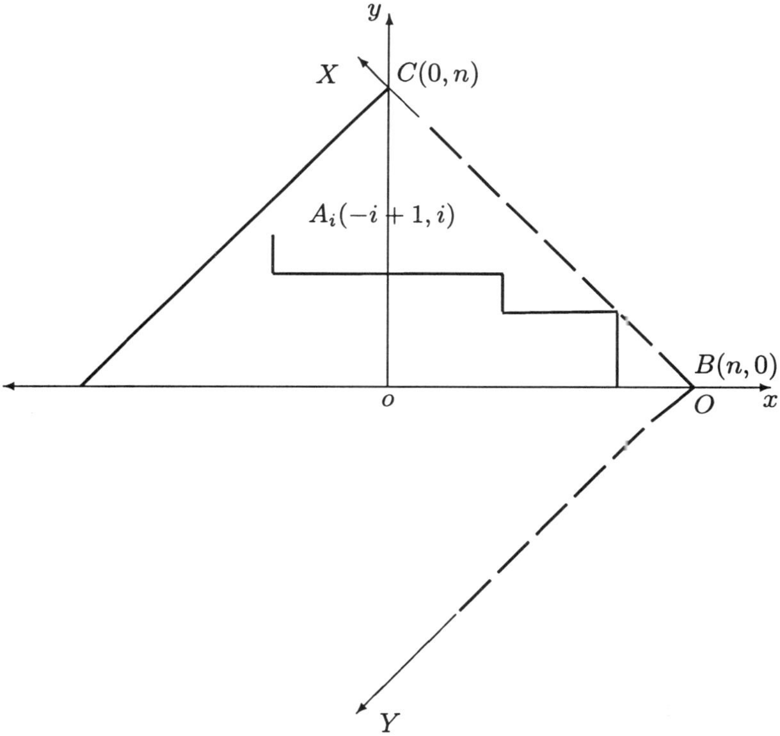

Fig. 2

It is easy to see that the number $N_1(i,t)$ coincides with the number of various $k = t - i$ scheduling (indistinguishable) rises on $l = i - 1$ distinguishable positions (cells) under the assumption that some of them may be empty. Thus, we have

$$N_1(i,t) = \binom{l+k-1}{k} = \binom{t-2}{t-i}. \tag{8}$$

Choose the new coordinate system (X, Y) with the origin at the point B, and coordinate axes as those indicated on Fig. 2 by dashed line. Then we can consider the above-mentioned paths $N_2(m,n)$ with rectangular steps from the point $A_m = (1,m)$ to the point B as the paths from the point B to the point A_m on the same lattice but with diagonal steps. Moreover, the coordinates of the point $A_m = (X_m, Y_m)$ in the coordinate system (X, Y) are $X_m = 2n - m - 1$, $Y_m = m - 1$. Therefore, from (6) it follows that

$$N_2(m,n) = \frac{m}{2n-m}\binom{2n-m}{n}, \tag{9}$$

and from (7)–(9) we obtain

$$d(D_n, K) = d(A^2 A_{n-2}, K) + \sum_{i=2}^{n-1}\sum_{t=i}^{n-1}\sum_{m=t+1}^{n}\binom{t-2}{t-i}\frac{m}{2n-m}\binom{2n-m}{n}$$

$$+d(A_{n-1} - d(A_{n-2}, K)). \tag{10}$$

3.2. The case $d(C_n, K)$. Due to the combinatorial description given in Section 1, we have

$$d(C_n, K) = \sum_{i=1}^{n}\sum_{m=1}^{n}(m-i+2)S(i,m,n),$$

where $S(i,m,n) = N_1(i,m) \times N_2(m,n)$. Here $N_1(i,m)$ (in suitable coordinate system (x,y); see Fig. 2) stands for the number of various paths with rectangular steps from points $A_i = (-i+1, i)$, $i \in \overline{1,n}$, to points $B_m = (0,m)$, $m \in \overline{i,n}$, and $N_2(m,n)$ denotes the number of various paths from the point $(1,m)$ to the point $(n,0)$ which are placed below or on the line L passing through the points $(n,0)$ and $(0,n)$.

In the same manner as in the case of $d(D_n, K)$ we obtain

$$d(C_n, K) = \sum_{i=1}^{n}\sum_{m=i}^{n}(m-i+2)\binom{m-1}{m-i}\frac{m}{2n-m}\binom{2n-m}{n}. \tag{11}$$

4. Computation of combinatorial sums

Further we apply the method of integral representation and calculation of combinatorial sums (the method of coefficients) which were developed in [2].

Lemma 6. *Let a sum S_n be given by*

$$S_n = \sum_{i=2}^{n-1}\sum_{t=i}^{n-1}\sum_{m=t+1}^{n} \binom{t-2}{t-i} \frac{m}{2n-m} \binom{2n-m}{n-m}, \quad n \geqslant 3.$$

Then

$$S_n = \binom{2n-2}{n-3} \tag{12}$$

Proof. First, by the algorithm of the method of coefficients [2, Chapter 1], we find the following integral representations for the numbers $\binom{t-2}{t-i}$ and $\frac{m}{2n-m}\binom{2n-m}{n-m}$:

$$\binom{t-2}{t-i} = \frac{1}{2\pi i} \int\limits_{|y|=\rho_1} (1+y)^{t-2} y^{-t+i-1}, \quad \rho_1 > 0;$$

$$\frac{m}{2n-m}\binom{2n-m}{n-m} = 2\binom{2n-m-1}{n-m} - \binom{2n-m}{n-m}$$

$$= -\frac{1}{2\pi i} \int\limits_{|x|=\rho_2} (1-x)^{-n-1} x^{-n+m-1} dx + \frac{2}{2\pi i} \int\limits_{|x|=\rho_2} (1-x)^{-n} x^{-n+m-1} dx$$

$$= \frac{1}{2\pi i} \int\limits_{|x|=\rho_2} (1-x)^{-n-1}(1-2x) x^{-n+m-1} dx, \quad 0 < \rho < 1.$$

Substituting the integral representations into (12) we obtain

$$S_n = \sum_{i=2}^{n-1}\sum_{t=i}^{n-1}\sum_{m=t+1}^{n} \frac{1}{(2\pi i)^2} \int\limits_{|x|=\rho_1}\int\limits_{|y|=\rho_2} (1+y)^{t-2} y^{-t+i-1}$$

$$\times (1-x)^{-n-1}(1-2x) x^{-n+m-1} dx \wedge dy$$

$$= \sum_{i=2}^{n-1} \sum_{t=i}^{n-1} \sum_{m=t+1}^{\infty} \cdots$$

[since the added summands equal 0]

$$= \sum_{i=2}^{n-1} \sum_{t=i}^{n-1} \frac{1}{(2\pi i)^2} \int\limits_{|x|=\rho_1} \int\limits_{|y|=\rho_2} (1+y)^{t-2} y^{-t+i-1}$$

$$\times (1-x)^{-n-1}(1-2x)x^{-n-1} \left(\sum_{m=t+1}^{\infty} x^m \right) dx \wedge dy$$

$$= \sum_{i=2}^{n-1} \sum_{t=i}^{n-1} \frac{1}{(2\pi i)^2} \int\limits_{|x|=\rho_1} \int\limits_{|y|=\rho_2} (1+y)^{t-2} y^{-t+i-1}$$

$$\times (1-x)^{-n-1}(1-2x)x^{-n-1} \frac{x^{t+1}}{1-x} dx \wedge dy$$

$$= \sum_{i=2}^{n-1} \sum_{t=i}^{\infty} \cdots$$

$$= \sum_{i=2}^{n-1} \frac{1}{(2\pi i)^2} \int\limits_{|x|=\rho_1} \int\limits_{|y|=\rho_2} \frac{(1+y)^{i-2}(1-x)^{-n-2}(1-2x)x^i}{x^n y(1-x(1+y)y^{-1})} dx \wedge dy$$

$$= \sum_{i=2}^{\infty} \cdots = \frac{1}{(2\pi i)^2} \int\limits_{|x|=\rho_1} \int\limits_{|y|=\rho_2} \frac{(1-x)^{-n-2}(1-2x)x^2}{x^n(y-(1+y)x)(1-x(1+y))} dx \wedge dy$$

$$= \frac{1}{(2\pi i)^2} \int\limits_{|x|=\rho_1} \int\limits_{|y|=\rho_2} \frac{(1-x)^{-n-3}(1-2x)x^2}{x^{n,-2}(y-x(1-x)^{-1})(1-x(1+y))} dx \wedge dy$$

[$y = x(1-x)^{-1}$ a simple pole]

$$= \frac{1}{2\pi i} \int\limits_{|x|=\rho_1} \frac{(1-x)^{-n-3}(1-2x)}{x^{n-2}(1-x(1+x(1-x)^{-1}))} dx$$

$$= \frac{1}{2\pi i} \int\limits_{|x|=\rho_1} (1-x)^{-n-2} x^{-n+2} dx$$

$$= \binom{n+2+n-3-1}{n-3} = \binom{2n-2}{n-3}.$$

Here we may summarize under the integration sign because the geometric progression converges uniformly on the sceleton

$$\gamma = \{(x,y) \in \mathbb{C}^2 : |x| = \rho_1, |y| = \rho_2,\ 0 < \rho_1 \ll \rho_2 < 1/2\}$$

and, consequently, the inequalities

$$|x| < 1, \quad |(1+y)xy^{-1}| < 1, \quad |x(1+y)| < 1$$

are valid on γ. Lemma 6 is proved.

Lemma 7. *Let n be a natural number. Then the following identity holds:*

$$\sum_{i=1}^{n} \sum_{m=i}^{n} (m-i+2) \binom{m-1}{m-i} \frac{m}{2n-m} \binom{2n-m}{n-m} = 4^{n-1} + \binom{2n-1}{n}. \tag{13}$$

Proof. The integral representations for binomial coefficients lead to the formulas

$$(m-i+2) = \frac{1}{2\pi i} \int\limits_{|z|=\rho_1} (1-z)^{-2} z^{-m+i-2} dz, \qquad \rho_1 < 1, \tag{14}$$

$$\binom{m-1}{m-i} = \binom{m-1}{i-1} = \frac{1}{2\pi i} \int\limits_{|y|=\rho_2} (1+y)^{m-1} y^{-i} dy, \quad \rho_2 > 0, \tag{15}$$

$$\frac{n}{2n-m} \binom{2n-m}{n-m} = \frac{1}{2\pi i} \int\limits_{|x|=\rho_3} (1-x)^{-n-1}(1-2x) x^{-n+m-1} dx, \quad \rho_3 > 0. \tag{16}$$

By (14)–(16), the sum S on the left-hand side of (13) can be presented as follows:

$$S = \sum_{i=1}^{n} \sum_{m=i}^{n} \frac{1}{(2\pi i)^3} \int\limits_{\gamma} \{(1-x)^{-n-1}(1-2x)(1-z)^{-2}\}$$

$$\times (1+y)^{m-1} y^{-i} z^{-m+i-2} x^{-n+m-1} dx \wedge dy \wedge dz,$$

where

$$\gamma = \{(x, y, z) \in \mathbb{C}^3 : |z| = \rho_1,\ |y| = \rho_2,\ |z| = \rho_3,\ 0 < \rho_1 < 1,\ \rho_2 > 0,\ 0 < \rho_3 < 1\}.$$

Continue the sum

$$S = \sum_{i=1}^{\infty} \sum_{m=i}^{\infty} \cdots$$

[by (14)–(16), the added summands equal 0]. Replacing $m - i$ by m, we find

$$\begin{aligned}
S &= \sum_{i=1}^{\infty} \sum_{m=0}^{\infty} \frac{1}{(2\pi i)^3} \int_{\gamma} \{\dots\}(1+y)^{m+i-1} y^{-i} z^{-m-2} x^{-n+m+i-1} dx \wedge dy \wedge dz \\
&= \frac{1}{(2\pi i)^3} \int_{\gamma} \{\dots\} z^{-2} x^{-n-1} (1+y)^{-1} \left[\sum_{i=1}^{\infty} \left(\frac{x(1+y)}{z} \right)^{i} \right] \\
&\qquad \times \left[\sum_{m=0}^{\infty} \left(\frac{x(1+y)}{y} \right)^{m} \right] dx \wedge dy \wedge dz \\
&= \frac{1}{(2\pi i)^3} \int_{\gamma} \{\dots\} z^{-2} x^{-n-1} (1+y)^{-1} \frac{x(1+y)}{z} \\
&\qquad \times \left(1 - \frac{x(1+y)}{z}\right)^{-1} \left(1 - \frac{x(1+y)}{y}\right)^{-1} dx \wedge dy \wedge dz \\
&= \frac{1}{(2\pi i)^3} \int_{\gamma} \frac{(1-x)^{-n-1}(1-2x)(1-z)^{-2} dx \wedge dy \wedge dz}{z x^n (z - x(1+y))(y - x(1+y))}.
\end{aligned} \tag{17}$$

Here we may summarize under the integration sign because the geometric progression converges uniformly on the sceleton γ. The latter is true if we choose ρ_1, ρ_2, and ρ_3 such that

$$0 < \rho_3 \ll \rho_2 \ll \rho_1 < 1.$$

Then the inequalities

$$|x(1+y)z^{-1}| < 1, \quad |x(1+y)y^{-1}| < 1$$

are valid on γ.

Applying the residue theorem on variable z at the simple poles $z_1 = 0$ and $z_2 = x(1+y)$, we obtain

$$J_1 = \frac{-1}{(2\pi i)^2} \int\limits_{|x|=\rho_3} \int\limits_{|y|=\rho_2} \frac{(1-x)^{-n-1}(1-2x)dx \wedge dy}{x^{n+1}(1+y)(y-x(1+y))}$$

$$[y = x(1-x)^{-1} \text{a simple pole}, \quad y = -1 \text{ an absent pole}]$$

$$= -\frac{1}{2\pi i} \int\limits_{|x|=\rho_3} \frac{(1-x)^{-n-2}(1-2x)}{x^{n+1}\left(1+\frac{x}{1-x}\right)} dx = -\frac{1}{2\pi i} \int\limits_{|x|=\rho_2} \frac{(1-x)^{-n-1}(1-2x)}{x^{n+1}} dx$$

$$= -\binom{2n}{n} + 2\binom{2n-1}{n-1} = 0$$

and

$$J_2 = \frac{1}{(2\pi i)^2} \int\limits_{|x|=\rho_3} \int\limits_{|y|=\rho_2} \frac{(1-x)^{-n-2}(1-2x)dx \wedge dy}{1-x(1+y))^2 \left(y - \frac{x}{1-x}\right) x^{n+1}(1+y)}$$

$$[y = x(1-x)^{-1} \text{ a simple pole}]$$

$$= \frac{1}{2\pi i} \int\limits_{|x|=\rho_3} \frac{(1-x)^{-n-2}(1-2x)dx}{\left(1+\frac{x}{1-x}\right) x^{n+1}(1-x(1+x(1-x)^{-1}))^2}$$

$$= \frac{1}{2\pi i} \int\limits_{|x|=\rho_3} \frac{(1-x)^{-n+1}}{(1-2x)x^{n+1}} dx$$

$$[\text{substituting } w = x(1-x) \text{ we have } 0 \to 0, \text{ hence } x = (1+(1-4w)^{1/2})/2]$$

$$= \frac{1}{2\pi i} \int\limits_{|w|=\rho_3} \frac{1+2(1-4w)^{1/2}}{4(1-4w)w^{n+1}} dw$$

$$= \frac{1}{4}\frac{1}{2\pi i} \int\limits_{|w|=0,2} (1-4w)^{-1} w^{-n-1} dw + \frac{1}{2}\frac{1}{2\pi i} \int\limits_{|w|=0,2} (1-4w)^{-1/2} w^{-n-1} dw$$

$$= 4^{n-1} + \frac{1}{2}\binom{2n}{n} = 4^{n-1} + \binom{2n-1}{n}.$$

Thus, $S = J_1 + J_2 = 4^{n-1} + \binom{2n-1}{n}$. Lemma 7 is proved.

References

[1] Levchuk, V. M., Automorphisms of unipotent subgroups of Lie type groups of small ranks. Algebra Logika 29:2 (1990), 141–161; 29:3 (1990), 315–338. English translation: Algebra Logic 29:2 (1990), 97–112; 29:3 (1990), 211–223.

[2] Egorychev, G. P., Integral Representation and the Computation of Combinatoric Sums (Math. Monogr. 59). 2nd ed. Am. Math. Soc., Providence, RI 1984.

[3] Tolasov, B. A., The number of normal subgroups of triangular group that are contained in the uni-triangular subgroup. In: Algebra and Number Theory 2, 122–126. Kabardino–Balkarsk. Gos. Univ., Nalchik 1977.

[4] Levchuk, V. M., Subgroups of the unitriangular group. Izv. Akad. Nauk SSSR , Ser. Mat. 38:6 (1974), 1202–1220.

[5] Levchuk, V. M., Chevalley groups and their unipotent subgroups. In: Proc. Int. Conf. on Algebra (Novosibirsk, 1989) (Contemp. Math. 131), pt 1, 227–241. Am. Math. Soc., Providence, RI 1992.

[6] Carter, R. W., Simple groups of Lie type (Pure Appl. Math. Ser. texts and monogr. 28). J. Wiley & Sons, London–New York–Sydney 1972.

[7] Hurley, J. F., Ideals in Chevalley algebras. Trans. Am. Math. Soc. 137 (1969), 245–258.

[8] Egorychev, G. P., Levchuk, V. M., Enumerate problems for the groups and algebras of Lie types. Dokl. Ross. Akad. Nauk 330:3 (1993), 291–293. English translation: Sov. Math., Dokl. (1993).

[9] Levchuk, V. M., Parabolic subgroups of some ABA-groups. Mat. Zametki 31:4 (1982), 509–525. English translation: Math. Notes 31:3–4 (1982), 259–267.

[10] Bourbaki, N., Groupes et algébres de Lie. Ch. IV–VI. Paris 1968.

[11] Feller, W., An Introduction to Probability Theory and its Applications. Mir, Moscow 1967. Original: V.1: Science Press, Peking 1964, 2nd ed.; V.2: J. Wiley & Sons, New York–London–Sydney 1966, 2nd ed.

Projectivity of absolute Galois groups of RC^*_ζ-fields

Yuriĭ L. Ershov

Abstract. We introduce and study a notion of a projective Δ-group which is close to a notion of projectivity with respect to a separable family of subgroups introduced by D. Haran earlier. We establish projectivity of the absolute Galois Δ-group $\mathbb{G}(F, W)$ for RC^*_ζ-fields (i.e. RC^*-fields such that the mapping $\zeta : W \to \operatorname{Subg}^* G(F)$ is continuous). We establish sufficient conditions for continuity of ζ.

1991 Mathematics Subject Classification: 12F10, 12J05, 13F05, 03C20, 08B25, 06F15.

In a significant paper [1] absolute Galois groups of PC_p-fields were characterized by means of a notion of projectivity. It turns out that an appropriate projectivity property for absolute Galois groups of RC^*-fields holds under very general assumptions (see [2]).

In the first section we introduce and study a notion of a projective Δ-group which is close to a notion of projectivity with respect to a separated family of subgroups introduced in [3].

In the second section we establish (Theorem 3) projectivity of an absolute Galois Δ-group $\mathbb{G}(F, W)$ for RC^*_ζ-fields (i.e. RC^*-fields such that the mapping $\zeta : W \to \operatorname{Subg}^* G(F)$ is continuous).

In the third section we establish sufficient conditions for continuity of ζ.

1.

Let G be a profinite group and $\operatorname{Subg} G$ be the set of all closed subgroups of G. A topology of a Boolean space on $\operatorname{Subg} G$ is defined in a natural way (see [1]). The group G acts (continuously) on the space $\operatorname{Subg} G$ by conjugation; denote by $\operatorname{Subg}^* G$ the corresponding factor space $\operatorname{Subg} G/G$. The corresponding projection $\pi_G : \operatorname{Subg} G \to \operatorname{Subg}^* G$ is an open continuous mapping.

A pair $\mathbb{G} = \langle G, \mathcal{D}\rangle$, where G is a profinite group and $\mathcal{D} \subset \operatorname{Subg} G$ is a closed subset such that $\Gamma^g \in \mathcal{D}$ for all $\Gamma \in \mathcal{D}$ and $g \in G$, and $\operatorname{Subg}\Gamma \subset \mathcal{D}$ for any $\Gamma \in \mathcal{D}$, is called a *Δ-group*. If $\mathbb{G}_0 = \langle G_0, \mathcal{D}_0\rangle$ and $\mathbb{G}_1 = \langle G_1, \mathcal{D}_1\rangle$ are Δ-groups, then a *homomorphism* φ from $\mathbb{G}_0$ to $\mathbb{G}_1$ is any (continuous) homomorphism $\varphi : G_0 \to G_1$ of profinite groups such that $\varphi(\Gamma) \in \mathcal{D}_1$ for every $\Gamma \in \mathcal{D}_0$. A homomorphism $\varphi : \mathbb{G}_0 \to \mathbb{G}_1$ is an *epimorphism* if $\varphi : G_0 \to G_1$ is an epimorphism and $\mathcal{D}_1 = \{\varphi(\Gamma) \mid \Gamma \in \mathcal{D}_0\}$. An epimorphism $\varphi : \mathbb{G}_0 \to \mathbb{G}_1$ is called *flat*, if $\Gamma \cap \operatorname{Ker}\varphi = \{e\}$ for all $\Gamma \in \mathcal{D}_0$.

A Δ-group $\mathbb{G}$ is called *projective* if for any pair of epimorphisms $\varphi_0 : \mathbb{G} \to \mathbb{G}_0$, $\psi : \mathbb{G}_1 \to \mathbb{G}_0$ such that $\mathbb{G}_1$ is finite and ψ is flat there exists a homomorphism $\varphi_1 : \mathbb{G} \to \mathbb{G}_1$ such that $\psi\varphi_1 = \varphi_0$.

Establish some properties of projective Δ-groups. Let $\mathbb{G} = \langle G, \mathcal{D}\rangle$ be a projective Δ-group.

Proposition 1. *If $\Gamma_0, \Gamma_1 \in \mathcal{D}$ and $\Gamma_0 \cap \Gamma_1 \neq \{e\}$, then there exists $\Gamma \in \mathcal{D}$ such that $\Gamma_0, \Gamma_1 \leqslant \Gamma$.*

Assume the converse and let $\Gamma_0, \Gamma_1 \in \mathcal{D}$ be such that $\Gamma_0 \cap \Gamma_1 \neq \{e\}$ and there is no $\Gamma \in \mathcal{D}$ such that $\Gamma_0, \Gamma_1 \leqslant \Gamma$. Prove that there exists an open normal subgroup $N \trianglelefteq G$ such that in the (finite) Δ-group

$$\mathbb{G}/N \leftrightharpoons \langle G/N, \mathcal{D}/N\rangle,$$

$$\mathcal{D}/N \leftrightharpoons \{\Gamma N/N \mid \Gamma \in \mathcal{D}\},$$

$$\Gamma_0 \cap \Gamma_1 \nleqslant N,$$

$$\Gamma_0' \leftrightharpoons \Gamma_0 N/N, \quad \Gamma_1' \leftrightharpoons \Gamma_1 N/N$$

are such that there is no $\Gamma' \in \mathcal{D}/N$ such that $\Gamma_0', \Gamma_1' \leqslant \Gamma'$. Indeed, for any open normal subgroup $N \trianglelefteq G$ the sets

$$F_{N,0} \leftrightharpoons \{\Gamma \mid \Gamma \in \mathcal{D}, \Gamma_0 N \leqslant \Gamma N\}, \quad F_{N,1} \leftrightharpoons \{\Gamma \mid \Gamma \in \mathcal{D}, \Gamma_1 N \leqslant \Gamma N\}$$

are closed and if for every open normal subgroup $N \trianglelefteq G$ $F_{N,0} \cap F_{N,1} \neq \varnothing$, then

$$\bigcap_N (F_{N,0} \cap F_{N,1}) = \Big(\bigcap_N F_{N,0}\Big) \cap \Big(\bigcap_N F_{N,1}\Big) \neq \varnothing$$

also. But it is easy to verify that

$$\bigcap_N F_{N,0} = \{\Gamma \mid \Gamma \in \mathcal{D}, \Gamma_0 \leqslant \Gamma\}, \quad \bigcap_N F_{N,1} = \{\Gamma \mid \Gamma \in \mathcal{D}, \Gamma_1 \leqslant \Gamma\},$$

and for

$$\Gamma \in \bigcap_N (F_{N,0} \cap F_{N,1})$$

we have $\Gamma_0, \Gamma_1 \leqslant \Gamma$ and $\Gamma \in \mathcal{D}$, which is impossible. Thus, there exists an open normal subgroup $N \leqslant G$ such that $F_{N,0} \cap F_{N,1} = \varnothing$. If $N_0 \leqslant N$, then $F_{N_0,0} \cap$

$F_{N_0,1} = \varnothing$; hence, we may assume that $\Gamma_0 \cap \Gamma_1 \nleqslant N$. Let $M_0, \dots, M_k$ be all maximal with respect to inclusion subgroups of $\mathcal{D}/N$ and

$$G_1 \leftrightharpoons \Big(\prod_{i\leqslant k}^{*} M_i\Big) * F_\omega$$

be the free profinite product of $M_i, i \leqslant k$, and the free profinite group of countable rank F_ω. If

$$\mathcal{D}_1 \leftrightharpoons \{M^{g_1} \mid M \leqslant M_i, g_1 \in G, i \leqslant k\}$$

then $\mathcal{D}_1$ is a closed subset of $\operatorname{Subg} G_1$ and $\mathbb{G}_1 \leftrightharpoons \langle G_1, \mathcal{D}_1\rangle$ is a Δ-group. Let $\mu_i \leftrightharpoons \mathrm{id}_{M_i} : M_i \to M_i$, $i \leqslant k$, and $\mu_\omega : F_\omega \to G/N$ be an arbitrary epimorphism. Then for the tuple $\langle \mu_0, \dots, \mu_k, \mu_\omega\rangle$ there exists a uniquely determined epimorphism $\mu : G_1 \to G/N$ which is easily seen to be a flat epimorphism of a Δ-group $\mathbb{G}_1$ onto a Δ-group $\mathbb{G}/N$. Using separability of $\mathbb{G}_1$ and projectivity of $\mathbb{G}$ one can build a homomorphism $\psi : \mathbb{G} \to \mathbb{G}_1$ of Δ-groups such that $\mu\psi : \mathbb{G} \to \mathbb{G}/N$ is a natural projection. Since $\psi(\Gamma_0'), \psi(\Gamma_1') \in \mathcal{D}_1$, there exist $i, j \leqslant k$ and $g_0, g_1 \in G_1$ such that

$$\psi(\Gamma_0) \leqslant M_i^{g_0}, \quad \psi(\Gamma_1) \leqslant M_j^{g_1};$$

then

$$\mu\psi(\Gamma_0) = \Gamma_0' \leqslant M_i^{\mu(g_0)}, \quad \mu\psi(\Gamma_1) = \Gamma_1' \leqslant M_j^{\mu(g_1)}.$$

We have $\mu\psi(\gamma) \neq e$ for $\gamma \in (\Gamma_0 \cap \Gamma_1) \setminus N$; hence, $\psi(\gamma) \neq e$ and $\psi(\gamma) \in M_i^{g_0} \cap M_j^{g_1}$. Since $M_i^{g_0} \cap M_j^{g_1} \neq \{e\}$ we have $i = j$ and $g_1 g_0^{-1} \in M_i$ (see [4]). Then

$$\mu(g_1 g_0^{-1}) = \mu(g_1)\mu(g_0)^{-1} \in M_i,$$

$$M_i^{\mu(g_0)} = M_i^{\mu(g_1)} = M_j^{\mu(g_1)},$$

$$\Gamma_0', \Gamma_1' \leqslant M_i^{\mu(g_0)} \in \mathcal{D}/N,$$

which is impossible. □

Since $\mathcal{D}$ is closed, the inclusion relation is inductive. Hence, every subgroup $\Gamma \in \mathcal{D}$ is contained in a maximal $\Gamma^* \in \mathcal{D}$. The proposition shows that for $\Gamma \neq \{e\}$ this Γ^* is unique. Denote by $\mathcal{D}^*$ the family of all maximal subgroups in $\mathcal{D}$. Then Proposition 1 implies that if $\Gamma_0 \neq \Gamma_1 \in \mathcal{D}^*$, then $\Gamma_0 \cap \Gamma_1 = \{e\}$.

Similarly to Proposition 1 one can prove (for projective Δ-groups $\mathbb{G}$) the following

Proposition 2. *If $\Gamma \in \mathcal{D}^*$, $g \in G$, and $\Gamma^g = \Gamma$, then $g \in \Gamma$.* □

A notion of projectivity of Δ-groups is closely related to a notion of projectivity with respect to a separated family of subgroups which was introduced and studied by D. Haran in [1].

Proposition 3. (a) *If a group G is projective with respect to a separated family of subgroups $\mathcal{D}'$, then*

$$\mathcal{D} \leftrightharpoons \{\Gamma^g \mid \text{there exists } \Gamma_0 \in \mathcal{D}', \Gamma \leqslant \Gamma_0, g \in G\}$$

is a closed family of subgroups, $\mathbb{G} = \langle G, \mathcal{D}\rangle$ is a projective Δ-group, and

$$\mathcal{D}^* = \{\Gamma^g \mid \Gamma \in \mathcal{D}', \Gamma \neq \{e\}, g \in \Gamma\}.$$

(b) *If $\mathbb{G} = \langle G, \mathcal{D}\rangle$ is a projective Δ-group and $\mathcal{D}^*$ is a separated family, then G is projective with respect to $\mathcal{D}^*$.*

Statement (a) follows easily from the following definitions.

Let ($\varphi : G \to A$, $\alpha : B \to A$, $\mathrm{Con}(B)$) be a finite $\mathcal{D}^*$-embedding problem in the sense of Haran (see [3, p. 277]). Without loss of generality one may assume that φ is an epimorphism (if it is not, let $A_0 \leftrightharpoons \varphi(G)(\leqslant A)$, $B_0 \leftrightharpoons \alpha^{-1}(A_0)$, $\alpha_0 \leftrightharpoons \alpha \restriction B_0$, $\mathrm{Con}(B_0) \leftrightharpoons \{\Gamma \mid \Gamma \in \mathrm{Con}(B), \Gamma \leqslant B_0\}$; then ($\varphi : G \to A_0$, $\alpha_0 : B_0 \to A_0$, $\mathrm{Con}(B_0)$) is also a $\mathcal{D}^*$-embedding problem).

Let $\mathbb{B} \leftrightharpoons \langle B, \mathrm{Con}(B)\rangle$, $\mathbb{A} \leftrightharpoons \langle A, \alpha(\mathrm{Con}(B))\rangle$; then α is an epimorphism of Δ-groups and φ is a homomorphism of a Δ-group $\mathbb{G}$ into $\mathbb{A}$ (which is an epimorphism of groups). By assumptions for every $\Gamma \in \mathcal{D}^*$ (hence, for every $\Gamma \in \mathcal{D}$) there exists a homomorphism $\psi_\Gamma : \Gamma \to B$ such that $\alpha\psi_\Gamma = \varphi \restriction \Gamma$ and $\psi_\Gamma(\Gamma) \in \mathrm{Con}(B)$. Let N_Γ be an open normal subgroup of G such that $N_\Gamma \leqslant \mathrm{Ker}\,\varphi$ and $N_\Gamma \cap \Gamma \leqslant \mathrm{Ker}\,\psi_\Gamma$. Put $V_\Gamma \leftrightharpoons \{\Gamma' \mid \Gamma' \in \mathrm{Subg}\,G, N_\Gamma\Gamma' = N_\Gamma\Gamma\}$, then $\Gamma \in V_\Gamma$ and V_Γ is open. Since $\mathcal{D}$ is closed and

$$\mathcal{D} \subseteq \bigcup_{\Gamma \in \mathcal{D}} V_\Gamma,$$

there exist $\Gamma_0, \dots, \Gamma_n \in \mathcal{D}$ such that

$$\mathcal{D} \subseteq \bigcup_{i \leqslant n} V_{\Gamma_i}.$$

Let

$$N \leftrightharpoons \bigcap_{i \leqslant n} N_{\Gamma_i}, \quad \mathbb{G}_0 \leftrightharpoons \mathbb{G}/N, \quad G_1 \leftrightharpoons B \times_A G_0,$$

$$\mathcal{D}_1' \leftrightharpoons \{\{\langle \overline{\psi}_{\Gamma_i}(\overline{\gamma}), \overline{\gamma}\rangle \mid \overline{\gamma} \in \Gamma' N\} \mid \Gamma' \in V_{\Gamma_i}, i \leqslant n\} \subseteq \mathrm{Subg}\, G_1,$$

where $\overline{\psi}_{\Gamma_i}$ is induced by the homomorphism ψ_{Γ_i}. Put

$$\mathcal{D}_1 \leftrightharpoons \bigcup_{g_1 \in G_1} (\mathcal{D}_1')^{g_1};$$

then $\mathbb{G}_1 \leftrightharpoons \langle G_1, \mathcal{D}_1\rangle$ is a Δ-group and its projection π_1 onto $\mathbb{G}_0$ is easily seen to be a flat epimorphism $\pi_1 : \mathbb{G}_1 \to \mathbb{G}_0$. Hence there exists a homomorphism $\psi : \mathbb{G} \to \mathbb{G}_1$ such that $\pi_1\psi$ is a natural projection of $\mathbb{G}$ onto $\mathbb{G}_0 = \mathbb{G}/N$. If π_0 is a projection of $\mathbb{G}_1$ onto $\mathbb{B}$, then $\pi_0\psi : \mathbb{G} \to \mathbb{B}$ is a homomorphism such that $\alpha\pi_0\psi = \varphi$, which was required for the solution of this $\mathcal{D}^*$-embedding problem. □

Remark 1. If for every projective Δ-group $\mathbb{G} = \langle G, \mathcal{D} \rangle$ the set $\mathcal{D}^* \cup \{\{e\}\}$, where $\mathcal{D}^*$ is the set of maximal subgroups, is separated, then the notions of a projective Δ-group and of a group projective with respect to a separated family are essentially the same. On the other hand, if there exist (unknown to the author) examples of projective Δ-groups $\langle G, \mathcal{D} \rangle$ such that $\mathcal{D}^* \cup \{\{e\}\}$ is not separated, then the latter notion is more general. But in either case The Subgroup Theorem is valid for projective Δ-groups.

Theorem 1. *If $\mathbb{G} = \langle G, \mathcal{D} \rangle$ is a projective Δ-group, $H \leqslant G$, and $\mathbb{H} \leftrightharpoons \langle H, \mathcal{D} \cap \operatorname{Subg} H \rangle$, then $\mathbb{H}$ is also a projective Δ-group.*

The proof of Theorem 3.1 from [3] is valid in this case also. □

2.

Let W be a Boolean family of valuation rings of a field F and F_0 be a Galois extension of F. We relate this extension to the Δ-group $\mathbb{G}(F_0/F, W) \leftrightharpoons \langle G(F_0/F), \mathcal{D}_{F_0}(W) \rangle$, where

$$\mathcal{D}_{F_0}(W) \leftrightharpoons \{\Gamma \mid \Gamma \leqslant G(F_0/F), \text{ there exists } R \in W \text{ such that } \Gamma \text{ is a subgroup of a decomposition group of the valuation ring } R \text{ in } F_0\}.$$

Obviously, $\mathcal{D}_{F_0}(W)$ is closed with respect to subgroups and conjugates. If F_0 is a finite extension, then obviously $\mathcal{D}_{F_0}(W)$ is closed in $\operatorname{Subg} G(F_0/F)$. In the case of an infinite extension F_0 the closedness of $\mathcal{D}_{F_0}(W)$ follows from the representation of $\mathcal{D}_{F_0}(W)$ as the limit of the inverse spectrum $\mathcal{D}_{F_0}(W) = \varprojlim \mathcal{D}_{F_1}(W)$ along finite Galois extensions $F_1 \leqslant F_0$. There exists a naturally defined mapping $\zeta_{F_0} : W \to \operatorname{Subg}^* G(F_0/F)$ such that

$$\pi^{-1}_{G(F_0/F)}(\zeta_{F_0}(R)) = \{\Gamma \mid \Gamma \leqslant G(F_0/F), \Gamma \text{ is a decomposition group of the valuation ring } R \text{ in } F_0\} \subseteq \mathcal{D}_{F_0}(W).$$

If $F_0 = \tilde{F}^s$ is the separable closure of F, then we will write $\mathbb{G}(F_1, W)$ and ζ instead of $\mathbb{G}(F_0/F, W)$ and ζ_{F_0} respectively.

Though ζ is a naturally defined mapping, in general it is not continuous, as the example in author's paper [5] shows. Nevertheless, the case of continuity of ζ is quite common, and in such a case the continuity of ζ can be used for the proof of projectivity of the Δ-group $\mathbb{G}(F, W)$ (see Theorem 3).

Theorem 2. *The class of all pairs $\langle F, R \rangle$ satisfying the conditions*

1) *R is a regular Preufer ring with the quotient field F (and W_R is the corresponding Boolean family of valuation rings);*

2) *W_R is independent;*

3) *F is dense in* $\mathbb{H}_{R'}(F)$ *for all* $R' \in W_R$;

4) *the mapping* $\zeta : W_R \to \operatorname{Subg}^* G(F)$ *is continuous,*

is axiomatizable in the language $\sigma_f \cup \langle R \rangle$.

Property 1 is elementary; independency of W_R is equivalent to the property BAP for $\langle F, R \rangle$ (see [2]); elementarity of the denseness property is established similarly to the proof of Theorem 3 in [6] (see Remark after that theorem).

Let $\langle F, R \rangle$ satisfy properties 1–3. If property 4 also holds, then for every finite Galois extension F_0 of F and any $\Gamma \leqslant G(F_0/F)$ the set

$$V_\Gamma \leftrightharpoons \{R' \mid R' \in W_R,\ \Gamma \text{ is a decomposition group of } R' \text{ in } F_0\}$$

is open in W_R. (It is easily seen then that V_Γ is also closed.)

Let $R' \in V_\Gamma$, then $F_0^\Gamma = F_0 \cap H_{R'}(F)$; for an appropriate integral over R element α we have $F_0^\Gamma = F(\alpha)$; let $f_\alpha(x) = x^n + a_1 x^{n-1} + \cdots + a_n$ be a minimal polynomial for α over F. Then $a_1, \ldots, a_n \in R \leqslant R'$. Since F is dense in $\mathbb{H}_{R'}(F)$, there exists $a \in R'$ such that

$$v_{R'^h}(a - \alpha) > v_{R'^h}(f'_\alpha(\alpha)^2) = 2v_{R'^h}(f'_\alpha(\alpha));$$

then

$$v_{R'}(f_\alpha(a)) = v_{R'^h}(f_\alpha(a) - f_\alpha(\alpha)) \geqslant v_{R'^h}(a - \alpha) > 2v_{R'^h}(f'_\alpha(\alpha));$$
$$v_{R'^h}(f'_\alpha(a)^2 - f'_\alpha(\alpha)^2) \geqslant v_{R'^h}(a - \alpha) > v_{R'^h}(f'_\alpha(\alpha)^2).$$

This implies $v_{R'}(f'_\alpha(a)^2) = v_{R'^h}(f'_\alpha(\alpha)^2) < v_{R'}(f_\alpha(a))$ and $f_\alpha(a) \cdot f'_\alpha(a)^{-2} \in m(R')$. Put

$$V^*_\alpha \leftrightharpoons (V_{f_\alpha(a) f'_\alpha(a)^{-2}} \setminus V_{f_\alpha(a)^{-1} f'_\alpha(a)^2}) \cap V_\Gamma.$$

The set V^*_α is closed open, $R' \in V^*_\alpha$ and for any $R'' \in V^*_\alpha$ we have $v_{R''}(f_\alpha(a)) > 2v_{R''}(f'_\alpha(a))$. Since W_R is Boolean, V^*_α is $V_{a_0^{-1}}$ for an appropriate $a_0 \in R$.

Let $W' \leftrightharpoons \{R' \mid R' \text{ is a valuation ring of } F_0^\Gamma,\ R' \cap F \in W_R\}$; then W' is a Boolean family. Consider a closed open subset

$$V' \leftrightharpoons V^{F_0^\Gamma}_{a_0^{-1}} \cap V_{(a-\alpha) f_\alpha(a)^{-1} f'_\alpha(a)} \cap V_{(a-\alpha)^{-1} f_\alpha(a) f'_\alpha(a)^{-1}}.$$

For every $R' \in V'$ we have $(R' \cap F) \leqslant R' \leqslant (R' \cap F)^h$ and Γ is a decomposition group of R' in F_0. Since V' is closed open, there exists $a_1 \in R_{W'}$ (= integral closure of R in F_0^Γ) such that $V' = V^{F_0^\Gamma}_{a_1^{-1}}$. Put

$$R_{a_1} \leftrightharpoons \cap\{R' \mid R' \in W', a_1^{-1} \in R'\} = \cap\{R' \mid R' \in V'\}.$$

If $R^*_{a_1}$ is an integral closure of R_{a_1} in F_0, then the rings $R_{a_1}/J(R_{a_1}) \leqslant R^*_{a_1}/J(R^*_{a_1})$ have the same idempotents (this is equivalent to the fact that every $R' \in V'$ has unique extension to a valuation ring of F_0). In view of compactness of W_R we can state that if condition 4 is satisfied, the following property (*) is also satisfied:

Property (*). *For every finite Galois extension F_0 of F there exist subgroups $\Gamma_0, \ldots, \Gamma_k \leqslant G(F_0/F)$, elements $a_0, \ldots, a_k, b_0, \ldots, b_k \in R$, and $\beta_0, \ldots, \beta_k, \gamma_0, \ldots, \gamma_k \in C_{F_0}(F)$ (= integral closure of R in F_0) such that the following five conditions hold:*

1) $[a_0, \ldots, a_k]$ *is a partition of* W_R;

2) $F(\beta_i) = F_0^{\Gamma_i}$, $\gamma_i \in F(\beta_i)$ *for* $i \leqslant k$;

   [let

   $W_i \leftrightharpoons \{R' \mid R'$ is a valuation ring of the field $F(\beta_i)$ such that $R' \cap F \in W_R\}$

   and $f_i(x) \in R[x]$ be the minimal polynomial of β_i over F, $i \leqslant k$]

3) $v_{R'}(f_i(b_i)) > 2v_{R'}(f_i'(b_i))$ *for all* $R' \in V^F_{a_i^{-1}}$, $i \leqslant k$;

4) $V^{F(\beta_i)}_{\gamma_i^{-1}} = \{R' \mid R' \in W_i, R' \cap F \in V^F_{a_i^{-1}}, v_{R'}(b_i - \beta_i) = v_{R'}(f_i(b_i)f_i'(b_i)^{-1})\}$, $i \leqslant k$.

   [let $R_i \leftrightharpoons \cap\{R' \mid R' \in V^{F(\beta_i)}_{\gamma_i^{-1}}\}$, $i \leqslant k$]

5) *if* $R_i^* = C_{F_0}(R_i)$ *is the integral closure of* R_i *in* F_0, *then the rings* $R_i/J(R_i) \leqslant R_i^*/J(R_i^*)$, $i \leqslant k$, *have the same idempotents.*

It is easily seen that property (*) can be expressed by a system of propositions in the language $\sigma_f \cup \langle R \rangle$. It is even easier to verify that if $\langle F, R \rangle$ satisfies conditions 1–3 of the theorem and the property (*), then condition 4 of the theorem also holds. □

Remark 2. Employing the technique used in the proof of Proposition 1 in [7] (or [8, p. 20, 21]), one can prove the following propositions.

1. *If $\mathbb{G} = \langle G, \mathcal{D} \rangle$ is a finite Δ-group, then one can build (effectively in $\mathbb{G}$) a proposition $\varphi_{\mathbb{G}}$ in the language $\sigma_f \cup \langle R \rangle$ such that for every pair $\langle F, R \rangle$ satisfying conditions 1–4 of the theorem the following assertions are equivalent:*

   (a) $\langle F, R \rangle \models \varphi_{\mathbb{G}}$;

   (b) *there exists a finite Galois extension F_0 of F such that $\mathbb{G}(F_0/F, W_R) \simeq \mathbb{G}$.*

2. *If $\alpha : \mathbb{G}_0 \to \mathbb{G}$ is an epimorphism of finite Δ-groups, $k = |G|$, then there exists a formula $\varphi_\alpha(x_1, \ldots, x_k)$ in the language $\sigma_f \cup \langle R \rangle$ such that for any pair $\langle F, R \rangle$ satisfying conditions 1–4 of the theorem and any $a_1, \ldots, a_k \in F$ the following conditions are equivalent:*

   (a) $\langle F, R \rangle \models \varphi_\alpha(a_1, \ldots, a_k)$;

(b) *the polynomial* $f(x) = x^k + a_1x^{k-1} + \cdots + a_k \in F[x]$ *is irreducible over* F, $F(\alpha_f)$ *is a Galois extension of* F *(here* α_f *is a root of* f*), there exist an isomorphism* $\varphi : \mathbb{G}(F(\alpha_f)/F, W_R) \xrightarrow{\sim} \mathbb{G}$, *a Galois extension* $F_0 \geqslant F$, *and an isomorphism* $\varphi_0 : \mathbb{G}(F_0/F, W_R) \xrightarrow{\sim} \mathbb{G}_0$ *such that* $F(\alpha_f) \leqslant F_0$ *and* $\alpha\varphi_0 = \varphi\rho$, *where* $\rho : G(F_0/F) \to G(F(\alpha_f)/F)$ *is the restriction epimorphism.*

A pair $\langle F, R\rangle$ is called a *RC^*_ζ-field,* if $\langle F, R\rangle$ is a RC^*-field (i.e. satisfies the conditions BAP and THR) and satisfies condition 4 of Theorem 1, i.e. the mapping $\zeta : W_R \to \operatorname{Subg}^* G(F)$ is continuous.

Corollary. *The class of all RC^*_ζ-fields is axiomatizable in the language* $\sigma_f \cup \langle R\rangle$.

Follows from Theorem 2 and the fact that the class of RC^*-fields is axiomatizable. □

Theorem 3. *If* $\langle F, R\rangle$ *is a RC^*_ζ-field, then the Δ-group* $\mathbb{G}(F, W_R)$ *is projective.*

Let $\alpha : \mathbb{H} \to \mathbb{G}$ be a flat epimorphism of finite Δ-groups and $\varphi : \mathbb{G}(F, W_R) \to \mathbb{G}$ be an epimorphism of Δ-groups. Let $F_0 \leftrightharpoons \tilde{F}^{s\ \operatorname{Ker}\varphi}$, then F_0 is a finite Galois extension and the epimorphism φ induces an isomorphism $\overline{\varphi} : \mathbb{G}(F_0/F, W_R) \xrightarrow{\sim} \mathbb{G}$. By the property (*) from the proof of Theorem 2 choose $\Gamma_0, \ldots, \Gamma_k \leqslant G(F_0/F)$, $a_0, \ldots, a_k, b_0, \ldots, b_k \in R$, and $\beta_0, \ldots, \beta_k, \gamma_0, \ldots, \gamma_k \in C_{F_0}(R)$ which satisfy conditions 1–5.

Since $\alpha : \mathbb{H} \to \mathbb{G} = \langle G, \mathcal{D}\rangle$ is a flat epimorphism and $\overline{\varphi}(\Gamma_0), \ldots, \overline{\varphi}(\Gamma_k) \in \mathcal{D}$, there exist $H_0, \ldots, H_k \in \mathcal{D}(\mathbb{H})$ such that $H_i \cap \operatorname{Ker}\alpha = \{e\}$ and $\alpha(H_i) = \overline{\varphi}(\Gamma_i)$ for $i \leqslant k$.

Considering elements of the group H as transcendental over F_0, define the field $F_1 \leftrightharpoons F_0(H)$ and define an action of H on F_1 by $a^{h_0} \leftrightharpoons \overline{\varphi}^{-1}\alpha(h_0)(a)$, $h^{h_0} \leftrightharpoons hh_0$ for $h_0, h \in H$, $a \in F_0$. Then F_1 is a Galois extension of the field $F_2 \leftrightharpoons F_1^H$ with Galois group $G(F_1/F_0)$ naturally isomorphic to H; note that F_1 is a regular extension of F.

Fix $i_0 \leqslant k$ and consider the fields $F^{i_0} \leftrightharpoons F_0^{\Gamma_{i_0}}$, $F_1^{i_0} \leftrightharpoons F_1^{H_{i_0}}$, $F^{i_0} \leqslant F_1^{i_0}$. Similarly to [1], prove

Lemma 1. *The field* $F_1^{i_0}$ *is purely transcendental extension of* F^{i_0}.

Let $\alpha \in F_0$ be such that $F_0 = F(\alpha)$ and let $H = h_0H_{i_0} \cup h_1H_{i_0} \cup \cdots \cup h_sH_{i_0}$, $h_0 = e$, be an expansion of H into left residue classes with respect to the subgroup H_{i_0}. Put

$$u_{ij} \leftrightharpoons \sum_{h\in H_{i_0}} (\alpha^i)^h h_j h, \quad i < |H_{i_0}|, j \leqslant s.$$

Since the determinant of the Vandermonde matrix

$$\left[(\alpha^i)^h\right]_{i<|H_{i_0}|, h\in H_{i_0}}$$

is different from zero, every element h in $H \subseteq F_1 = F_0(H)$ is uniquely representable as a linear combination of elements u_{ij} over the field F_0. Moreover, since the matrix $((\alpha^i)^h)$ is block diagonal, $h \in h_j H_{i_0}$ is a linear combination of u_{ij}, $i < |H_{i_0}|$. It is easily seen that $u_{ij}^h = u_{ij}$ for any $h \in H_{i_0}$. Hence,

$$F^{i_0}(\{u_{ij} \mid i < |H_{i_0}|, j \leqslant s\}) \leqslant F_1^{i_0}.$$

Further,

$$F_0 F^{i_0}(\{u_{ij} \mid i < |H_{i_0}|, j \leqslant s\}) = F_0(\{u_{ij}\}) \geqslant F_0(H) = F_1;$$

hence $F_0 F^{i_0}(\{u_{ij}\}) = F_1$ and

$$[F_1 : F^{i_0}(\{u_{ij}\})] \leqslant [F_0 : F^{i_0}] = |H_{i_0}|;$$

but $[F_1 : F_1^{i_0}] = |H_{i_0}|$, it follows $F_1^{i_0} = F^{i_0}(\{u_{ij}\})$. To complete the proof of the lemma it is sufficient to notice that the set $\{u_{ij}\}$ is algebraically independent over F^{i_0}. The transcendental degree of F_1 over F_0 (and hence, over F^{i_0}) is $|H| = |H_{i_0}| \cdot (s+1)$. The field F_1 is algebraic over $F^{i_0}(\{u_{ij}\})$ and

$$|\{u_{ij} \mid i < |H_{i_0}|, j \leqslant s\}| = |H_{i_0}| \cdot (s+1) = |H|.$$

It implies that the set $\{u_{ij}\}$ is algebraically independent. $\square$

We will use the notation introduced in the proof of the lemma. On the set $\{u_{ij}\}$ we define a linear order which is induced by the lexicographical order on indices:

$$u_{ij} \leqslant u_{i'j'} \leftrightharpoons i < i' \text{ or } i = i' \text{ and } j \leqslant j'.$$

For this order there is a uniquely determined valuation ring R_{i_0} of F_1 such that $F_0 \leqslant R_{i_0}$, $F_0 = F_{R_{i_0}}$, and, if we denote by v_{i_0} the valuation determined by the ring R_{i_0}, then $0 < v_{i_0}(u_{ij}) \ll v_{i_0}(u_{i'j'})$ for $u_{ij} < u_{i'j'}$, where $\gamma \ll \gamma'$ in a linearly ordered group Γ means that $0 \leqslant \gamma \leqslant \gamma'$ and there exists a convex subgroup $\Gamma' < \Gamma$ such that $\gamma \in \Gamma'$ and $\gamma' \notin \Gamma'$. In the group $\Gamma_{i_0} \leftrightharpoons \Gamma_{R_{i_0}}$ $(= \Gamma_{v_{i_0}})$ there exists a convex subgroup Δ (generated by the set $\{v_{i_0}(u_{0j}) \mid j \leqslant s\}$) such that $v_{i_0}(u_{ij}) \notin \Delta$ for $0 < i < |H_{i_0}|$. Note that for any $h \in H \subseteq F_1 = F_0(H)$ the following equivalence holds:

$$v_{i_0}(h) \in \Delta \iff h \in H_{i_0}.$$

This follows from the fact that if $h \in h_j H_{i_0}$, then h has a representation $h = \sum_{i<|H_{i_0}|} \alpha_i u_{ij}$, $\alpha_i \in F_0$ and $v_{i_0}(h) = \min\{v_{i_0}(u_{ij}) \mid i < |H_{i_0}|\ \alpha_i \neq 0\}$. The equivalence implies that H_{i_0} is a decomposition group of the valuation ring $\overline{R}_{i_0} \leftrightharpoons R_{i_0} \cap F_2$ of the field F_2 in F_1. Indeed, if $h \in H_{i_0}$, then $R_{i_0}^h = R_{i_0}$; if $h \notin H_{i_0}$, then $v_{i_0}(e^h) = v_{i_0}(h) \notin \Delta$, but $v_{i_0}(e) \in \Delta$, hence $R_{i_0}^h \neq R_{i_0}$.

Let

$$W_{i_0} \leftrightharpoons \{\overline{R}_{i_0} \circ R' \mid R' \in V^F_{a_{i_0}^{-1}}\},$$

$$W_2^* \leftrightharpoons \bigcap_{i_0 \leqslant k} W_{i_0};$$

the mapping $\pi^* : R^* \mapsto R^* \cap F$, $R^* \in W_2^*$ is a one-one mapping of W_2^* onto W_R. It is easy to verify that it is continuous. Moreover, W_2^* is weakly Boolean, hence, π^* is a homomorphism and W_2^* is Boolean.

Notice that $\mathbb{G}(F_1/F_2, W_2^*)$ is naturally isomorphic to a Δ-subgroup of a Δ-group $\mathbb{H}$. This follows from $G(F_1/F_2) = H$ and the fact that for every $\overline{R}_{i_0} \circ R' \in W_{i_0}$ the subgroup H_{i_0} is a decomposition group of this valuation ring $\overline{R}_{i_0} \circ R'$. Verify the last statement. Let R'' be a valuation ring of the field F_0 such that $R'' \cap F = R'$ and $\Gamma_{i_0} = \{g \mid g \in G(F_0/F), (R'')^g = R''\}$. If $R_* \leftrightharpoons R_{i_0} \circ R''$, then it is easily verified that $R_*^h = R_*^h$ for $h \in H_{i_0}$, but $R_*^h \neq R_*$ for $h \in H \setminus H_{i_0}$, since $v_{R_*}(e) \ll v_{R_*}(e^h) = v_{R_*}(h)$.

Let

$$R_2 \leftrightharpoons \bigcap \{R' \mid R' \in W_2^*\}.$$

By Proposition 1 in [9], $\langle F, R\rangle \leqslant_1 \langle F_2, R_2\rangle$ (the condition $\mathbb{H}_{R' \cap F}(F) \leqslant_1 \mathbb{H}_{R'}(F_2)$ for $R' \in W_2^*$ is easily verified from the definition of W_2^*). Then there exists $\langle F_*, R_*\rangle$ such that $\langle F, R\rangle \preccurlyeq \langle F_*, R_*\rangle$ ($\langle F_*, R_*\rangle$ is an elementary extension of $\langle F, R\rangle$) and $\langle F_2, R_2\rangle \leqslant \langle F_*, R_*\rangle$. Now establish that $\mathbb{G}(F_1F_*/F_*, W_{R_*})$ is naturally isomorphic to a Δ-subgroup of a Δ-group $\mathbb{G}(F_1/F_2, W_2^*)$. There is a natural embedding $\rho : G(F_1F_*/F_*) \to G(F_1/F_2) = H$, namely the restriction homomorphism. Let $R^* \in W_{R_*}$ and $\Gamma^* \leqslant G(F_1F_*/F_*)$ be a decomposition group of R^* in F_1F_*. Since $R^* \geqslant R_* \geqslant R_2$, then $R^* \cap F_2 \geqslant R'$ for an appropriate $R' \in W_2^*$. Since $(R^* \cap F_2) \cap F = R^* \cap F \in W_R$ and $R' \cap F \in W_R$, it implies that $R^* \cap F = R' \cap F$, and since R' is $\overline{R}_{i_0} \circ (R' \cap F)$ for an appropriate i_0, it follows $R^* \cap F_2 = R'$. Since Γ_{i_0} is a decomposition group of $R^* \cap F$ (and of all $R'' \in V^F_{a_{i_0}^{-1}}$) in F_0, the elementarity of extension $\langle F_*, R_*\rangle \succcurlyeq \langle F, R\rangle$ implies that Γ_{i_0} is also a decomposition group of R^* in $F_0F_* \leqslant F_1F_*$. But a decomposition group of R^* in F_1F_* should be (can be chosen) a subgroup of the decomposition group H_{i_0} of the ring $R' = R^* \cap F_2$. Since this subgroup H_{i_0} is mapped onto Γ_{i_0}, it should coincide with H_{i_0}. Thus, we have an embedding

$$\mathbb{G}(F_1F_*/F_*, W_{R_*}) \leqslant \mathbb{G}(F_2/F_1, W_2^*) \leqslant \mathbb{H}$$

(under natural identifications). Moreover, the diagram

$$\begin{array}{ccc}
\mathbb{G}(F_1F_*/F_*, W_{R_*}) & \xrightarrow{\quad i \quad} & \mathbb{H} \\
& {\scriptstyle \rho'} \searrow \qquad \swarrow {\scriptstyle \overline{\varphi}\alpha} & \\
& \mathbb{G}(F_0/F, W_R) &
\end{array}$$

where ρ' is the restriction homomorphism, is commutative. Let $F_0 = F(\alpha)$ and $f_\alpha = x^n + c_1x^{n-1} + \cdots + c_n$ be the minimal polynomial of α over F. According to the remark after Theorem 1, there exists a formula $\varphi_{\rho'}(x_1, \ldots, x_n)$ in the language $\sigma_f \cup \langle R \rangle$ such that for any pair $\langle F', R' \rangle$ satisfying conditions 1–4 of the theorem, and any $c_1', \ldots, c_n' \in F'$ the following conditions are equivalent:

a) $\langle F', R' \rangle \vDash \varphi_{\rho'}(c_1', \ldots, c_n')$;

b) the polynomial $f = x^n + c_1'x^{n-1} + \cdots + c_n'$ is irreducible over F', $F'(\alpha_{f'})$ is a Galois extension of F' and there exist an isomorphism

$$\varphi' : \mathbb{G}(F'(\alpha_{f'})/F', W_{R'}) \xrightarrow{\sim} \mathbb{G}(F_0/F, W_R)$$

and a Galois extension $F_0' \geqslant F'$ such that $F_0' \geqslant F'(\alpha_{f'})$, and an isomorphism

$$\varphi_0' : \mathbb{G}(F_0'/F', W_{R'}) \xrightarrow{\sim} \mathbb{G}(F_1F_*/F_*, W_{R_*})$$

such that $\rho'\varphi_0' = \varphi'\overline{\rho}$, where

$$\overline{\rho} : \mathbb{G}(F_0'/F', W_{R'}) \to \mathbb{G}(f'(\alpha_{f'})/F', W_{R'})$$

is the restriction epimorphism.

For this formula we have $\langle F_*, R_* \rangle \vDash \varphi_{\rho'}(c_1, \ldots, c_n)$; then $\langle F, R \rangle \vDash \varphi_{\rho'}(c_1, \ldots, c_n)$ and hence there exist a Galois extension F_0' such that $F_0' \geqslant F_0 = F(\alpha_f)$ and an isomorphism

$$\varphi_0' : \mathbb{G}(F_0'/F, W_R) \xrightarrow{\sim} \mathbb{G}(F_1F_*/F_*, W_{R_*})$$

such that $\rho'\varphi_0' = \overline{\rho}$, where

$$\overline{\rho} : \mathbb{G}(F_0'/F, W_R) \to \mathbb{G}(F_0/F, W_R)$$

is the restriction epimorphism. The isomorphism

$$\varphi_0' : \mathbb{G}(F_0'/F, W_R) \xrightarrow{\sim} \mathbb{G}(F_1F_*/F_*, W_{R_*})$$

with the embedding

$$\iota : \mathbb{G}(F_1F_*/F_*, W_{R_*}) \to \mathbb{H}$$

induces the homomorphism $\varphi_0 : \mathbb{G}(F, W_R) \to \mathbb{H}$ such that the diagram

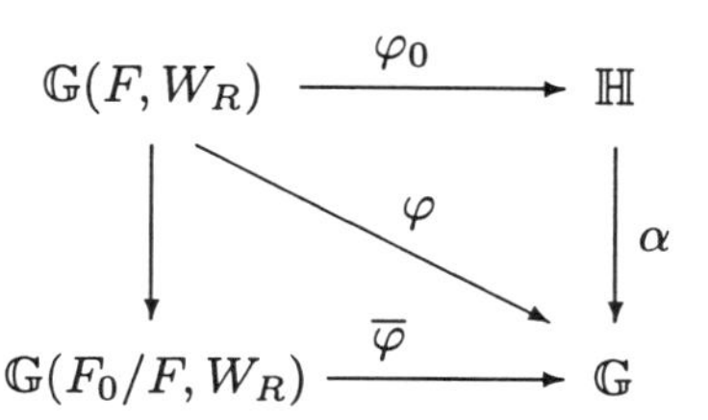

commutes. Hence, $\mathbb{G}(F, W_R)$ is a projective Δ-group. $\square$

3.

In this section we establish sufficient conditions for the continuity of the mapping ζ.

We start with general considerations. Let W be a Boolean family of valuation rings of a field F and $R_W = \cap\{R \mid R \in W\}$. For every $R \in W$ there exists an epimorphism $\lambda_R : R_W \to F_R$ having as the kernel that maximal ideal $\mathfrak{m} < R_W$ for which $R = (R_W)_\mathfrak{m}$. The system of epimorphisms λ_R, $R \in W$, defines a homomorphism $\lambda_W : R_W \to \prod\limits_{R \in W} F_R$. It is easily seen that the kernel of this homomorphism is the Jacobson radical $J(R_W)$ of the ring R_W, which is the intersection of all maximal ideals of R_W. If we denote $R_W/J(R_W)$ by $\overline{R}_W$, then λ_W induces an embedding $\overline{\lambda}_W : \overline{R}_W \to \prod\limits_{R \in W} F_R$ of the ring $\overline{R}_W$ as a subdirect product of the ring $\prod\limits_{R \in W} F_R$. Actually this embedding defines $\overline{R}_W$ as a Boolean product in the sence of the following definition (see [10]).

Let X be a Boolean space and $\mathfrak{A}_\xi$, $\xi \in X$, be a family of algebraic systems in some fixed language. Let $\lambda : \mathfrak{A} \to \prod\limits_{\xi \in X} \mathfrak{A}_\xi$ be an embedding of $\mathfrak{A}$ as a subdirect product. We say that λ *defines* $\mathfrak{A}$ *as a Boolean product* ($\exists$*-product, elementary product*), if the following conditions hold:

1. If $f, g \in \lambda(\mathfrak{A})$ and Y is a closed open subset of X, then $\langle f, g, Y\rangle \in \lambda(\mathfrak{A})$, where
$$\langle f, g, Y\rangle(\xi) \leftrightharpoons \begin{cases} f(\xi) & \text{for } \xi \in Y; \\ g(\xi) & \text{for } \xi \in X \setminus Y. \end{cases}$$

2. If $\varphi(x_0, \dots, x_n)$ is a quantifier free formula ($\exists$-formula, arbitrary formula) and $f_0, \dots, f_n \in \lambda(\mathfrak{A})$, then the set $\{\xi \mid \mathfrak{A}_\xi \vDash \varphi(f_0(\xi), \dots, f_n(\xi))\}$ if closed open in X.

Proposition 4. *The embedding* $\overline{\lambda}_W : \overline{R}_W \to \prod\limits_{R \in W} F_R$ *defines* $\overline{R}_W$ *as a Boolean product.*

Let $\overline{a}, \overline{b} \in \overline{R}_W$ and Y be a closed open subset of W, then there exists $c \in R_W$ such that $Y = V_{c^{-1}}$. Since $\overline{R}_W$ is regular, there exists $\overline{d} \in \overline{R}_W$ such that $\overline{c}^2\overline{d} = \overline{c}$, where $\overline{c}$ is the image of c in $\overline{R}_W$. The element $\overline{e} \leftrightharpoons \overline{c}\overline{d}$ is an idempotent; put $\overline{f} \leftrightharpoons \overline{a}\overline{e} + \overline{b}(1 - \overline{e})$. It is easy to verify that $\overline{\lambda}_W(\overline{f}) = \langle \overline{\lambda}_W(\overline{a}), \overline{\lambda}_W(\overline{b}), Y\rangle$. Condition 1 is verified.

To verify condition 2 it is sufficient to check that for formulas of the form $p(x) = 0$ and $p(x) \neq 0$, where $p(x) \in Z[x_0, \dots, x_n]$ and for $a_0, \dots, a_n \in R_W$ the sets

$$A \leftrightharpoons \{R \mid F_R \vDash p(\lambda_R(a_0), \dots, \lambda_R(a_n)) = 0\}$$

and

$$B \leftrightharpoons \{R \mid F_R \vDash p(\lambda_R(a_0), \dots, \lambda_R(a_n)) \neq 0\}$$

are open. If $p(a_1, \dots, a_n) = 0$, then $A = W$ is open. If $p(\overline{a}) \neq 0$ and $R \in A$, then $p(\overline{a}) = p(a_0, \dots, a_n) \in \mathfrak{m}(R)$ and $R \in V_{p(\overline{a})} \setminus V_{p(\overline{a})^{-1}} \subseteq A$; the set A is open. Let $R \in B$, then $R \in V_{p(\overline{a})^{-1}} \subseteq B$ and B is open. □

An example from [5] shows that $\overline{\lambda}_W$ not always defines $\overline{R}_W$ as $\exists$-product.

We say that W satisfies the property (*), if for any unitary polynomial $f(x) = x^n + a_1x^{n-1} + \dots + a_n \in R_W[x]$ the set

$$V_f \leftrightharpoons \{R \mid \text{ the polynomial } \lambda_R f \leftrightharpoons x^n + \lambda_R(a_1)x^{n-1} + \dots + \lambda_R(a_n) \in F_R[x] \text{ is irreducible over } F_R\}$$

is open.

Corollary 1. *If W satisfies property (*), then for any unitary $f(x) \in R_W[x]$ the set V_f is closed open.*

Show that $W \setminus V_f$ is open. Let $R \in W \setminus V_f$, then there exist unitary $g, h \in R_W[x] \setminus R_W$ such that $\lambda_R(f) = \lambda_R(g) \cdot \lambda_R(h)$, but $\{R' \mid F_{R'} \vDash \lambda_{R'}(f) = \lambda_{R'}(g) \cdot \lambda_{R'}(h)\}$ is open and is contained in $W \setminus V_f$. □

Corollary 2. *If $\overline{\lambda}_W$ determines $\overline{R}_W$ as $\exists$-product, then W satisfies condition* (*). □

We associate with a Boolean family W of valuation rings of a field F the following partially ordered abelian group $\langle \Gamma_W, P_W \rangle$: $\Gamma_W \leftrightharpoons F^*/U(R_W)$, where $U(R_W)$ is the multiplicative of units (group invertible elements) of the ring R_W and the cone P_W is defined to be the set $(R_W \setminus \{0\})/U(R_W) \subseteq \Gamma_W$. The notation $\gamma_0 \leqslant_W \gamma_1$ for $\gamma_0, \gamma_1 \in \Gamma_W$ is used to designate that $\gamma_1 - \gamma_0 \in P_W$ (the operation in the group Γ_W will be written additively). Denote by v_W the natural epimorphism of F^* onto F_W.

Remark 3. The mapping $v_W : F^* \to \Gamma_W$ extended to F in such a way that $v_W(0) = \infty$ is a semi-valuation (*Demi-Bewertung*) in the sence of the thesis [11]. The proof of the following proposition can be found in that thesis; we give another proof for the convenience of the reader.

Proposition 5. *A partially ordered group $\langle \Gamma_W, P_W \rangle$ is a lattice ordered (l-) group.*

It is sufficient to show that for any $\gamma_0, \gamma_1 \in \Gamma_W$ there exists the greatest lower bound $\gamma_0 \wedge \gamma_1$ with respect to the order $\leqslant_W$. Let $\gamma_0 = a_0U(R_W)$, $\gamma_1 = a_1U(R_W)$, $a_0, a_1 \in F^*$. Since R_W is a Bézout ring (see Proposition 3 in [12]), there exists $b \in F$ such that $a_0R_W + a_1R_W = bR_W$. It is clear that $b \neq 0$, $a_iR_W \subseteq bR_W$, $i = 0, 1$. Check that for any $c, d \in F^*$

$$v_W(c) \leqslant_W v_W(d) \Longleftrightarrow dR_W \subseteq cR_W$$

Indeed, $v_W(c) \leqslant_W v_W(d)$ is equivalent ($\Longleftrightarrow$) to $v_W(d) - v_W(c) = v_W(dc^{-1}) \in P_W \Longleftrightarrow dc^{-1} \in R_W \Longleftrightarrow dR_W \subseteq cR_W$.

Thus, $v_W(b) \leqslant \gamma_i = v_W(a_i)$, $i = 0,1$. If $\gamma_0, \gamma_1 \geqslant v_W(d)$, then $a_i R_W \subseteq dR_W$, $i = 0,1$; $bR_W = a_0 R_W + a_1 R_W \subseteq dR_W$, $v_W(d) \leqslant_W v_W(b)$, that is $v_W(b) = \gamma_0 \wedge \gamma_1$. □

For any $R \in W$ the restriction of the valuation v_R to F^* is an epimorphism $v_R : F^* \to \Gamma_R$. The system of epimorphisms v_R, $R \in W$, defines a homomorphism $\lambda_W^* : F^* \to \prod\limits_{R \in W} \Gamma_R$ with the kernel, as it is easy to verify, $U(R_W)$. Then λ_W^* induces the embedding $\lambda_W : \Gamma_W \to \prod\limits_{R \in W} \Gamma_R$. If we consider a lattice ordered group $\langle \Gamma_W, P_W \rangle$ and linearly ordered groups Γ_R in the language $\sigma_\ell = \langle 0, +, -, \wedge \rangle$ (here $\wedge$ is the operation of greatest lower bound), then λ_W is embedding in the language σ_ℓ too.

Proposition 6. *The embedding $\lambda_W : \Gamma_W \to \prod\limits_{R \in W} \Gamma_R$ determines Γ_W as a Boolean product if and only if W is independent.*

Let W be independent. Verify condition 1 in the definition of a Boolean product.

Let $a_0, a_1 \in F^*$, $Y = V_a$, $a \in F^*$. Choose $0 \neq \varepsilon \in J(R_W)$ such that $v_R(a_i) > v_R(\varepsilon)$, $i = 0,1$, for all $R \in W$. By the property BAP which is equivalent to independence of W, there exists $b \in F$ such that $v_R(b - a_0) > v_R(\varepsilon)$ for $R \in V_a$ and $v_R(b - a_1) > v_R(\varepsilon)$ for $R \in W \setminus V_a$. The choice of ε implies that $v_R(b) = v_R(a_0)$ for $R \in V_a$ and $v_R(b) = v_R(a_1)$ for $R \in W \setminus V_a$. But then $\lambda_W v_W(b) = \langle \lambda_W v_W(a_0), \lambda_W v_W(a_1), Y \rangle$, that is condition 1 is satisfied.

Now show that if condition 1 is satisfied, then W satisfies the property BAP. Condition 1 is equivalent to the following property:

For any $a_0, a_1 \in F^*$ and any closed open $Y \subseteq W$ there exists $b \in F^*$ such that $v_R(b) = v_R(a_0)$ for $R \in Y$ and $v_R(b) = v_R(a_1)$ for $R \in W \setminus Y$.

To establish the BAP property it is sufficient to prove the property:

For any $a_0, a_1 \in R_W$, any closed open $Y \subseteq W$ and any $0 \neq \varepsilon \in J(R_W)$ there exists b ($\in R_W$) such that $v_R(b - a_0) > v_R(\varepsilon)$ for $R \in Y$ and $v_R(b - a_1) > v_R(\varepsilon)$ for $R \in W \setminus Y$.

By property 1 there exist b_0, b_1 ($\in R_W$) such that $v_R(b_0) = 0, v_R(b_1) = 2v_R(\varepsilon)$ for $R \in Y$ and $v_R(b_0) = 2v_R(\varepsilon), v_R(b_1) = 0$ for $R \in W \setminus Y$.

Put $b \leftrightharpoons b_0 + b_1 \in U(R_W)$, $c_0 \leftrightharpoons b_0 b^{-1}$, $c_1 \leftrightharpoons b_1 b^{-1} \in R_W$. Let $d \leftrightharpoons a_0 c_0 + a_1 c_1$, $R \in Y$, then

$$v_R(d - a_0) = v_R(a_0(c_0 - 1) + a_1 c_1);$$
$$c_0 - 1 = b^{-1} b_0 - b^{-1} b = b^{-1}(b_0 - b) = -b^{-1} b_1;$$

$$v_R(c_0 - 1) = v_R(b^{-1}) + v_R(b_1) = 0 + v_R(b_1) = 2v_R(\varepsilon);$$
$$v_R(a_1c_1) = v_R(a_1) + v_R(c_1) \geqslant v_R(b^{-1}b_1) = v_R(b_1) = 2v_R(\varepsilon);$$

hence, $v_R(d - a_0) \geqslant 2v_R(\varepsilon) > v_R(\varepsilon)$. Similarly one can verify that if $R \in W \setminus Y$, then $v_R(d - a_1) > v_R(\varepsilon)$.

Thus, if condition 1 of the definition of a Boolean product holds for λ_W, then W is independent.

Let W be independent. Verify condition 2 of the definition. It is sufficient to verify that for any $m_i \in Z$, $i = 0, \dots, n$, and $a_0, \dots, a_n \in F^*$ the sets

$$A \leftrightharpoons \{R \mid \Gamma_R \vDash \sum m_i v_R(a_i) = 0\}$$

and

$$B \leftrightharpoons \{R \mid \Gamma_R \vDash \sum m_i v_R(a_i) > 0\}$$

are open, but it is easy to see that

$$A = V_{\prod\limits_{i\leqslant n} a_i^{m_i}} \cap V_{(\prod\limits_{i\leqslant n} a_i^{m_i})^{-1}},$$

$$B = V_{\prod\limits_{i\leqslant n} a_i^{m_i}} \setminus V_{(\prod\limits_{i\leqslant n} a_i^{m_i})^{-1}},$$

which are open. □

Let us say that W satisfies *property* (**) if for any $n > 1$ and $a \in F^*$ the set

$$V_{[a,n]} \leftrightharpoons \{R \mid \Gamma_R \vDash \forall x(nx \neq v_R(a))\}$$

is open.

Corollary 3. *If λ_W determines Γ_W as $\exists$-product, then W satisfies the property (**).* □

Remind that a valued field $\langle F, R\rangle$ is *algebraically complete*, if for every extension $\langle F, R\rangle \leqslant \langle F_0, R_0\rangle$ such that F_0 is a finite algebraic extension of F the equality

$$[F_0 : F] = [F_{R_0} : F_R] \cdot [\Gamma_{R_0} : \Gamma_R]$$

holds.

Theorem 4. *Let W be a Boolean family of valuation rings of a field F such that W is independent, satisfies conditions (*) and (**) and for every $R \in W$ the field F is dense in $\mathbb{H}_R(F)$, F_R is perfect, and $\mathbb{H}_R(F) = \langle q(R^h), R^h\rangle$ is algebraically complete. Then the mapping $\zeta : W \to \mathrm{Subg}^* G(F)$ is continuous.*

To prove the theorem it is sufficient to establish the following proposition:

For every finite Galois extension $F_0 \geqslant F$ and $R \in W$, if $\Gamma \leqslant G(F_0/F)$ is a decomposition group of R in F_0, then there exists a neighbourhood V of the ring R in W such that for any $R' \in V$, Γ is a decomposition group of R' in F_0.

Let R_0 be a valuation ring of the field F_0 such that $R_0 \cap F = R$ and

$$\Gamma = \{g \mid g \in G(F_0/F), R_0^g = R_0\}.$$

Let $F_1 \leftrightharpoons F_0^\Gamma$ and $R_1 \leftrightharpoons R_0 \cap F_1$; then $R \leqslant R_1 \leqslant R^h$ and $H_R(F) \|_{F_1} F_0$ ($H_R(F)$ and F_0 are linearly disjoint over F_1). In view of algebraic completeness of $\mathbb{H}(F)$ we have the equality

$$[F_0 : F_1] = [F_{R_0} : F_{R_1}] \cdot [\Gamma_{R_0} : \Gamma_{R_1}];$$

notice that $F_{R_1} = F_R$ and $\Gamma_{R_1} = \Gamma_R$. Let $F_1 = F(\alpha)$ for some α integral over R_W and let $f_\alpha(x) = x^k + a_1 x^{k-1} + \cdots + a_k \in R_W[x]$ be the minimal polynomial for α over F. By denseness of F in $\mathbb{H}_R(F)$ there exists $a \in R$ such that $v_{R^h}(a - \alpha) > 2v_{R^h}(f'_\alpha(\alpha))$, then $v_R(f(a)) > 2v_R(f'_\alpha(a))$. If

$$V_0 \leftrightharpoons V_{f_\alpha(a) f'_\alpha(a)^{-2}} \setminus V_{f_\alpha(a)^{-1} f'_\alpha(a)},$$

then $R \in V_0$; if

$$V_1^{F_1} \leftrightharpoons V^{F_1}_{(a-\alpha) f_\alpha(a)^{-1} f'_\alpha(a)} \cap V^{F_1}_{(a-\alpha)^{-1} f_\alpha(a) f'_\alpha(a)},$$

then for any $R' \in V_0$ there exists unique $R'_1 \in V_1^{F_1}$ such that $R'_1 \cap F = R'$, $R' \leqslant R'_1 \leqslant (R')^h$, and $v_{R'_1}(a - \alpha) = v_{R'_1}(f_\alpha(a)) - v_{R'_1}(f'_\alpha(a))$.

Let $\Gamma' \leqslant \Gamma$ be the inertia group of the valuation ring R_0, that is

$$\Gamma' = \{g \mid g \in \Gamma \leqslant G(F_0/F), \forall r \in R_0 (r - r^g \in \mathfrak{m}(R_0))\},$$

then Γ' is a normal subgroup of Γ and Γ/Γ' is isomorphic to the Galois group of the Galois extension F_{R_0} of F_R (this extension is normal and separable, since F_R is perfect). Let $F_2 \leftrightharpoons F^{\Gamma'}$, $R_2 \leftrightharpoons R_0 \cap F_2$; then

$$[F_2 : F_1] = [F_{R_2} : F_{R_1}] = [F_{R_0} : F_R].$$

Let $\overline{\beta} \in F_{R_2}$ be such that $F_{R_2} = F_{R_1}(\overline{\beta})$ and $\beta \in R_1$ be such that $\overline{\beta} = \beta + \mathfrak{m}(R_1)$. Then obviously $F_2 = F_1(\beta)$. Let $f_\beta(x) = x^m + b_1 x^{m-1} + \cdots + b_m$ be the minimal polynomial for β over F_1, then $f_\beta(x) \in R_1[x]$ and $\lambda_{R_1}(f_\beta) = x^m + \lambda_{R_1}(b_1)x^{m-1} + \cdots + \lambda_{R_1}(b_m)$ is the minimal polynomial for $\overline{\beta}$ over $F_{R_1} = F_R$. Since F is dense in $\mathbb{H}_R(F)$, $\langle F, R\rangle \leqslant \langle F_1, R_1\rangle \leqslant \mathbb{H}_R(F)$, then, as in the proof of Proposition 8 in Section 1 of [13], one can find β such that $f_\beta \in R[x]$. By property (*) V_{f_β} is open and $R \in V_{f_\beta}$. Let

$$V_1 \leftrightharpoons V_0 \cap V_{f_\beta} \cap \left(\bigcap_{i=1}^{m} V_{b_i}\right);$$

V_1 is open and for any $R' \in V_1$, if $R'_1 \in V_1^{F_1}$ is such that $R'_1 \cap F = R'$, then R'_1 has a unique extension to a valuation ring R'_2 of the field F_2 and $[F_{R'_2} : F_{R'_1}] = [F_2 : F_1]$ $(= m)$.

Let the finite abelian group Γ_{R_0}/Γ_R be presented as direct sum $\sum_{(i,j) \in S} Z_{p_i^{n_{ij}}}$ of primary cyclic groups; for every pair $(i, j) \in S$ let $\gamma_{ij} \in \Gamma_{R_0}$ be such that its image in Γ_{R_0}/Γ_R is the generator of the group $Z_{p_i^{n_{ij}}}$; let $c_{ij} \in F_0^*$ be such that

$v_{R_0}(c_{ij}) = \gamma_{ij}$; since $p_i^{n_{ij}}\gamma_{ij} \in \Gamma_R$, there exist $d_{ij} \in F^*$ such that $v_R(d_{ij}) = p_i^{n_{ij}}\gamma_{ij}$. Notice that $v_R(d_{ij}) = v_{R_0}(c_{ij}^{p_i^{n_{ij}}})$. Let

$$V_{ij}^{F_0} \leftrightharpoons V^{F_0}_{d_{ij}c_{ij}^{-p_i^{n_{ij}}}} \cap V^{F_0}_{d_{ij}^{-1}c_{ij}^{p_i^{n_{ij}}}}, \qquad V_2^{F_0} \leftrightharpoons \bigcap_{(i,j)\in S} V_{ij}^{F_0};$$

then $R_0 \in V_2^{F_0}$; $V_2^{F_1} \leftrightharpoons V_1^{F_1} \cap \pi_0(V_2^{F_0})$, where π_0 is the projection $R'_0 \mapsto R'_0 \cap F_1$, $R'_0 \in W_0$, and

$$W_0 \leftrightharpoons \{R'_0 \mid R'_0 \text{ is a valuation ring of } F_0 \text{ such that } R'_0 \cap F \in W\},$$
$$W_1 \leftrightharpoons \{R \mid R \text{ is a valuation ring of } F_1 \text{ such that } R \cap F \in W\}.$$

By Proposition 1 from [12] the projection π_0 is an open mapping; so $V_2^{F_1}$ is an open neighbourhood of ring R_1. Let

$$V_2 \leftrightharpoons V_1 \cap \pi_1(V_2^{F_1}) \cap \bigcap_{(i,j)\in S} V_{[d_{ij},p_i]},$$

where $\pi_1 : W_1 \to W$ is a natural projection and V_2 is an open neighbourhood of R. Let $R' \in V_2$, $R'_1 \in W_1$ be such that $R'_1 \cap F = R'$, $R'_1 \in V_1^{F_1}$, $R'_0 \in W_0$ be such that $R'_0 \cap F_1 = R'_1$ and $R'_0 \in V_2^{F_0}$. Then we have $R' \leqslant R'_1 \leqslant (R')^h$; $R'_2 \leftrightharpoons R'_0 \cap F_2$ is such that

$$[F_{R'_2} : F_{R'_1}] = [F_2 : F_1] = [F_{R_2} : F_{R_1}],$$

that is $\langle F_2, R'_2\rangle$ is a totally unramified extension of $\langle F_1, R'_1\rangle$. Further, $R' \in V_{[d_{ij},p_i]}$ implies that $v_{R'}(d_{ij})$ is not divisible by p_i in $\Gamma_{R'} = \Gamma_{R'_1}$. On the other hand, for $(i,j) \in S$

$$v_{R'}(d_{ij}) = v_{R'_0}(d_{ij}) = v_{R'_0}(c_{ij}^{p_i^{n_{ij}}}) = p_i^{n_{ij}} v_{R'_0}(c_{ij});$$

hence

$$[\Gamma_{R'_0} : \Gamma_{R'}] = [\Gamma_{R'_2} : \Gamma_{R'_1}] = [\Gamma_{R'_0} : \Gamma_{R'_1}] \geqslant \prod_{(i,j)\in S} p_i^{n_{ij}} = [\Gamma_{R_0} : \Gamma_{R_1}].$$

It follows

$$[F_0 : F_1] = [F_{R_0} : F_{R_1}] \cdot [\Gamma_{R_0} : \Gamma_{R_1}] = [F_{R_2} : F_{R_1}] \cdot [\Gamma_{R_0} : \Gamma_{R_1}]$$
$$\leqslant [F_{R'_0} : F_{R'_1}] \cdot [\Gamma_{R'_0} : \Gamma_{R'_1}] \leqslant [F_0 : F_1].$$

Then R'_0 is the unique extension of R'_1 and hence F_1 is a decomposition field for R' and $\Gamma = G(F_0/F_1)$ is a decomposition group of R' over F_0. □

Remark 4. The conclusion of Theorem 4 remains valid if we omit the assumption that fields F_R are perfect for all $R \in W$, but replace condition (*) by a stronger assumption that $\overline{\lambda}_W$ determines $\overline{R}_W$ as $\exists$-product.

References

[1] Haran, D., Jarden, M., The absolute Galois group of a pseudo p-adically closed field. J. Reine Angew. Math. 383 (1988), 147–206.

[2] Ershov, Yu. L., RC^*-fields. Algebra Logika 33:4 (1994), 367–386. English translation: Algebra Logic 33:4 (1994).

[3] Haran, D., On closed subgroups of free products of profinite groups. Proc. Lond. Math. Soc., III. Ser. 55:2 (1987), 266–298.

[4] Herfort, W., Ribes, L., Torsion elements and centralizers in free products of profinite groups. J. Reine Angew. Math. 358 (1985), 155–161.

[5] Ershov, Yu. L., On Galois groups of RC^*_ζ-fields. Dokl. Ross. Akad. Nauk 334:3 (1994), 275–277. English translation: Sov. Math., Dokl. (1994).

[6] Ershov, Yu. L., Relative regular closedness and π-valuations. Algebra Logika 31:6 (1992), 592–623. English translation: Algebra Logic 31:6 (1992), 342–359.

[7] Ershov, Yu. L., RRC-fields with small Galois groups. Ann. Pure Appl. Logic 43:3 (1989), 197–208.

[8] Ershov, Yu. L., PC_p-fields with universal Galois group. Sib. Adv. Math. 1:4 (1991), 1–26.

[9] Ershov, Yu. L., Model-theoretic properties of RC^*-fields. Dokl. Ross. Akad. Nauk 335:2 (1994), 138–141. English translation: Sov. Math., Dokl. (1994).

[10] Burris, S., Werner, H., Sheaf constructions and their elementary properties. Trans. Am. Math. Soc. 248:2 (1979), 269–309.

[11] Heinemann, B., Über mehrfach bewertete "Henselsche" Körper. Diss. Konstanz 1982.

[12] Ershov, Yu. L., Boolean families of valuation rings. I, II. Algebra Logika 31:3 (1992), 274–294; 32:3 (1993), 261–266. English translation: Algebra Logic 31:3 (1992), 170–181; 32:3 (1993), 139–141.

[13] Ershov, Yu. L., Multiply valued fields. Usp. Mat. Nauk 37:3 (1982), 55–93. English translation: Russ. Math. Surv. 37:3 (1982), 63–107.

On finite basedness of abstract T-spaces

Aleksandr V. Grishin

Abstract. Let F be a free countably generated associative algebra over a field k of characteristic 0, and let T be the semigroup of endomorphisms of F. The notion of a T-subspace of F is generalized to an abstract T-space over k, which is an arbitrary unitary right kT-module. **Theorem:** *If F/I is a relatively free algebra such that the T-ideal I contains some standard polynomial, then F/I is a Noetherian kT-module.* This is equivalent to the assertion that every T-space in F/I is finitely based. The theorem implies that T-ideals are finitly based (Kemer's Theorem) as well as many other T-spaces are, for example, the T-space of polynomials in F with values in the center of an algebra satisfying a standard identity.

1991 Mathematics Subject Classification: 16A33, 16A05, 16A38.

Introduction

Let $F = k\langle x_1, \dots, x_i, \dots\rangle$ be a free countably generated associative algebra over a field k of characteristic zero, and let T be the semigroup of endomorphisms (substitutions) of F. By *substitution τ of type* $(x_l, \dots, x_i, \dots) \mapsto (g_l, \dots, g_i, \dots)$, where $g_i \in F$, we mean an endomorphism of F which is defined by the above mapping. Denote by f^τ the image of a polynomial f from F under the action of the substitution τ. For the sake of convenience, we sometimes write other letters with single or double indices to denote some variables of F.

Following [1], by a *T-space* of the algebra F we mean a linear subspace of F stable under the action of the semigroup T. In particular, a T-ideal of F is exactly an ideal of F which is a T-space.

The semigroup T acts from the right on any relatively free algebras. Hence we may introduce the notions of a T-space and a T-ideal on these algebras in analogous way. Any T-space of a relatively free algebra can be equipped with the structure of a unitary right kT-module (kT stands for the semigroup k-algebra). This leads to the following substantial generalization of the definition of a T-space.

By a (*abstract*) *T-space* over a field k we mean any unitary right kT-module. *Morphisms of T-spaces* are defined as homomorphisms of T-spaces considered as kT-modules. Denote by S^T a T-space generated by a subset S of a T-space.

For wide classes of T-spaces, it is possible to prove that T-spaces are finitely generated as kT-modules.

Theorem. *Let F/I be a relatively free algebra such that the T-ideal I contains some standard polynomial. Then F/I is a Noetherian kT-module.*

The theorem implies the finite basedness of T-ideals (this fact was first proved by A. R. Kemer [2]) and of many other T-spaces, for example, of the T-space constituted by the polynomials in F with values in the center of an algebra satisfying a standard identity. It is clear that the theorem is equivalent to the assertion that any T-space in F/I is finitely based.

Below, we present a series of T-spaces from the proof of the theorem and give a sketch of the proof.

1. Matrix T-spaces

Let k_r be an algebra of $r \times r$-matrices. Denote by M_r the T-ideal of identities of k_r in the algebra F, and let $GM(r, X) \cong F/M_r$ be the algebra consisting of general $r \times r$-matrices embedded in a usual way into the algebra $k[X]_r$ of $r \times r$-matrices over the ring with identity of commutative polynomials $k[X]$ with a countable set of variables $X = \{x_i^{(\alpha,\beta)} \mid 1 \leqslant \alpha, \beta \leqslant r,\, i = 1, 2, \dots\}$, $\overline{x}_i = \sum x_i^{(\alpha,\beta)} e_{\alpha\beta}$, where $e_{\alpha\beta}$ stand for matrix units of k_r and $\overline{x}_i$ denotes the image of the variable x_i (see [1]). Consider the subalgebra $\mathrm{Tr}\,(r, X) = k[\mathrm{tr}\,(GM(r, X))]$ of $k[X]$ generated by the identity and the traces of all general matrices, and the subalgebra $\mathrm{Tr}\, GM(r, X)$ of $k[X]_r$ generated by algebras $\mathrm{Tr}\, GM(r, X)$, $\mathrm{Tr}\,(r, X)$, and the matrix unit. The algebra $\mathrm{Tr}\, GM(r, X)$ is endowed with a structure of a T-space in a natural way.

Consider s natural numbers $r_1, \dots, r_s$ (not necessarily different) and s countable sets of commutative variables $X_1, \dots, X_s$ such that $X_i \cap X_j = \varnothing$ for $i \neq j$. Consider the subalgebra R of $k[X]$ generated by all algebras $\mathrm{Tr}\,(r_i, X_i)$ where $X = \bigcup_{i=1}^{s} X_i$. Algebras $\mathrm{Tr}\, GM(r_i, X_i)$ can be considered as subalgebras of $k[X]_r$.

Let $\Phi_i = RGM(r_i, X_i)$ be an algebra generated by R and $\mathrm{Tr}\, GM(r_i, X_i)$. We may introduce T-structure on the algebra Φ_1 so that algebras $GM(r_i, X_i)$ and $\mathrm{Tr}\, GM(r_i, X_i)$ become T-subspaces of Φ_1.

In the sequel, we use an R-algebra equipped with the T-structure

$$\Phi(r_1, \dots, r_s) = \Phi_1 \oplus \dots \oplus \Phi_s.$$

2. Algebra of quasipolynomials

Let $\overline{F} = F_1^e \oplus \dots \oplus F_s^e$ be a direct sum of algebras $F_i^e = F_i + ke_i$ where F_i is a relatively free algebra; e_i denotes the unit augmented to the algebra F_i, if the latter is not an algebra of general matrices, or the corresponding matrix unit otherwise. Let $\overline{x}_i = x_i^{(1)} + \cdots + x_i^{(s)}$ where $x_i^{(\alpha)}$ is the image of the variable x_i in the algebra F_i, and let $k\langle\theta_1, \ldots, \theta_i, \ldots\rangle$ be a free countably generated associative algebra.

The free product

$$Q(\overline{F}) = \overline{F} *_k k\langle\theta_1, \ldots, \theta_i, \ldots\rangle$$

over k is called an *algebra of $\overline{F}$-quasipolynomials with coefficients in the field k.*

Every element (quasipolynomial) in $Q(\overline{F})$ is a linear combination of the following expressions (quasimonomials)

$$u_0\theta_{i_1}u_1 \ldots \theta_{i_n}u_n \tag{1}$$

where u_i is a monomial of one of the algebras $F_1, \ldots, F_s$ or one of the units $e_1, \ldots, e_s$. The number n is called the *θ-degree* of the quasimonomial (1) and the number $n + \sum_{i=0}^{n} \deg u_i$ is called the *degree* of (1).

For any quasipolynomial in $Q(\overline{F})$, the notions of θ-degree and degree are defined in an obvious way.

We say that *a variable x_i occurs in the quasimonomial* (1) *with degree* $m = p+q$ if the summary degree of occurrence of the images of x_i in $u_0, \ldots, u_n$ equals p and exactly q elements among $\theta_{i_1}, \ldots, \theta_{i_n}$ coincide with θ_i. By the homogenity in each variable in identities in the algebra $Q(\overline{F})$, the above definition can be transferred onto any non-zero polynomials homogeneous in x_i.

Denote by $\Theta(\overline{F})$ the ideal of the algebra $Q(\overline{F})$ which is generated by all elements θ_i. The space generated by all monomials of θ-degree n is identified with the quotient algebra $Q_n(\overline{F}) = \Theta(\overline{F})^n/\Theta(\overline{F})^{n+1}$ in an obvious way. Moreover, we have $\Theta(\overline{F})^0 = Q(\overline{F})$ and $Q_0(\overline{F}) = \overline{F}$.

Put $x_i = \overline{x}_i + \theta_i$. It is evident that elements x_i generate a free associative algebra that is a subalgebra of $Q(\overline{F})$. Thus, the algebra $F = k\langle x_1, \ldots, x_i, \ldots\rangle$ is embedded into the algebra $Q(\overline{F})$. Introduce the action of the semigroup T on $Q(\overline{F})$ so that it continues the action of T on algebras F and $\overline{F}$. It suffices to define the action on elements θ_i only. Let τ be a substitution of type $x_i \mapsto g_i$. Set

$$\theta_i^\tau = x_i^\tau - \overline{x}_i^\tau = g_i(x_1, \ldots, x_d) - g_i(\overline{x}_1, \ldots, \overline{x}_d),$$

i.e., θ_i^τ is $g_i(x_1, \ldots, x_d)$ without a component of zero θ-degree. The definition of the action of the semigroup T implies that the ideal $\Theta(\overline{F})$ and all its powers are T-ideals of the algebra $Q(\overline{F})$.

3. Algebra of matrix quasipolynomials

Denote by $Q(\Phi(r_1, \ldots, r_s))$ the algebra of $\Phi(r_1, \ldots, r_s)$-quasipolynomials with coefficients in the field k and the action of the semigroup T as above: $\theta_i^\tau = x_i^\tau - \overline{x}_i^\tau$.

As above, we have the decomposition of quasipolynomials in $Q(\Phi(r_1, \ldots, r_s))$ into quasimonomials (1). We can introduce the notations of degree, θ-degree and degree of occurrence of a variable x_i in the quasimonomial (1), the ideals $(\Theta(\Phi(r_1,\dots,r_s)))^m$, and T-spaces $Q_n(\Phi(r_1,\dots,r_s))$. It is evident that, for $\overline{F} = GM(r_1,X_1)^e \oplus \cdots \oplus GM(r_s,X_s)^e$, the algebra of quasipolynomials $Q(\overline{F})$ is a T-subalgebra in $Q(\Phi(r_1,\dots,r_s))$ and the objects of the type Θ^m, Q_n of the first algebra are T-spaces of similar objects of the second one.

The algebra of quasipolynomials $Q(GM(r,X)^e)$ is of particular interest. It was considered in [1] where the following proposition related to the inclusion $F \subset Q(GM(r,X)^e)$ was proved.

Proposition 1. $F \cap (\Theta(GM(r,X)^e))^m = M_r^m$.

Corollary 1. *F/M_r^m is a T-space in $Q(GM(r,X)^e)/(\Theta(GM(r,X)^e))^m$.*

Corollary 2. *M_r^n/M_r^{n+1} is a T-space in $Q_n(GM(r,X)^e)$.*

4. Algebra of matrix quasipolynomials with a trace

Consider the above-introduced algebra of traces $R = k[\mathrm{Tr}\,(r_1,X_1),\dots,\mathrm{Tr}\,(r_s,X_s)]$ and the R-algebra

$$Q_R(\Phi(r_1,\dots,r_s)) = \Phi(r_1,\dots,r_s) *_R R\langle\theta_1,\dots,\theta_i,\dots\rangle$$

on which the notions of degree, θ-degree, action of semigroups T, and T-spaces $\Theta^R(\Phi(r_1,\dots,r_s))^m$ and $Q_n^R(\Phi(r_1,\dots,r_s))$ are introduced as above. In contrast to $Q(\Phi(r_1,\dots,r_s))$, in the algebra $Q^R(\Phi(r_1,\dots,r_s))$ traces commute with the elements θ_i.

To each quasipolynomial of the first algebra, we associate exactly the same polynomial expression in elements $x_i^{(j)}$, θ_i and traces from R, which is, however, considered in the second algebra. It is obvious that this correspondence defines the surjective homomorphism

$$\pi : Q(\Phi(r_1,\dots,r_s)) \mapsto Q^R(\Phi(r_1,\dots,r_s)).$$

5. Complete space of quasipolynomials of θ-degree n

Let $\overline{F} = F^{(1)} \oplus F^{(2)}$ where $F^{(1)} = \oplus_{i=1}^p F_i^e$, $F^{(2)} = \oplus_{i=1}^q F_{p+i}^e$, and F_i are relatively free algebras. An element of the T-space $Q_n(F^{(1)})$ is represented as a linear combination of expressions (1) where u_i is a monomial of one of the algebras

$F_1^e, \dots, F_p^e$. But the same linear combinations form a T-subspace of the space $Q_n(\overline{F})$. Thus, we can say that $Q_n(F^{(1)})$ and $Q_n(F^{(2)}$ considered as T-subspaces are embedded into the T-space $Q_n(\overline{F})$. This allows us to consider a T-space $Q_n = \varinjlim Q_n(\overline{F})$ where the inductive limit is taken over all finite direct sums $\oplus_{i=1}^s F_i^e$. It is convenient to regard all T-spaces $Q_n(\overline{F})$ considered below as embedded into Q_n.

6. Elementary spaces

Let $F_{i_0}^e, \dots, F_{i_n}^e$ be algebras from the set of the direct summands $F_1^e, \dots, F_s^e$ that form the algebra $\overline{F}$. Note that numbers $i_0, \dots, i_n$ are not necessarily pairwise distinct and do not necessarily contain all the natural numbers from 1 to s. By an *elementary space* we mean the following T-subspace of the space $Q_n(\overline{F})$:

$$Q_n\langle F_{i_0}^e, \dots, F_{i_n}^e\rangle = \oplus_{(j_1,\dots,j_n)} F_{i_0}^e \theta_{j_1} F_{i_1}^e \dots \theta_{j_n} F_{i_n}^e.$$

It follows from the definition that $Q_n(\overline{F})$ is a finite direct sum of elementary spaces, i.e.,

$$Q_n(\overline{F}) = \oplus_{(i_0,\dots,i_n)} Q_n\langle F_{i_0}^e, \dots, F_{i_n}^e\rangle.$$

Let $V_{i_0}, \dots, V_{i_n}$ be T-ideals in algebras $F_{i_0}^e, \dots, F_{i_n}^e$ (maybe, coinciding with F_{i_α} or $F_{i_\alpha}^e$, or zero). It is evident that the space

$$Q_n\langle V_{i_0}, \dots, V_{i_n}\rangle = \oplus_{(j_1,\dots,j_n)} V_{i_0} \theta_{j_1} V_{i_1} \dots \theta_{j_n} V_{i_n}$$

is a T-subspace of the corresponding elementary space.

Denote by $Q_n\langle F_{i_0}^e, \dots, F_{i_n}^e\rangle(d,c)$ the linear subspace in $Q_n\langle F_{i_0}^e, \dots, F_{i_n}^e\rangle$ which is generated by quasimonomials depending only on variables $x_1, \dots, x_d, \dots, x_{d+c}$ and such that the variables $x_{d+1}, \dots, x_{d+c}$ occur in them with degree ≤ 1. As above, we define linear spaces $Q_n\langle V_{i_0}, \dots, V_{i_n}\rangle(d,c)$. In precisely the same manner we can introduce matrix elementary spaces $Q_n(\Phi_{i_0}, \dots, \Phi_{i_n})$, $Q_n^R(\Phi_{i_0}, \dots, \Phi_{i_n})$, their subspaces

$$V = Q_n\langle GM(r_{i_0}, X_{i_0})^e, \dots, GM(r_{i_n}, X_{i_n})^e\rangle,$$

$$V^R = Q_n^R\langle GM(r_{i_0}, X_{i_0})^e, \dots, GM(r_{i_n}, X_{i_n})^e\rangle,$$

and their analogs of the type $V(d,c)$ and $V^R(d,c)$.

Proposition 2. *If sets $X_{i_1}, \dots, X_{i_n}$ are pairwise distinct then the restriction of the mapping π* (see item 4) *to the T-space V is an isomorphism of T-spaces V and V^R.*

The proof is direct verification.

Proposition 2 allows us to employ the following schema of proving the finite basedness for T-spaces. It suffices to prove the finite basedness of any T-space in

$$V = Q_n\langle GM(r_{i_0}, X_{i_0})^e, \dots, GM(r_{i_n}, X_{i_n})^e\rangle$$

where $X_{i_\alpha} \neq X_{i_\beta}$ for $\alpha \neq \beta$, since there is a surjective morphism onto an analogous T-space with repeating sets X_{i_α}. By Proposition 2, the T-space is isomorphic to the corresponding T-space V^R belonging to $Q_n^R\langle \Phi_{i_0}, \dots, \Phi_{i_n}\rangle$. For the latter, one can effectively employ the technique of spaces.

Denote by $[GM(r_\alpha, X_\alpha)^e]_{r_\alpha}$ the algebra of $r_\alpha^2 \times r_\alpha^2$-matrices of the type

$$[f^{(\alpha)}]_{r_\alpha} = \begin{pmatrix} f^{(\alpha)} & & 0 \\ & \ddots & \\ 0 & & f^{(\alpha)} \end{pmatrix}$$

where $f^{(\alpha)}$ stands for the image of the polynomial f under a mapping from $F^e = k\langle 1, x_1, \dots, x_i, \dots\rangle$ to $GM(r_\alpha, X_\alpha)^e$. It is obvious that algebras $[GM(r_\alpha, X_\alpha)^e]_{r_\alpha}$ and $GM(r_\alpha, X_\alpha)^e$ are isomorphic and may be identified in a natural way. The same is true for R-algebras $R[GM(r_\alpha, X_\alpha)^e]_{r_\alpha}$ and $RGM(r_\alpha, X_\alpha)^e$. The unit matrices in these algebras will be identified and denoted by e_α.

Let Δ be a subalgebra of the direct sum

$$[GM(r_{i_0}, X_{i_0})^e]_{r_{i_0}} \oplus \dots \oplus [GM(r_{i_n}, X_{i_n})^e]_{r_{i_n}}$$

consisting of the elements of the form $[f^{(i_0)}]_{r_{i_0}} + \dots + [f^{(i_n)}]_{r_{i_n}}$. It is clear that

$$[GM(r_\alpha, X_\alpha)^e]_{r_\alpha} = e_\alpha\, \Delta;$$

moreover, the algebra Δ is isomorphic to the algebra $GM(r_{i_m}, X_{i_m})^e$ where r_{i_m} is the greatest of the numbers $r_{i_0}, \dots, r_{i_n}$. On the other hand, the algebra Δ can be viewed as an algebra of diagonal block $r \times r$-matrices of the form

$$\widetilde{f} = \begin{pmatrix} [f^{(i_0)}]_{r_{i_0}} & & 0 \\ & \ddots & \\ 0 & & [f^{(i_n)}]_{r_{i_n}} \end{pmatrix}$$

where $r = r_{i_0}^2 + \dots + r_{i_n}^2$. Consider the subalgebra $L = k[\mathrm{tr}\,(\Delta)]$ of R generated by the traces of the matrices $\widetilde{f}$. The generators of L have the form

$$\sum_{\alpha=0}^{n} \mathrm{tr}\, r_{i_\alpha} f^{(i_\alpha)}.$$

Denote by $L(d)$ the algebra s similar to L but consisting of elements depending only on variables $x_1, \dots, x_d$. We write $\Delta(d, c)$ for the subspace in Δ composed of elements depending on variables $x_1, \dots, x_d, \dots, x_{d+c}$ such that variables $x_{d+1}, \dots, x_{d+c}$ occur in these elements with degree ≤ 1. It is easy to see that $L(d)$ is a finitely generated k-algebra and

$$H_{d,c} = L(d) Q_n^R\langle GM(r_{i_0}, X_{i_0})^e, \dots, GM(r_{i_n}, X_{i_n})^e\rangle(d, c)$$

is a Noetherian $L(d)$-module. Indeed, it suffices to prove the assertion for each of $L(d)$-modules

$$L(d)\, e_{i_0}\, \Delta(d,c)\, \theta_{j_1}\, \Delta(d,c)\ldots\theta_{j_n} e_{i_n} \Delta(d,c).$$

The standard argument related to the Shirshov Theorem on height and the Hamilton–Cayley type identities (see [1]) shows that $L(d)$ is a finitely generated k-algebra and $L(d)\Delta(d,c)$ is a finitely generated $L(d)$-module. Consequently, the $L(d)$-module $H_{d,c}$ is Noetherian.

In what follows, we need

Proposition 3. *Let $\overline{x}_1, \ldots, \overline{x}_{r^2}, \overline{y}$ be the images of the variables $x_1, \ldots, x_{r^2}, y$ of the algebra F in the algebra of general matrices $GM(r,X)$. Then*

$$\operatorname{tr}\overline{y}\, \operatorname{tr}\overline{x}_1 \ldots \operatorname{tr}\overline{x}_{r^2} \in \{\operatorname{tr}\overline{x}_1, \operatorname{tr}\overline{x}_1 \operatorname{tr}\overline{x}_2, \ldots, \operatorname{tr}\overline{x}_1 \ldots \operatorname{tr}\overline{x}_{r^2}\}^T.$$

The proof of Proposition 3 (we omit it here) is based on the important property of Capelli polynomials c_n, i.e., polynomials of F having the form

$$c_n(x_1, \ldots, x_n, y_0, \ldots, y_n) = \sum_{\sigma \in \mathrm{Sym}(n)} (-1)^\sigma y_0 x_{\sigma(1)} y_1 \ldots x_{\sigma(n)} y_n.$$

The following property of Capelli polynomials is well known (see [3, p. 25]): in the algebra $\operatorname{Tr} GM(r,X)$ the following identity holds:

$$\sum_{i=1}^{r^2} c_{r^2}(\overline{x}_1, \ldots, \overline{x}_i \overline{z}, \ldots, \overline{x}_{r^2}, \overline{y}_0, \ldots, \overline{y}_{r^2}) = r\ \operatorname{tr}\overline{z}\ c_{r^2}(\overline{x}_1, \ldots, \overline{x}_{r^2}, \overline{y}_0, \ldots, \overline{y}_{r^2}). \tag{2}$$

Corollary 1. *The T-space $k[\operatorname{tr}(GM(r,X))]$ is generated by*

$$\operatorname{tr}\overline{x}_1, \operatorname{tr}\overline{x}_1 \operatorname{tr}\overline{x}_2, \ldots, \operatorname{tr}\overline{x}_1 \ldots \operatorname{tr}\overline{x}_{r^2}.$$

Corollary 2. *Let $\widetilde{x}_j$ and $\widetilde{y}$ be diagonal block matrices of the form*

$$\widetilde{x}_j = \begin{pmatrix} [x_j^{(i_0)}]_{r_{i_0}} & & 0 \\ & \ddots & \\ 0 & & [x_j^{(i_n)}]_{r_{i_n}} \end{pmatrix}, \quad \widetilde{y} = \begin{pmatrix} [y_j^{(i_0)}]_{r_{i_0}} & & 0 \\ & \ddots & \\ 0 & & [y_j^{(i_n)}]_{r_{i_n}} \end{pmatrix}$$

where $x_j^{(i_\alpha)}$ and $y^{(i_\alpha)}$ denote the images of variables x_j and y of the algebra F in the algebra $GM(r_{i_\alpha}, X_{i_\alpha})$. Then

$$\operatorname{tr}\widetilde{y}\, \operatorname{tr}\widetilde{x}_1 \ldots \operatorname{tr}\widetilde{x}_{r^2} \in \{\operatorname{tr}\widetilde{x}_1, \operatorname{tr}\widetilde{x}_1 \operatorname{tr}\widetilde{x}_2, \ldots, \operatorname{tr}\widetilde{x}_1 \ldots \operatorname{tr}\widetilde{x}_{r^2}\}^T$$

where $r = r_{i_0}^2 + \cdots + r_{i_n}^2$.

In order to prove the last assertion, we apply Proposition 3 to general $r \times r$-matrices, vanish some of commutative variables, and set equal some of the rest commutative variables to obtain the matrices $\widetilde{x}_j$, $\widetilde{y}$ from general $r \times r$-matrices.

7. Capelli spaces

Let $r \geq 2$ and $C(r, X)$ denote the T-ideal in the algebra $GM(r, X)$ generated by the Capelli polynomial c_{r^2} (Capelli ideal). As is well-known (see [3]), this ideal is distinct from zero, and the relatively free algebra $GM(r, X)/C(r, X)$ is covered by the algebra F/M_{r-1}^m for some m (its radical is nilpotent as was shown in [4]). By definition, set $C(1, X) = 0$.

By a *Capelli space* in the elementary space

$$Q_n\langle GM(r_{i_0}, X_{i_0})^e, \dots, GM(r_{i_n}, X_{i_n})^e\rangle$$

we mean its T-subspace of the form $V = Q_n\langle V_0, \dots, V_n\rangle$ where V_α stands for the T-ideal $C(r_{i_\alpha}, X_{i_\alpha})$ of the algebra $GM(r_{i_\alpha}, X_{i_\alpha})$ if $r_{i_\alpha} \geq 2$ or the algebra $GM(r_{i_\alpha}, X_{i_\alpha})^e$. Note that an elementary space is a particular case of a Capelli space when all the V_α coincide with algebras $GM(r_{i_\alpha}, X_{i_\alpha})^e$. The number $r(V)$ equal the maximum of all r_{i_α} such that V_α coincides with the algebra $GM(r_{i_\alpha}, X_{i_\alpha})^e$ is called the *complexity* of Capelli space. We set $r(V) = 0$ if there are no such spaces V_α. Thus, the inequality $r(V) \leqslant 1$ implies that each space V_α is either a Capelli ideal or a ring of commutative polynomials. Note that the space $Q_0(V_0)$ is either $C(r_{i_0}, X_{i_0})$ or $GM(r_{i_0}, X_{i_0})^e$.

Since the T-space $Q_n(GM(r, X))$ containing the T-space M_r^n/M_r^{n+1} (see Proposition 1, Corollary 2) is elementary and, consequently, a Capelli space, to prove the main result it suffices to establish the finite basedness of a T-space in any Capelli space.

After suitable renumbering, it suffices to consider an elementary space of the form

$$Q_n\langle GM(r_0, X_0)^e, \dots, GM(r_n, X_n)^e\rangle$$

where $X_i \neq X_j$ if $i \neq j$. Furthermore, we may assume that the Capelli space V is represented as

$$V = Q_n\langle GM(r_0, X_0)^e, \dots, GM(r_p, X_p)^e,\ C(r_{p+1}, X_{p+1}), \dots, C(r_n, X_n)\rangle$$

where $0 \leqslant p \leqslant n$. Then $r(V) = \max\{r_0, \dots, r_p\}$ or (formally set $p = -1$)

$$V = Q_n\langle C(r_0, X_0), \dots, C(r_n, X_n)\rangle$$

if $r(V) = 0$. Taking into account the fact that all sets X_i are pairwise distinct, we may assume that V is a T-subspace in $Q_n^R(\Phi(r_0, \dots, r_n))$.

Consider the matrix

$$Z = \begin{pmatrix} z_{11} & \dots & z_{1r^2} \\ & \ddots & \\ z_{N1} & \dots & z_{Nr^2} \end{pmatrix}$$

composed by the variables of the free algebra F which are numbered for convenience by double indices, where $r = r_0^2 + \dots + r_n^2$ and N is a natural number.

If $r(V) \leqslant 1$ and N is greater than some number depending on $r_0, \dots, r_n$ and d, then there exists a mapping $f \mapsto f^Z$ acting from $V(d, C)$ into $V(d, Nr^2)$ (the construction of this mapping is based on (2), we omit details) such that

1) $f^Z = \sum_{i=1}^{N} \operatorname{tr} \widetilde{z}_{i1} \dots \operatorname{tr} \widetilde{z}_{ir^2} f^{[i]}$ *where*

$$\widetilde{z}_{ij} = \begin{pmatrix} [z_{ij}^{(0)}]_{r_0} & & 0 \\ & \ddots & \\ 0 & & [z_{ij}^{(n)}]_{r_n} \end{pmatrix},$$

   $z_{ij}^{(\alpha)}$ *denotes the image of the variable z_{ij} in the algebra of general matrices $GM(r_\alpha, X_0)$, and the quasipolynomial $\operatorname{tr} \widetilde{z}_{i1} \dots \operatorname{tr} \widetilde{z}_{ir^2} f^{[i]}$ is linear in each of Nr^2 variables z_{ij} and $f^{[i]} \mid_{z_{ij}=1} = tf$ for some natural t;*

2) $f^Z \in \{f\}^T$ *and any quasipolynomial obtained from a quasipolynomial belonging to the T-space $\{f^Z\}^T$ by the specialization of any of the variables in the set $\{z_{ij}\}$ to unit, belongs to the T-space $\{f\}^T$.*

Note that the specialization of the variable $x_i = \overline{x}_i + \theta_i$ to the unit means the result of replacement of $\overline{x}_i$ and θ_i by e and 0.

8. The proof of the main result

To prove the theorem we need the following lemmas.

Lemma 1. *Let $r(V) \leqslant 1$. Then any T-space in the space V is finitely based.*

Proof. On the algebra $Q^R(\Phi(r_0, \dots, r_s))/(\Theta^R(\Phi(r_0, \dots, r_n)))^{n+1}$, a Capelli identity of some degree is valid. Consequently, with the help of the representation theory of symmetric groups, the proof is reduced to the case of quasipolynomials in d variables for some natural number d, i.e., the space $V(d, 0)$.

For any subset S of $V(d, 0)$ we put

$$S^Z = \{f^Z \mid f \in S\}, \quad S^{[i]} = \{f^{[i]} \mid f \in S\}.$$

By the properties of the mapping $f \mapsto f^Z$, the inclusion

$$V(d,0)^Z \subset \sum_{i=1}^{N} \operatorname{tr} \widetilde{z}_{i1} \dots \operatorname{tr} \widetilde{z}_{ir^2}\, V(d,0)^{[i]}$$

is valid. Therefore, for any subset $S \subset V(d,0)$ we have

$$S^Z \subset \sum_{i=1}^{N} \operatorname{tr} \widetilde{z}_{i1} \dots \operatorname{tr} \widetilde{z}_{ir^2}\, S^{[i]} \subset \sum_{i=1}^{N} \operatorname{tr} \widetilde{z}_{i1} \dots \operatorname{tr} \widetilde{z}_{ir^2}\, S_i$$

where S_i is the subspace of $V(d, Nr^2)$ obtained from the set $S^{[i]}$ by means of all substitutions of the type $z_{\alpha\beta} \mapsto g_{\alpha\beta}$, where $\alpha \neq i$ and $g_{\alpha\beta} \in k\langle 1, x_1, \dots, x_d, \{z_{\alpha\beta} \mid \alpha \neq i\}\rangle$, such that the result of a substitution is again a quasipolynomial linear in all variables $z_{\alpha\beta}$, $\alpha \neq i$. Since the $L(d)$-module H_{d,Nr^2} is Noetherian (see Section 6), the module $L(d)\operatorname{tr} \widetilde{z}_{i1} \dots \operatorname{tr} \widetilde{z}_{ir^2} S_i$ is also Noetherian. Then, by Proposition 3, the multiplication by elements of the algebra $L(d)$ may be replaced by substitutions in variables $z_{i1}, \dots, z_{ir^2}$ and linear actions. Hence, for any subset P_i of the space $\operatorname{tr} \widetilde{z}_{i1} \dots \operatorname{tr} \widetilde{z}_{ir^2} S_i$, there exists a finite subset $G_i \subset P_i$ such that each element of P_i can be obtained from elements of the set G_i by means of substitution of variables $z_{i1}, \dots, z_{ir^2}$ and linear actions. It is clear that in this case there exists a finite subset G of the set S^Z which generates S^Z with the help of the same actions. It remains to observe that, after specialization $z_{ij} \mapsto 1$, any element f^Z of S^Z turns into an element divisible by f. Thus, any subset $S \subset V(d,0)$ generates a finitely based T-space. Lemma 1 is proved.

Remark 1. It is easy to see that specialization of some of the variables z_{ij} to the unit is needed to obtain the products $\operatorname{tr} \widetilde{z}_{i1} \dots \operatorname{tr} \widetilde{z}_{ir^2-1}, \dots, \operatorname{tr} \widetilde{z}_{i1}$, along with the products $\operatorname{tr} \widetilde{z}_{i1} \dots \operatorname{tr} \widetilde{z}_{ir^2}$ and, further, to obtain from them the product of r^2+1 traces.

Remark 2. The mapping $f \mapsto f^Z$ allows us to provide the fact that the $L(d)$-module $H_{d,c}$ is Noetherian without using multiplication by traces, but by means of substitutions and linear operations only, i.e., by means of the algebra kT.

Lemma 2. *Any T-space in the Capelli space V is finitely based.*

Proof. We proceed by induction on $(r(V),\ p+1)$. For the base of induction see Lemma 1. Let $r(V) > 1$. Consequently, $p+1 > 0$. Consider the morphism of T-spaces

$$\varphi :\ V \to Q_n\langle GM(r_0, X_0)^e / C(r_0, X_0), \dots, GM(r_p, X_p)^e / C(r_p, X_p)$$

$$C(r_{p+1}, X_{p+1}), \dots, C(r_n, X_n)\rangle$$

induced by canonical homomorphisms

$$\varphi_i :\ GM(r_i, X_i) \to GM(r_i, X_i)/C(r_i, X_i).$$

It is clear that

$$\ker\varphi = \sum_{i=1}^{p} Q_n\langle GM(r_0,X_0)^e,\dots,C(r_i,X_i),\dots,GM(r_p,X_p)^e,$$

$$C(r_{p+1},X_{p+1}),\dots,C(r_n,X_n)\rangle,$$

i.e., $\ker\varphi$ can be represented as the sum of Capelli spaces for which the number $\rho+1$ is by 1 less than that for the space V and hence the assertion of the lemma is true in $\ker\varphi$ by the induction assumption. On the other hand, in view of the existence of surjective morphisms of T-spaces

$$\psi_i:\ F^e/M_{s_i}^{m_i}\to GM(r_i,X_i)^e/C(r_i,X_i)$$

where

$$s_i=\begin{cases} r_i-1 & \text{if } r_i>1,\\ 1 & \text{if } r_i=1,\end{cases}$$

and m_i is some natural number (see Section 7), there exists a surjective morphism of T-spaces

$$\psi:\ Q_n\langle F^e/M_{s_0}^{m_0},\dots,F^e/M_{s_p}^{m_p},C(r_{p+1},X_{p+1}),\dots,C(r_n,X_n)\rangle$$

$$\to Q_n\langle GM(r_0,X_0)^e/C(r_0,X_0),\dots,GM(r_p,X_p)^e/C(r_p,X_p),$$

$$C(r_{p+1},X_{p+1}),\dots,C(r_n,X_n)\rangle.$$

Thus, it suffices to prove the finite basedness of any T-subspace in any of the T-spaces

$$\Lambda=Q_n\langle\Lambda_0,\dots,\Lambda_p,C(r_{p+1},X_{p+1}),\dots,C(r_n,X_n)\rangle$$

where $\Lambda_i = GM(1,X_i)$ if $r_i=1$ and $\Lambda_i=M_{r_i-1}^{t_i}/M_{r_i-1}^{t_i+1}$ if $r_i>1$; here $0\leqslant t_i<m_i$. Since, for $r_i>1$, the T-space Λ is embeddable into $Q_{t_i}\langle GM(r_i-1,Y_i)^e,\dots,GM(r_i-1,Y_i)^e\rangle$ (see Proposition 1, Corollary 2), Λ is embeddable (as a T-subspace) into the T-space

$$W=Q_{n+1}\langle\underbrace{GM(s_0,Y_0)^e,\dots,GM(s_0,Y_0)^e}_{t_0+1},\dots,\underbrace{GM(s_p,Y_p)^e,\dots,GM(s_p,Y_p)^e}_{t_p+1},$$

$$C(r_{p+1},X_{p+1}),\dots,C(r_n,X_n)\rangle$$

where $t=t_0+\dots+t_p$ and $t_i=0$ if $r_i=1$. It remains to observe that $r(W)=r(V)-1$ and employ the induction assumption. Lemma 2 is proved.

The proof of the theorem. Since the radical of the algebra F/I is nilpotent (see [4]), it suffices to prove the theorem for the algebra F/M_r^m for some r and m. But this case is reduced to a similar assertion for $M_r^n/M_r^{n+1}\subset Q_n(GM(r,X))$, i.e., to elementary spaces. Hence the assertion of the theorem is reduced to Lemma 2.

Remark 3. Since the kT-module F/I is Noetherian, the theorem remains valid if we replace the ground field k by an arbitrary commutative k-algebra of finite type.

Remark 4. The methods and results of this article lead to the essential generalization of the results of [1] related to finite basedness of systems of generalized polynomials.

Remark 5. Employing the methods analogous to the exposed above, one can prove that a tensor product over a field k of finite number of algebras F/I (considered as a module over a tensor product over a field k of the same number of algebras kT) is Noetherian. This fact and Proposition 3 imply finite basedness of any T-space in the algebra $\operatorname{Tr} GM(r, X)$. Indeed, by Proposition 3, this algebra considered as a kT-module is generated by the following finite set of its elements:

$$\operatorname{tr}\overline{x}_1, \dots, \operatorname{tr}\overline{x}_1 \dots \operatorname{tr}\overline{x}_{r^2},\ (\operatorname{tr}\overline{x}_1)\,\overline{y}, \dots, (\operatorname{tr}\overline{x}_1 \dots \operatorname{tr}\overline{x}_{r^2})\,\overline{y},\ \overline{y};$$

here $\overline{x}_i$ and $\overline{y}$ are images of variables of the algebra F.

Remark 6. As in [1], the results obtained allow us to give a positive answer to the question of representability of the algebra F/I.

References

[1] Grishin, A. V., On finitely based systems of generalized polynomials. Izv. Akad. Nauk SSSR, Ser. Mat. 54:5 (1990), 899–927. English translation: Math. USSR, Izv. 37:2 (1991), 243–272.

[2] Kemer, A. R., Finite basability of identities of associative algebras. Algebra Logika 26:5 (1987), 597–641. English translation: Algebra Logic 26:5 (1987), 362–397.

[3] Rowen, L. H., Polynomial identities in ring theory. Academic Press, New York 1980.

[4] Brown, A., The radical in a finitely generated PI-algebra. Bull. Am. Math. Soc., New Ser. 7:2 (1982), 385–386.

Non-cancellation phenomena in a class of finitely generated nilpotent groups

Peter Hilton and Christopher Schuck

Abstract. A subclass of finitely-generated nilpotent groups with finite commutator subgroup is defined. The recent results of the authors are applied to obtain non-cancellation theorems for direct products of groups in this class. Here "non-cancellation" includes "non-uniqueness of decomposition as a direct product of indecomposable."

1991 Mathematics Subject Classification: 20F18.

Recently, we have made a systematic study of the abelian group $\mathcal{G}(N^k)$, where N is a finitely-generated (fg) nilpotent group belonging to the class $\mathcal{N}_1$ defined below, N^k is the direct product of k copies of N, $k \geq 1$, and $\mathcal{G}$ is the Mislin genus [1, 2].

Let TN be the torsion subgroup of N and FN the torsionfree quotient, $FN = N/TN$, so that

$$TN \rightarrowtail N \twoheadrightarrow FN \tag{0.1}$$

is a short exact sequence naturally associated with N. We assume N to be fg, so that TN is finite. We say that $N \in \mathcal{N}_0$ if FN is *commutative* (equivalently, if the commutator subgroup $[N, N]$ is finite). Then Mislin showed in [1] that $\mathcal{G}(N)$ may be given the structure of a finite abelian group with N as zero element. We recall that $\mathcal{G}(N)$ is the set of isomorphism classes of fg nilpotent groups M such that $M_p \cong N_p$ for all primes p. By abuse of language we write $M \in \mathcal{G}(N)$ if the isomorphism class of M belongs to $\mathcal{G}(N)$.

We define a subclass $\mathcal{N}_1$ of $\mathcal{N}_0$ by three requirements (see [3]). Thus $N \in \mathcal{N}_1$ if

(i) TN is commutative;

(ii) the extension (0.1) is a split extension for the action $\omega : FN \rightarrow \operatorname{Aut} TN$;

(iii) $\omega(FN)$ lies in the center of $\operatorname{Aut} TN$.

We remark

Proposition 0.1. *The fg nilpotent group N belongs to $\mathcal{N}_1$ if TN and FN are cyclic.*

Starting in [3], and continuing in [4–6], the group $\mathcal{G}(N^k)$ has been completely calculated for $N \in \mathcal{N}_1$. It should be noted that N^k will, in general, fail to belong to $\mathcal{N}_1$ if $k \geq 2$, since condition (iii) is not inherited by direct products. It turns out that $\mathcal{G}(N^k)$ is completely determined by a certain invariant $t = t(N)$, defined as the *height* of $\ker \omega$ in FN. Thus we have, as a consequence of the main theorem of [5],

Theorem 0.2. *$\mathcal{G}(N^k) = 0$, $k \geq 1$, if FN is not cyclic.*

Theorem 0.3. *If FN is cyclic, then $\mathcal{G}(N) \cong (\mathbb{Z}/t)^*/\{\pm 1\}$, where $(\mathbb{Z}/t)^*$ is the multiplicative group of units in $\mathbb{Z}/t$. Moreover, the isomorphism may be chosen so that the group N_m corresponding to the class $[m]$, for m prime to t, is obtained by replacing the action ω by the action $FN \xrightarrow{m} FN \xrightarrow{\omega} \operatorname{Aut} TN$.*

Notice that, with FN cyclic, t is the order of $\omega(\xi)$ in $\operatorname{Aut} TN$, where $FN = \langle \xi \rangle$. Let us illustrate Theorem 0.3 with an example.

Example 0.1. Let $N = \langle x, y \mid x^{25} = 1, yxy^{-1} = x^6 \rangle$. Obviously N is nilpotent of class 2 and $N \in \mathcal{N}_1$ by Proposition 0.1, since $TN = \langle x \rangle$ and FN is cyclic, say, $FN = \langle \xi \rangle$. Then it is easy to verify that $\ker \omega = \langle \xi^5 \rangle$, so that $t = 5$ and $\mathcal{G}(N) \cong (\mathbb{Z}/5)^*/\{\pm 1\} \cong C_2$, where C_r is a cyclic group of order r. As generator of $(\mathbb{Z}/5)^*/\{\pm 1\}$ we may take the class [2]. Then N_2 is obtained by replacing the second relation in the presentation of N by $yxy^{-1} = x^{6^2}$. Thus

$$N_2 = \langle x, y \mid x^{25} = 1, yxy^{-1} = x^{11} \rangle.$$

Remarks. (i) Notice that, in Theorem 0.3, $(\mathbb{Z}/t)^*/\{\pm 1\}$ is a multiplicative group while $\mathcal{G}(N)$ is written additively. Thus, for m_1, m_2 prime to t, we have

$$N_{m_1} + N_{m_2} = N_{m_1 m_2} \tag{0.2}$$

(ii) We will exploit Proposition 0.1 so that all our explicit examples will have TN cyclic; of course, Theorem 0.2 obliges us to take FN cyclic to avoid triviality. However, we emphasize that we are always able to find examples of the phenomena discussed in this paper with TN non-cyclic.

(iii) It is critical to our arguments that, as shown in [3], t is a T-number, where $T = T(N)$ is the set of primes p such that N has p-torsion. In fact, given t, we may always choose N so that t is divisible by *every* prime such that N has p-torsion. This is fully explained in Section 1.

Now let $\rho : \mathcal{G}(N) \to \mathcal{G}(N^k)$, $k \geq 2$, be given by $\rho(M) = M \times N^{k-1}$. It is shown in [3] that, for $N \in \mathcal{N}_0$, ρ is a surjective homomorphism. Thus, if $N \in \mathcal{N}_1$, we determine $\mathcal{G}(N^k)$, $k \geq 2$, by calculating $\ker \rho$.

Theorem 0.4 [4, 6]. *Let $N \in \mathcal{N}_1$ with FN cyclic. Then* $\ker \rho$ *consists of those groups N_m such that*

$$m \equiv \pm 1 \mod p_i^{l_i},\ i = 1, 2, \dots, \lambda \tag{0.3}$$

where

$$t = \prod_{i=1}^{\lambda} p_i^{l_i}. \tag{0.4}$$

Our object in this paper is to exploit Theorems 0.3 and 0.4 to obtain non-cancellation theorems for direct products of groups in $\mathcal{N}_1$. Here we stretch the meaning of "non-cancellation" to include "non-uniqueness of decomposition as a direct product of indecomposables." In addition to Theorems 0.3 and 0.4, we also exploit a result from [2] which applies to groups in $\mathcal{N}_0$. This crucial result (Theorem 1.1) relates the addition in $\mathcal{G}(N)$ to the direct product operation; and we exploit Theorem 0.4 to show that its converse is false. We also show in Section 1 how to ensure that groups in $\mathcal{N}_1$ are indecomposable.

We lay stress on the fact that our results depend only on elementary arithmetical properties of the invariant $t(N)$. Thus the non-cancellation results which we establish for N will apply equally well to any other group $\overline{N}$ in $\mathcal{N}_1$ such that $t(\overline{N}) = t(N)$.

Finally, we make a point about notation. If we write, for example, $L_i \in \mathcal{G}(N)$, the subscript i is merely a numerical index. However, if we write N_m $(N_\mu, N_\nu) \in \mathcal{G}(N)$, then the subscript is an integer prime to t and N_m is constructed according to Theorem 0.3 and relation (0.2) holds.

1. On the genus and direct products

We collect together in this section some basic results not explicitly related to the calculation of $\mathcal{G}(N^k)$. The first result is to be found in [2].

Theorem 1.1. *Let $N \in \mathcal{N}_0$, and let $L_1, L_2 \dots, L_k$; $M_1, M_2, \dots, M_k \in \mathcal{G}(N)$ with*

$$L_1 + L_2 + \cdots + L_k = M_1 + M_2 + \cdots + M_k.$$

Then

$$L_1 \times L_2 \times \cdots \times L_k \cong M_1 \times M_2 \times \cdots \times M_k.$$

The proof proceeds as follows. We consider the function

$$\Phi : \mathcal{G}(N)^k \to \mathcal{G}(N^k),\ k \geq 2,$$

given by

$$\Phi(L_1, L_2, \ldots, L_k) = L_1 \times L_2 \times \cdots \times L_k. \tag{1.1}$$

Then it is shown in [2] that Φ is a homomorphism, from which it follows that

$$\Phi(L_1, L_2, \ldots, L_k) = \Phi(L_1 + L_2 + \cdots + L_k, N, \ldots, N),$$

so that

$$L_1 \times L_2 \times \cdots \times L_k \cong (L_1 + L_2 + \cdots + L_k) \times N^{k-1}. \tag{1.2}$$

Now if $\rho : \mathcal{G}(N) \to \mathcal{G}(N^k)$ is given by $\rho(L) = L \times N^{k-1}$ then, of course, ρ is also a homomorphism. It is shown in [3] that ρ is surjective, so that Φ is certainly surjective. Moreover, Theorem 0.4 is obtained from Theorem 0.3 by analysing $\ker \rho$, when $N \in \mathcal{N}_1$, [4, 6]. It is a striking fact that $\ker \rho$ is independent of $k \geq 2$, so that $\mathcal{G}(N^k)$ is also independent of $k \geq 2$.

We now confine attention to groups $N \in \mathcal{N}_1$ with *TN cyclic*; this simplifies our descriptions and is adequate to exemplify all the phenomena we wish to describe. Since we require that $\mathcal{G}(N)$ should be non-trivial, we are compelled by Theorem 0.2 to take *FN* cyclic. Thus N may be described as follows.

$$N = \langle x, y \mid x^n = 1, yxy^{-1} = x^u \rangle. \tag{1.3}$$

It is easy to show

Proposition 1.2. *The group N, given by* (1.3), *is nilpotent if and only if $p | u - 1$ for every prime divisor p of n.*

Moreover, given that N is nilpotent, then certainly $N \in \mathcal{N}_1$. The calculation of t as a function of n and u is made as follows.

Let

$$n = \prod_{i=1}^{\lambda} p_i^{s_i},\ s_i \geq 1, \tag{1.4}$$

so that

$$u = 1 + c \prod_{i=1}^{\lambda} p_i^{m_i},\ p_i \nmid c,\ m_i \geq 1. \tag{1.5}$$

We assume the primes p_i enumerated in increasing order. Then, unless we are in the exceptional case $p_1 = 2$, $m_1 = 1$, $s_1 \geq 3$, we have

$$t = \prod_{i=1}^{\lambda} p_i^{l_i}, \tag{1.6}$$

with l_i given by

$$l_i = \max(0, s_i - m_i). \tag{1.7}$$

In the exceptional case we write $u = -1 + c'2^{m'}$, with c' odd, and replace l_1 in (1.6) by

$$l_1' = \max(1, s_1 - m'). \tag{1.7'}$$

We now discuss the question of the indecomposability of N, given by (1.3). Recall that, of course, $N \in \mathcal{N}_1$.

We partition $I = (1, 2, \ldots, \lambda)$ as $I_1 \cup I_2$, where

$$I_1 = \{i \in I \mid m_i < s_i\},\ I_2 = \{i \in I \mid m_i \geq s_i\}, \tag{1.8}$$

and set

$$n_1 = \prod_{i \in I_1} p_i^{s_i}, \quad n_2 = \prod_{i \in I_2} p_i^{s_i},$$

so that $n = n_1 n_2$. Now let

$$N_1 = \langle x, y | x^{n_1} = 1, yxy^{-1} = x^u\rangle,\ T_2 = \langle z | z^{n_2} = 1\rangle. \tag{1.9}$$

Theorem 1.3. *$N \cong N_1 \times T_2$ and $\mathcal{G}(N_1) \cong \mathcal{G}(N)$ under the mapping $L_1 \mapsto L_1 \times T_2$, $L_1 \in \mathcal{G}(N_1)$.*

Proof. The torsion subgroup of $N_1 \times T_2$ is $TN_1 \times T_2$. Since TN_1 is cyclic of order n_1, generated by x, and T_2 is cyclic of order n_2, generated by z, and since n_1, n_2 are mutually prime with $n = n_1 n_2$, it follows that $TN_1 \times T_2$ is cyclic of order n, generated by (x, z). Now $yxy^{-1} = x^u$ and $yzy^{-1} = z = z^u$, since $z^{n_2} = 1$ and $p_i^{s_i} | u - 1$ for each $i \in I_2$ by (1.5) and (1.8), so that $n_2 | u - 1$. This shows that $N_1 \times T_2 \cong N$. The second statement of the theorem is just a special case of Theorem 5.1 of [4]. □

In the light of this theorem we may confine ourselves, in drawing conclusions about non-cancellation from the calculation of $\mathcal{G}(N)$, to the case in which $m_i < s_i$ for *all* $i \in I$. *This we now assume.* In the light of (1.7), (1.7′), this is equivalent to assuming

$$l_i \geq 1,\ i \in I = (1, 2, \ldots, \lambda). \tag{1.10}$$

We now prove, for N given by (1.3), with n, u given by (1.4), (1.5),

Theorem 1.4. *Under the assumption* (1.10) *every group in $\mathcal{G}(N)$ is indecomposable.*

Proof. Since t is, of course, a genus invariant, it suffices to prove that N is itself indecomposable. Now if $N = N' \times N''$ then $FN = FN' \times FN''$. But FN is cyclic infinite; thus we may assume $FN' = FN$, $FN'' = 0$. Then $TN = TN' \times N''$. But

TN is cyclic of order n, so we may partition I as $I' \cup I''$, where N'' is cyclic of order

$$n'' = \prod_{i \in I''} p_i^{s_i}.$$

If $n = n'n''$ then $N'' = \langle x^{n'} \rangle$ and $yx^{n'}y^{-1} = x^{n'}$. Hence $un' \equiv n' \mod n$ or $u \equiv 1 \mod n''$. But if $i \in I''$ then the p_i-exponent of $u-1$ is m_i, whereas the p_i-exponent of n'' is $s_i > m_i$. Thus I'' must be empty, showing that N'' is trivial and N is indecomposable. □

We will henceforth assume that N is given by (1.3), with n, u given by (1.4), (1.5) respectively, subject to $m_i < s_i$ or, equivalently, (1.10).

It is now easy to prove

Theorem 1.5. *Given any positive integer t, we may choose n, u in* (1.3) *so that N is a nilpotent group in $\mathcal{N}_1$ such that $t(N) = t$. Moreover, N may be chosen to be indecomposable.*

Proof. Let $t = \prod_{i=1}^{\lambda} p_i^{l_i}$ with $l_i \geq 1$. If $p_1 \neq 2$, set

$$n = \prod_{i=1}^{\lambda} p_i^{l_i+1}, \quad u = 1 + \prod_{i=1}^{\lambda} p_i.$$

If $p_1 = 2$, set

$$n = 2^{l_1+2} \prod_{i=2}^{\lambda} p_i^{l_i+1}, \quad u = 1 + 4 \prod_{i=2}^{\lambda} p_i.$$

Then the formula (1.7) guarantees that $t(N) = t$; Proposition 1.2 guarantees that N is nilpotent, and Theorem 1.4 guarantees that N is indecomposable. □

Remarks. (a) Of course, all groups in the genus of N are indecomposable.

(b) Clearly there are infinitely many ways of constructing a suitable group N with $t(N) = t$. For all such groups, n and t will have precisely the same prime divisors. For any such prime divisor p, the p-exponent of t will be strictly less than the p-exponent of n. Thus n must be such that, for any prime divisor p, its p-exponent satisfies $p\text{-}\exp n \geq 2$.

(c) There is an obvious generalization of Theorem 1.3 to the case in which TN is not cyclic (but, of course, $N \in \mathcal{N}_1$). We still have quantities n, u, t, as in (1.4)-(1.7′) and we may again partition I as in (1.8). We then define

$$N_1 = \langle TN_p,\, p \in I_1,\, y \rangle,\; T_2 = \langle TN_p, p \in I_2 \rangle, \tag{1.11}$$

and prove

Theorem 1.6. *$N \cong N_1 \times T_2$ and $\mathcal{G}(N_1) \cong \mathcal{G}(N)$ under the mapping $L_1 \mapsto L_1 \times T_2$, $L_1 \in \mathcal{G}(N_1)$.*

However, care must be exercised in generalizing Theorem 1.4.

2. The non-cancellation theorems

We will assume (1.3); thus

$$N = \langle x, y | x^n = 1, yxy^{-1} = x^u \rangle, \tag{2.1}$$

where (n, u) are chosen so that N is nilpotent; specifically,

$$n = \prod_{i=1}^{\lambda} p_i^{s_i}, \; s_i \geq 2 \tag{2.2}$$

and

$$\mathcal{G}(N) \cong (\mathbb{Z}/t)^* / \{\pm 1\} \tag{2.3}$$

with

$$t = \prod_{i=1}^{\lambda} p_i^{l_i}, \; 1 \leq l_i < s_i. \tag{2.4}$$

Thus, in particular, every group in $\mathcal{G}(N)$ is indecomposable. Then Theorem 1.1, together with Theorem 0.3 and (0.2), provides us with a theorem on non-uniqueness of decomposition; however, this theorem will be improved later (Theorem 2.3).

Theorem 2.1. *Let N be as above, and let $\mu_1, \mu_2, \ldots, \mu_k; \nu_1, \nu_2, \ldots, \nu_k$ be prime to t such that $\mu_1 \mu_2 \ldots \mu_k \equiv \pm \nu_1 \nu_2 \ldots \nu_k \mod t$. Then*

$$N_{\mu_1} \times N_{\mu_2} \times \cdots \times N_{\mu_k} \cong N_{\nu_1} \times N_{\nu_2} \times \cdots \times N_{\nu_k}, \tag{2.5}$$

with all groups N_{μ_i}, N_{ν_j} indecomposable. In particular,

$$N_\mu^k \cong N^k,$$

for any μ prime to t and $k = \frac{1}{2}\Phi(t)$.

Example 2.1. Choose N so that $t = 15$. To be specific, we may take $n = 225$, $u = 16$. Then $(\mathbb{Z}/15)^*/\{\pm 1\} \cong C_4$, generated by [2]. Thus the genus of N consists of N, N_2, N_4, N_8 and we have

$$N \times N_2 \cong N_4 \times N_8,$$

$$N \times N_4 \cong N_2 \times N_2 \cong N_8 \times N_8,$$

$$N \times N_8 \cong N_2 \times N_4,$$

$$N \times N \cong N_4 \times N_4,$$

$$N^4 \cong N_2^4 \cong N_4^4 \cong N_8^4.$$

We now move to our second non-cancellation result. Let N, N' be two groups as in (2.1)-(2.4), such that $T(N) \cap T(N') = \emptyset$. Then $N \times N' \in \mathcal{N}_1$, but $t(N \times N') = 1$ (see [3]). Thus

$$\mathcal{G}(N \times N') = 0.$$

We conclude

Theorem 2.2. *Let N, N' be as above. Then the isomorphism class of $M \times M'$ is independent of the choice of M, M' within $\mathcal{G}(N)$, $\mathcal{G}(N')$ respectively.*

Example 2.2. Let N be as in Example 2.1 and choose N' so that $t' = 16$. To be specific, we may take $n' = 64$, $u' = 5$. Then $T(N) = \{3, 5\}$, $T(N') = \{2\}$, so that $\mathcal{G}(N \times N') = 0$. We know that $\mathcal{G}(N)$ consists of $N(= N_1)$, N_2, N_4, N_8. Likewise $\mathcal{G}(N') \cong (\mathbb{Z}/16)^*/\{\pm 1\} \cong C_4$, generated by [3], so that $\mathcal{G}(N')$ consists of $N'(= N'_1)$, N'_3, N'_9, N'_{11}. We infer that all 16 groups $N_i \times N'_j$, $i = 1, 2, 4, 8$, $j = 1, 3, 9, 11$, are isomorphic. Here we have examples of non-uniqueness of decomposition which include examples of genuine non-cancellation.

We now invoke our results on $\mathcal{G}(N^k)$, $N \in \mathcal{N}_1$, to obtain a third class of non-cancellation result. It is to be noted that this result implies non-cancellation *within the genus*, a phenomenon which had not previously been detected.

Let N be as before with $\mathcal{G}(N) \cong (\mathbb{Z}/t)^*/\{\pm 1\}$, for a given $t = \prod_{i=1}^{\lambda} p_i^{l_i}$, $l_i \geq 1$.

Theorem 2.3. *With n as above and $k \geq 2$, let $\mu_1, \mu_2, \ldots, \mu_k$, $\nu_1, \nu_2, \ldots, \nu_k$ be prime to t. Then*

$$N_{\mu_1} \times N_{\mu_2} \times \cdots \times N_{\mu_k} \cong N_{\nu_1} \times N_{\nu_2} \times \cdots \times N_{\nu_k}$$

if and only if

$$\mu_1\mu_2 \ldots \mu_k \equiv m\nu_1\nu_2 \ldots \nu_k \mod t,$$

with m satisfying (0.3).

Proof. Set $\mu = \mu_1\mu_2 \ldots \mu_k$, $\nu = \nu_1\nu_2 \ldots \nu_k$. Then, by (0.2) and (1.2),

$$N_{\mu_1} \times N_{\mu_2} \times \cdots \times N_{\mu_k} \cong N_\mu \times N^{k-1} = \rho(N_\mu),$$

$$N_{\nu_1} \times N_{\nu_2} \times \cdots \times N_{\nu_k} \cong N_\nu \times N^{k-1} = \rho(N_\nu).$$

Now $\mu \equiv m\nu \mod t$ for some m prime to t. Thus $N_\mu = N_{m\nu} = N_m + N_\nu$, and $\rho(N_\mu) = \rho(N_\nu)$ if and only if $N_m \in \ker\rho$. Thus Theorem 2.3 follows from Theorem 0.4. □

Notice that Theorem 2.3 strengthens Theorem 2.1.

Example 2.3. Let N be chosen so that $t = 60$. Thus we may take $n = 3600$, $u = 61$ in (1.3). Then $\mathcal{G}(N) \cong (\mathbb{Z}/60)^*/\{\pm 1\} \cong C_4 \times C_2$, and the groups in the genus of N are $N_1 (= N)$, N_7, N_{11}, N_{13}, N_{17}, N_{19}, N_{23}, N_{29}. The groups N_m in the kernel of ρ are given, from (0.3), by $m = 1, 11, 19, 29$, so $\mathcal{G}(N^k) \cong C_2$, with generator $N_7 \times N^{k-1}$, $k \geq 2$. We have the non-cancellation phenomena, for example,

$$M \times N \cong M \times N_{11} \cong M \times N_{19} \cong M \times N_{29}, \text{ for any } M \in \mathcal{G}(N), \tag{2.6}$$

with N, N_{11}, N_{19}, N_{29} pairwise non-isomorphic. Similarly,

$$M \times N_7 \cong M \times N_{17} \cong M \times N_{13} \cong M \times N_{23}, \text{ for any } M \in \mathcal{G}(N), \tag{2.7}$$

with N_7, N_{17}, N_{13}, N_{23} pairwise non-isomorphic.

Obviously, Theorem 2.3 shows that, within $\mathcal{G}(N)$, we may have

$$L_1 \times L_2 \times \cdots \times L_k \cong M_1 \times M_2 \times \cdots \times M_k$$

without the equality

$$L_1 + L_2 + \cdots + L_k = M_1 + M_2 + \cdots + M_k$$

holding in $\mathcal{G}(N)$. Indeed, in Example 2.3, no two of the groups $M + N$, $M + N_{17}$, $M + N_{19}$, $M + N_{29}$ are isomorphic, despite (2.6); a similar remark applies to (2.7).

References

[1] Mislin, G., Nilpotent groups with finite commutator subgroups. Springer Lecture Notes in Mathematics 418 (1974), 103–120.

[2] Hilton, P., Mislin, G., On the genus of a nilpotent group with finite commutator subgroup. Math. Z. 146 (1976), 201–211.

[3] Casacuberta, C., Hilton, P., Calculating the Mislin genus for a certain family of nilpotent groups. Comm. in Alg. 19:7 (1991), 2051–2069

[4] Schuck, C., Some Contributions to the Study of a Class of Nilpotent Groups. Ph.D. Dissertation, State University of New York at Binghamton (1992).

[5] Hilton, P., Schuck, C., On the structure of nilpotent groups of a certain type (to appear).

[6] Hilton, P., Schuck, C., Calculating the genus of certain nilpotent groups, Bolletín de la Sociedad Matemática Mexicana (to appear).

On certain classes of categories and monoids constructed from abstract Mal'cev clones. IV

Hans-Jürgen Hoehnke

Abstract. A structure theory of primal algebras in the total case was initiated by A. L. Foster essentially. Using the functorial language, we introduce the analogous concept (in the heterogeneous case) of a primal partial algebra $\underline{F} = (\,F|, F) : \underline{M} \to \mathrm{Par}$ of a Mal'cev theory $\underline{M}$. We show that it is equationally maximal, too. The Galois connection between congruences on a Mal'cev theory $\underline{M}$ and $\underline{M}$-subvarieties is established. The close relation between Mal'cev clones and Mal'cev theories may serve as a background for a categorical study of hyperidentities.

1991 Mathematics Subject Classification: 08C05, 08A40, 08A55, 08B05.

Introduction

A structure theory of primal algebras in the total case was initiated essentially by A. L. Foster [77]. Using the functorial language, we introduce the analogous concept (in the heterogeneous case) of a primal partial algebra $\underline{F} = (|F|, F)$: $\underline{M} \to \mathrm{Par}$ of a Mal'cev theory $\underline{M}$ (which is a generalization of an algebraic theory in the sense of F. W. Lawvere to the case of partial algebras). We show that again it is equationally maximal, i.e. that its Mal'cev theory $\underline{M} \cong \mathrm{Ary}^0\, \underline{M}F$ is congruence-simple. In non-functorial terms this result (in homogeneous case and using the concept of a strong regular identity) was also (independently) obtained by K. Denecke [75]. An advantage of using categories as theories is that non-regular identities are not excluded.

The Galois connection between congruences on a Mal'cev theory $\underline{M}$ and $\underline{M}$-subvarieties is established. The Mal'cev clone-part of a Mal'cev theory $\underline{M}$, $\mathrm{Mcl}\,\underline{M}$, is concerned with axiomatic Mal'cev clones (called abstract Mal'cev clones in [74]). The close connection between Mal'cev clones and Mal'cev theories may serve as a background for a categorical study of hyperidentities and for a simple proof of their varieties to be generated by the concrete representatives [26, 79, 80]. The notions of a Mal'cev clone and of a Mal'cev theory are chosen in order to indicate topics which belong to the theory of superposition of partial operations, e.g. A. I. Mal'cev [64, 78], differing from [17]; cf. also [26].

In order to avoid confusion of references, we continue with the numbering scheme of [Pt. 1–Pt. 3].

23. Schreckenberger's equational characterization of dht-symmetric categories

The author's original concept of a dht-symmetric category as a certain symmetric monoidal category [23] was used by J. Schreckenberger [49] to admit a purely equational axiomatic characterization, which is used in the sequel.

Let $\underline{K} = (\underline{K}, \otimes, I, a, s, r, l)$ be a *symmetric monoidal category* [30] with the object class $|\underline{K}|$, where

$$\otimes : \underline{K} \times \underline{K} \to \underline{K} \text{ is a bifunctor (the } \textit{monoidal product}),$$
$$I \in |\underline{K}| \text{ is a distinguished object (the } \textit{identity object} \text{ of } \underline{K}),$$

and a, s, r, l are natural isomorphisms,

$$a : - \otimes (- \otimes -) \to (- \otimes -) \otimes - \text{ ("associativity" of } \underline{K}),$$
$$s : - \otimes - \to s^x(- \otimes -) \text{ ("symmetry" of } \underline{K}), \text{ with}$$
$$s^x : \underline{K} \times \underline{K} \to \underline{K} \times \underline{K} : (f, f') \mapsto (f', f),$$
$$r : - \otimes I \to \mathrm{Id}_{\underline{K}} \text{ ("right identity" of } \underline{K}),$$
$$l : I \otimes - \to \mathrm{Id}_{\underline{K}} \text{ ("left identity" of } \underline{K}).$$

The naturality conditions are expressed as "equations" (i.e. identities if $\underline{K}$ is considered as a heterogeneous total algebra with the Hom-sets $\underline{K}(A, B)$ $(A, B \in |\underline{K}|)$ as the carrier sets).

Between these data certain additional "coherence" axioms are assumed to be valid. In the "strict" case (for simplicity, we restrict our considerations here by the condition) one supposes

$$a = 1, \quad r = 1, \quad l = 1. \tag{1}$$

This is equivalent $(\underline{K}, \otimes, 1_I)$ and $(|\underline{K}|, \otimes, I)$ to be monoids (generally, 1_A designates the identity morphism of $\underline{K}$ from the object A of $\underline{K}$). In this case the additional axioms are reduced just to the following equations $(A, B, C \in |\underline{K}|)$:

$$s(ss^x) = 1, \tag{2}$$
$$s_{A \otimes B, C} = (1_A \otimes s_{BC})(s_{AC} \otimes 1_B), \tag{3}$$
$$s_{AI} = 1_A. \tag{4}$$

23.1. Definition (cf. [50, 1.2.5, p. 20]). Let $\underline{K}$ be a symmetric (strict) monoidal category as above. Then $(\underline{K}, O, d, t, o)$ is said to be (*strict*) dht-*symmetric* in the case of $O \in |\underline{K}|$ being a distinguished object (the *zero object* of $\underline{K}$), $o : I \to O \in \underline{K}$ a distinguished morphism (the *zero morphism*), and $d = (d_A)_{A \in |\underline{K}|}$, $t = (t_A)_{A \in |\underline{K}|}$ distinguished families of morphisms of $\underline{K}$ such that

$$d : \mathrm{Id}_{\underline{K}} \to d^x \otimes \text{ (the } \textit{diagonal} \text{ of } \underline{K}),\ d^x : \underline{K} \to \underline{K} \times \underline{K} \text{ acting as } f \mapsto (f, f),$$

is a natural transformation, and

$$t : \mathrm{Id}_{\underline{K}} \to I \quad \text{(the } \textit{lax terminal} \text{ of } \underline{K})$$

is a natural transformation which satisfies the conditions posed below.

These data are connected by certain additional conditions reduced in the strict case to the following list:

$$\forall f : A \to O \in \underline{K} : t_A o = f, \tag{5}$$
$$\forall g : O \to B \in \underline{K} : 1_B \otimes t_O = g, \tag{6}$$
$$t_A \otimes t_B = t_{A \otimes B}, \tag{7}$$
$$d_A(1_A \otimes t_A) = d_A(t_A \otimes 1_A) = 1_A, \tag{8}$$
$$d_{A \otimes B}(1_A \otimes t_B \otimes t_A \otimes 1_B) = 1_{A \otimes B}, \tag{9}$$
$$d_I = t_I = 1_I. \tag{10}$$

These conditions imply that t is a lax natural transformation, i.e.

$$\forall f : A \to B, \quad g : B \to A \in \underline{K} : fg = 1_A \Rightarrow ft_B = t_A, \tag{11}$$

and that $(\underline{K}, \otimes, 1_I, 1_O)$ and $(|\underline{K}|, \otimes, I, O)$ are monoids with zero.

The symbol $(\leqq)$ designates the *canonical partial order* defined on the morphism sets $\underline{K}(A, B)$ $(A, B \in |\underline{K}|)$ of $\underline{K}$ by $f \leqq g \Leftrightarrow f = (\alpha f)g$, where $\alpha f = d_A(1_A \otimes f)p_{AB}$ for $p_{AB} = 1_A \otimes t_B$.

A congruence $\varkappa$ on $\underline{K}$ is given by a family of equivalence relations (one on each set $\underline{K}(A, B)$) which is compatible with the operations of $\underline{K}$; therefore, $\underline{K}$ is considered also as a heterogeneous total algebra.

A (strict) dht-symmetric category $\underline{K}$ *is modified* to (strict) dt-symmetric if in its axiomatics O, o are omitted and t is natural, i.e. (11) is replaced by

$$\forall f : A \to B \in \underline{K} : ft_B = t_A. \tag{11'}$$

In this case (with $p'_{AB} = t_A \otimes 1_B; A, B \in |\underline{K}|$) a categorical product in $\underline{K}$ is defined by

$$A \overset{p_{AB}}{\longleftarrow} A \otimes B \overset{p'_{AB}}{\longrightarrow} B.$$

A (strict) dht-symmetric category $\underline{M}$ is called a *Mal'cev theory* if the monoid $(|\underline{M}|, \otimes, I, O)$ with zero O is freely generated by a set J. As for the dt-symmetric case, one gets the notion of a *Lawvere theory*.

In the sequel, all monoidal categories are, without mentioning it, assumed to be strict monoidal.

24. Schreckenberger's characterization of Par

Recall [Pt. 3, Section 17] that a functor $\underline{F} = (|F|, F) : \underline{K} \to \underline{K}'$ between dht-symmetric categories is said to be *monoidal* dht-*symmetric* if there exist a family $\widetilde{F} = (\widetilde{F}_{AB})_{A,B\in|\underline{K}|}$ of morphisms

$$\widetilde{F}_{AB} : A|F| \otimes B|F| \to (A \otimes B)|F| \in \underline{K}'$$

and morphisms

$$F^1 : I' \to I|F|, \;\; F^0 : O' \to O|F| \in \underline{K}'$$

(here I' and O' are the distinguished objects of $\underline{K}'$ corresponding to I and O of $\underline{K}$ respectively), which are equivalent under certain conditions to:

$$\begin{aligned}
&\underline{F} \;\; \text{preserves} \;\; \alpha;\\
&\widehat{F}_{AB} := d_{(A\otimes B)|F|} \cdot (p_{AB}F \otimes p'_{AB}F) \in \operatorname{Iso}\underline{K}' \quad (A, B \in |\underline{K}|);\\
&I|F| \cong I', \quad O|F| \cong O'
\end{aligned}$$

(here, $\operatorname{Iso}\underline{K}'$ is the class of all isomorphisms of $\underline{K}'$, and the sign $(\cong)$ between objects means them being isomorphic).

Let MON designate the category of dht-symmetric categories and monoidal dht-symmetric functors between them as morphisms.

A functor $\underline{F} = (|F|, F) : \underline{K} \to \underline{K}'$ is said to be *representative* if for each object $A' \in |\underline{K}'|$ there exists an object $A \in |\underline{K}|$ such that $A|F| \cong A'$.

According to J. Schreckenberger [50, 4.1.1, p. 138, and 4.1.2, p. 141], the categories which are embeddable into the category Par of sets and mappings between sets as morphisms (with the usual relational products as the composition of morphisms) can be described as follows.

24.1. Proposition. *For each* dht-*symmetric category* $\underline{K}$ *there exists a full and faithful monoidal dht-symmetric functor* $\underline{K} \to$ Par *iff* $\underline{K}$ *satisfies*

(i) *I is a generator of $\underline{K}$ (i.e. $\forall f \neq g : A \to B \in \underline{K}, \exists u : I \to A \in \underline{K}$ such that $uf \neq ug$);*

(ii) $\operatorname{Card}\underline{K}(I, I) = 2$ (Card... *is the cardinal number of* ...);

(iii) $\forall f, g : A \to B \in \underline{K} \exists f \wedge g = \operatorname{Inf}\{f, g\} : A \to B \in \underline{K}$ *(the infimum of f and g in $\underline{K}$ is taken with respect to the canonical partial order $(\leqq)$ in $\underline{K}$) and $\forall h : X \to A \in \underline{K}$ we have $h(f \wedge g) = hf \wedge hg$;*

(iv) $\forall A, B \in |\underline{K}|$, $\forall (f_j)_{j\in L}$ *(where $f_j : A \to B \in \underline{K}$, $j \in L$, L an index set) one has the implication* $(\forall j_1, j_2 \in L : (f_{j_1} \neq f_{j_2}) \Rightarrow (\alpha f_{j_1} \wedge \alpha f_{j_2} = o_{AA})) \Rightarrow \exists \bigvee_{j\in L} f_j = \operatorname{Sup}\{f_j\}_{j\in L}$;

(v) $\forall A \in |\underline{K}|$, $\forall (e_j)_{j\in L}$, $\forall e$ *(where $e_j \leqq A$, $j \in L$, and $e \leqq A$ is an abbreviation for $e \leqq 1_A$ and means that e is a subidentity of A in $\underline{K}$) the suprema* $\bigvee_{j\in L} e_j$

and $\bigvee_{j\in L} e\wedge e_j$ *(with respect to* $(\leqq)$*) in* $\underline{K}$ *exist and the following distributive law holds:*

$$e\wedge(\bigvee_{j\in L} e_j)=\bigvee_{j\in L}(e\wedge e_j).$$

The necessity of the conditions (i)–(v) is obvious. Their sufficiency can be proved (cf. [50, p. 143–150]) by showing the monoidal dht-symmetric functor (cf. [Pt. 3.11])

$$\underline{F}_I:=\underline{F}_{\{1_I\}}=\underline{H}_{\underline{K}}\cdot \text{Part Sub}(I,-):\underline{K}\to \text{Part Set},$$

to be full and faithful, where Part Set = Par and $\underline{H}_{\underline{K}}:\underline{K}\to$ Part Sub$\underline{K}$ is given as in Theorem 3.2 of [Pt. 1] (cf. also [Pt. 2, Lemma 11.2] and [Pt. 3, Lemma 11.3 (b)]). By definition, for all $A\in|\underline{K}|$ one has $A|F_I|=\{f\in\underline{K}(I,A);\quad ft_A=1_I\}$ and faithfulness implies that $ft_A=1_I\Leftrightarrow f\neq o_{IA}$ (here, $o_{AB}\in\underline{K}$ is the unique zero morphism $A\to B$ represented by $o:I\to O$ as $o_{AB}=t_A\cdot o\cdot(t_O\otimes 1_B)$).

For the abstract characterization of Par one has to apply:

24.2. Proposition (cf. [50, 4.2.3, p. 151]). *For any* dht-*symmetric category* $\underline{K}$, *any full and faithful monoidal* dht-*symmetric functor* $\underline{F}:\underline{K}\to$ Par, *it is holds that* $\underline{F}$ *is representative iff*

(vi) *for each set* L *a coproduct* $\coprod_{j\in L} I$ *of the family* $(I)_{j\in L}$ *in* $\underline{K}$ *exists.*

Then one obtains

24.3. Theorem (J. Schreckenberger, [50, S. 4.2.4, p. 153]). *For each* dht-*symmetric category* $\underline{K}$ *there exists a full and faithful representative monoidal* dht-*symmetric functor* $\underline{K}\to$ Par *iff* $\underline{K}$ *satisfies* 24.1, (i)–(v) *and* 24.2, (vi).

24.4. Theorem. *For each Mal'cev theory* $\underline{M}$ *and a free generating set* J *of the free object monoid* $(|\underline{M}|,\otimes,I,O)$ *with zero there exists a full and faithful monoidal* dht-*symmetric functor* $\underline{F}:\underline{M}\to$ Par *such that for each set* $A|F|$ $(A\in J)$ *one has* $2\leqq$ Card $A|F|<\infty$ *and* $A|F|\neq B|F|$ *for* $A\neq B$ $(A,B\in J)$ *iff* $\underline{M}$ *satisfies* 24.1, (i)–(v) *and*

(vii) $3\leqq$ Card $\underline{M}(I,A)<\infty$ *for all* $A\in J$.

Proof. Necessity of the conditions is obvious. As for sufficiency, $\underline{F}_I=\underline{M}\to$ Par maps $A\in J$ to $A|F_I|=\underline{M}(I,A)\setminus\{o_{IA}\}$.

24.5. Corollary. *Each Mal'cev theory* $\underline{M}$ *satisfies the conditions* 24.1, (i)–(v) *and* 24.4, (vii) *iff* $\underline{M}$ *is isomorphic to the full subcategory* Par $K_J^{\supset}$ *of* Par *which belongs* *to the object monoid* $K_J^0 = (K_J^0, \otimes = x, I = \{\emptyset\}, O = \emptyset)$ *with zero generated by a* J*-indexed set of finite different sets of* Card $\geqq 2$.

25. Congruences on Par K_J^0

Each partial mapping can be represented as total mapping via the so-called one-point completion. For the purpose, we consider the category Par^+ of all pointed sets (A, ∞) (i.e. all sets A with fixed distinguished element ∞) as objects and all pointed mappings $f : (A, \infty) \to (B, \infty)$ (i.e. total mappings $f : A \to B \in$ Set, where Set is the category of all sets and all mappings between them, keeping the element ∞ fixed). Composition of pointed mappings in Par^+ is induced by the ordinary composition of these mappings in Set. Since Par^+ is equivalent to the category of algebras of the form (A, a) $(a \in A)$ (i.e. with a nullary operation as the only one), the category is well-known to admit limits and colimits. Especially, Par^+ has a categorical product Π given by

$$(A, \infty)\Pi(B, \infty) = (((A \otimes B) \setminus \{(\infty, \infty)\})\dot{\cup}\{\infty\}, \infty)$$

and for $f : (A, \infty) \to (C, \infty)$ and $g : (B, \infty) \to (D, \infty)$, the *categorical product* $f\Pi g : (A, \infty)\Pi(B, \infty) \to (C, \infty)\Pi(D, \infty)$ acts as follows

$$\begin{cases} (x, y) \mapsto (xf, yg) & \text{if } (xf, yg) \neq (\infty, \infty), \\ (x, y) \mapsto \infty & \text{otherwise}, \\ \infty \mapsto \infty. \end{cases}$$

Contrary to this we introduce another *monoidal product*, $\otimes^+$, in Par^+ as follows. If $A \in |\mathrm{Par}|$ we set $A^+ = A\dot{\cup}\{\infty\}$ and if also $B \in |\mathrm{Par}|$ we set

$$A^+ \otimes^+ B^+ = (A \otimes B)\dot{\cup}\{\infty\}.$$

Then $(^+) : (|\mathrm{Par}|, \otimes, I, O) \to (|\mathrm{Par}^+|, \otimes^+, I^+, O^+)$, mapping A to A^+, is an isomorphism of monoids with zero.

For $f : A \to B \in \mathrm{Par}$ we define its *completion* or *totalization* $f^+ : A^+ \to B^+$ in Par^+ as follows

$$x \mapsto \begin{cases} xf & \text{if } x \in \mathrm{Dex}\, f, \\ \infty & \text{otherwise}, \end{cases}$$

where $\mathrm{Dex}\, f$ designates the domain of existence of f. Then, for $f : A \to C$ and $g : B \to D \in \mathrm{Par}$ we set

$$f^+ \otimes^+ g^+ = (f \otimes g)^+ : (A \otimes B)^+ \to (C \otimes D)^+.$$

For $g : A^+ \to B^+ \in \mathrm{Par}^+$, $g^- : A \to B \in \mathrm{Par}$ acts as follows

$$x \mapsto \begin{cases} xg & \text{if } xg \in B, \\ * \text{ (non-existing)} & \text{otherwise.} \end{cases}$$

Then one has $f^{+-} = f : A \to B$, $g^{-+} = g : A^+ \to B^+$. With these notions one has:

25.1. Proposition. *There is an isomorphism of* dht-*symmetric categories* $(^+)$: Par $\to$ Par$^+$, *where* $(\mathrm{Par}^+, \otimes^+, I^+, O^+, s^+, d^+, t^+, o^+)$ *is considered as a* dht-*symmetric category.*

Proof. 1. Suppose that $x \in \mathrm{Dex}\, fg = \mathrm{Dex}\,(f(\alpha g)) \subseteqq \mathrm{Dex}\, f$. Then $xf \in \mathrm{Dex}\, g$, $x(fg)^+ = x(fg)$ and $x(f^+g^+) = (xf)g^+ = x(fg)^+$.

2. Suppose that $x \notin \mathrm{Dex}\, fg$, i.e. $x(fg) = \infty$. Then one has

$$x(f^+g^+) = \begin{cases} \infty g^+ = \infty & \text{if } x \notin \mathrm{Dex}\, f, \\ (xf)g^+ = \infty & \text{if } x \in \mathrm{Dex}\, f,\ xf \notin \mathrm{Dex}\, g, \end{cases}$$

i.e. $x(f^+g^+) = \infty$ and again $x(f^+g^+) = x(fg)^+$. Therefore $f^+g^+ = (fg)^+$. Thus $(^+)$ is a homomorphism of categories. Since $f^{+-} = f$, $(^+)$ is one-to-one.

25.2. Theorem (cf. [64, p. 10, Proof a)]). *Let K_J^0 be the object monoid with zero, which is generated by a J-indexed family of non-empty sets. Then the* dht-*symmetric category* Par $K_J^0 = \underline{M}$ *is a congruence-simple Mal'cev theory, i.e. $\underline{M}$ has only the two congruences*

$\underline{O_M}$ *(the least congruence on $\underline{M}$),*

$\underline{1_M} = \bigcup_{A,B \in |\underline{M}|} \underline{M}(A,B) \times \underline{M}(A,B)$ *(the largest congruence on $\underline{M}$).*

Proof. The category $\underline{M}$ is isomorphically mapped onto a dht-symmetric subcategory $\underline{M}^+$ of Par$^+$ by $(^+)$. Therefore, it is sufficient to prove that $\underline{M}^+$ is congruence-simple. Thus, suppose that $\varkappa \neq \underline{O}_{\underline{M}^+}$ is a congruence on $\underline{M}^+$, i.e. there are mappings $f_1^+, f_2^+ : A^+ \to B^+$ such that $(f_1^+, f_2^+) \in \varkappa$ and $f_1^+ \neq f_2^+$. (Observe that $A^+, B^+ \in |\underline{M}^+|$.) By assumption, there are $u \in A^+$, $v_1, v_2 \in B^+$ $(v_1 \neq v_2)$, such that $f_i(u) = v_i$ $(i = 1, 2)$. Define mappings $(v \in B^+)$

$$\begin{cases} q_v : B^+ \to B^+, \\ t_v^\infty : A^+ \to B^+ \quad (t_u^\infty : A^+ \to A^+), \\ e^\infty : (B \otimes A)^+ \to A^+, \end{cases} \tag{1}$$

according to the following list $(x \in A^+, y \in B^+)$:

$$q_v(y) = \begin{cases} v & y = v, \\ \infty & y \neq v, \end{cases} \qquad q_v^-(y) = \begin{cases} v & y = v \neq \infty, \\ * & v = \infty,\ y \neq \infty, \\ * & v \neq \infty,\ v \neq y \neq \infty, \end{cases}$$

$$t_v^\infty(x) = \begin{cases} v & x \neq \infty, \\ \infty & x = \infty, \end{cases} \qquad t_v^{\infty-}(x) = \begin{cases} v & v \neq \infty,\ x \neq \infty, \\ * & v = \infty,\ x \neq \infty, \end{cases}$$

$$e^\infty(z) = \begin{cases} x & z = (y,x) \in B \otimes A, \\ \infty & z = \infty, \end{cases} \qquad e^{\infty-}(y,x) = x\ (\neq \infty) \text{ for } y \neq \infty.$$

The mappings (1) obviously belong to $\underline{M}^+$, since

$$(q_v^-)^+ = q_v, \quad (t_v^{\infty-})^+ = t_v^\infty, \quad (e^{\infty-})^+ = e^\infty.$$

As $v_1 \neq v_2$ one of them, say v_1, is different from ∞. Firstly, we prove that $(t_{v_1}^\infty, t_\infty^\infty) \in \varkappa$. Since $(f_1, f_2) \in \varkappa$, the mappings $t_u^\infty f_i = t_{v_i}^\infty \quad (i = 1, 2)$ are congruent modulo $\varkappa$. This implies

$$t_{v_1}^\infty = t_{v_1}^\infty q_{v_1} \equiv t_{v_2}^\infty q_{v_1} = t_\infty^\infty : A^+ \to B^+(\varkappa).$$

Hence, the following mappings (which belong to $\underline{M}^+$)

$$(d_A(t_{v_1}^{\infty-} \otimes 1_A))^+ e^\infty = 1_{A^+} = (1_A)^+,$$
$$(d_A(t_\infty^{\infty-} \otimes 1_A))^+ e^\infty = t_\infty^\infty : A^+ \to A^+$$

are congruent modulo $\varkappa$. Thus, for every mapping $f^+ : A^+ \to B^+ \in \underline{M}^+$ we get

$$f^+ = 1_A^+ f^+ \equiv t_\infty^\infty f^+(\varkappa),$$

i.e. each f^+ is congruent modulo $\varkappa$ to the constant mapping $A^+ \to \{\infty\}$.

26. Subvarieties and congruences

Let $\underline{M}$ be a Mal'cev theory, $\underline{K}$ a dht-symmetric category. Then the monoidal dht-symmetric functors $\underline{F} : \underline{M} \to \underline{K}$ can be considered as the natural functorial description of the notion of an $\underline{M}$-algebra over $\underline{K}$. Thus, especially for $\underline{K} = \text{Par}$, one obtains the natural functorial description of partial $\underline{M}$-algebra.

The class MON $(\underline{M}, \underline{K})$ is called the *variety of all $\underline{M}$-algebras over $\underline{K}$*. As usual, one can introduce a certain Galois connection between subsets $P \subseteqq \underline{1}_{\underline{M}}$ of "equiations" $(f, f') \in P$ in $\underline{M}$ and subclasses $\underline{\underline{F}} \subseteqq \text{MON}(\underline{M}, \underline{K})$ by declaring that $\underline{\underline{F}} \models P$ if for all $\underline{F} \in \underline{\underline{F}}$ and all $(f, f') \in P$ one has $fF = f'F$.

In this case one also says the equations $(f, f') \in P$ to be valid in $\underline{\underline{F}}$. (In the case of $\underline{K} = \text{Par}$, the equality $fF = f'F$ is equivalent the identity $f = f'$ to be valid in the partial $\underline{M}$-algebra, which is functorially described by $\underline{F}$.)

One sets

$$P\,\mathrm{Fun} = \{\underline{F} \in \mathrm{MON}\,(\underline{M}, \underline{K}); \underline{F} \models P\},$$
$$\underline{\underline{F}}\,\mathrm{Eq} = \{(f, f') \in \underline{1}_M; \quad \underline{\underline{F}} \models (f, f')\}.$$

Then the pair (Fun, Eq) indeed is a Galois connection between $\mathrm{Sb}\,\underline{1}_M$ and $\mathrm{Sb}\,\mathrm{MON}\,(\underline{M}, \underline{K})$ ($Sb(-)$ the class of all subclasses of $(-)$). The subclass $P\,\mathrm{Fun}$ is called the *subvariety of* $\mathrm{MON}\,(\underline{M}, \underline{K})$, *defined by the set* P. For $\underline{\underline{F}} \subseteq \mathrm{MON}\,(\underline{M}, \underline{K})$ one has

$$\underline{\underline{F}}\,\mathrm{Eq} = \bigcap_{\underline{F} \in \underline{\underline{F}}} \mathrm{Ker}\,\underline{F} \in \mathrm{Con}\underline{M},$$

where $\mathrm{Ker}\,\underline{F} = \{(f, f') \in \underline{1}_M; \underline{F} \models (f, f')\}$ (the "kernel" of $\underline{F}$) is the congruence on $\underline{M}$ induced by $\underline{F}$. Here, $\underline{M}$ is again considered as a heterogeneous total algebra, and $\mathrm{Con}\,\underline{M}$ is its lattice of congruences on $\underline{M}$. We get:

26.1. Lemma. *With respect to the Galois connection* (Fun, Eq), *the closed subclasses of* $\mathrm{MON}\,(\underline{M}, \underline{K})$ *are (by definition) the subvarieties of* $\mathrm{MON}\,(\underline{M}, \underline{K})$, *and the closed subclasses of* $\underline{1}_M$ *are the congruences on* $\underline{M}$. *The pair* (Fun, Eq) *establishes a dual isomorphism between the complete lattices of subvarieties on* $\mathrm{MON}\,(\underline{M}, \underline{K})$ *and* $\mathrm{Con}\,\underline{M}$.

26.2. Definition. For any functorial $\underline{M}$-algebra over $\underline{K}$, $\underline{F} : \underline{M} \to \underline{K}$, we introduce the *arity-interpretation* of $\underline{M}F \subseteq \underline{K}$, which is again a Mal'cev theory, denoted by $\mathrm{Ary}^0\underline{M}F$, with the same object class as $\underline{M}$ and with

$$\mathrm{Ary}^0\underline{M}F(A, B) = \{(A, fF, B) : A \to B; f \in \underline{M}(A, B)\}.$$

Composition in $\mathrm{Ary}^0\underline{M}F$ between (A, fF, C) and (B, gF, D) (for $f : A \to C, g : B \to D \in \underline{M}$) is defined iff $B = C$, yielding $(A, fF, C)(B, gF, D) = (A, (fg)F, D)$. This is the main difference between $\underline{M}F$ and $\mathrm{Ary}^0\underline{M}F$. The other operations and distinguished families of morphisms of $\mathrm{Ary}^0\underline{M}F$ are properly defined.

The difference between $\underline{M}F$ (which in general needs not to be a category) and $\mathrm{Ary}^0\underline{M}F$ vanishes iff the object mapping

$$|F| : (|\underline{M}|, \otimes, I, O) \to (|\underline{K}|, \otimes, I, O)$$

is proved to be an isomorphism "into." For example, it is the case in the situations of Theorem 24.4 and Corollary 24.5. In general, given $\underline{F} \in \mathrm{MON}\,(\underline{M}, \mathrm{Par})$, one can embed $\mathrm{Ary}^0\underline{M}F$ into a full subcategory $\mathrm{Par}K_J^0$, where $K_J^0 = K_J^0(\underline{F})$ is the object monoid with zero, which is generated by the set $\{\{A\} \times A|F|; \quad A \in J\}$ simply by identifying A and $\{A\} \times A|F|$ $(A \in J)$, such that one obtains an inclusion $\mathrm{Ary}^0\underline{M}F \subseteq \mathrm{Par}\,K_J^0$. Therefore Mal'cev theories of this form, $\mathrm{Ary}^0\underline{M}F$, are called *concrete* (over Par).

The preceding remarks justify the choice of the symbol K_J^0 for the indication of the object class type of a Mal'cev theory $\underline{M}$. The category of Mal'cev theories (considered as heterogeneous total algebras) of fixed type K_J^0 (together with their homomorphisms) is designated by Mth_J.

26.3. Proposition (cf. [26, Proposition 9.2]). *For any Mal'cev theory* $\underline{M}$ *let* $\underline{F} : \underline{M} \to \mathrm{Par}$ *be a functorial* $\underline{M}$*-algebra,* $\underline{\underline{M}} \subseteq |\mathrm{Mth}_J|$ *any class of Mal'cev theories of type* K_J^0. *Then*

(i) $\mathrm{Ary}^0 \underline{M}F$ *is a Mal'cev theory,* $\mathrm{Ary}^0 \underline{M}F \cong \underline{M}/\mathrm{Ker}\, \underline{F}$ ($\mathrm{Ker}\, \underline{F} = \underline{F}\,\mathrm{Eq}$) *and the inclusion functor* $\mathrm{Ary}^0 \underline{M}F \overset{\subseteq}{\to} \mathrm{Par}\, K_J^0(\underline{F}) \overset{\subseteq}{\to} \mathrm{Par}$ *is an* $\mathrm{Ary}^0 \underline{M}F$*-algebra;*

(ii) *the variety* $\mathrm{MON}\,(\mathrm{Ary}^0 \underline{M}F, \mathrm{Par})$ *is essentially the same as the subvariety* $\mathrm{MON}_{\mathrm{Ker}\, \underline{F}}\,(\underline{M}, \mathrm{Par})$ *of* $\mathrm{MON}\,(\underline{M}, \mathrm{Par})$ *generated by* $\underline{F}$*; the subvariety consists of all* $\underline{M}$*-algebras* $\underline{G} : \underline{M} \to \mathrm{Par}$ *such that* $\mathrm{Ker}\, \underline{F} \subseteq \mathrm{Ker}\, \underline{G}$*, i.e. of* $\underline{M}$*-algebras which factorize according to* $\underline{G} : \underline{M} \to \underline{M}/\mathrm{Ker}\, \underline{F} \to \mathrm{Par}$ *through* $\mathrm{Ker}\, \underline{F} \in \mathrm{Con}\, \underline{M}$;

(iii) *the identities, which are valid in all Mal'cev theories* $\mathrm{Ary}^0 \underline{M}G$ *for* $\underline{G} \in \mathrm{MON}_{\mathrm{Ker}\, \underline{F}}(\underline{M}, \mathrm{Par})$*, are the same as ones valid in* $\mathrm{Ary}^0 \underline{M}F$;

(iv) *the identities, which are valid in all concrete Mal'cev theories* $\mathrm{Ary}^0 \underline{M}F$ *for* $\underline{M} \in \underline{\underline{M}}$ *and* $\underline{F} \in \mathrm{MON}\,(\underline{M}, \mathrm{Par})$*, are the same as ones valid in* $\underline{\underline{M}}$.

Proof. The proof of (i) and (ii) is straightforward, (iii) is a consequence of (iv). The last expresses the fact that each subvariety of $|\mathrm{Mth}_J|$ is generated by its concrete Mal'cev theories, which was already stated in [50, 3.2.3, p. 134; 3.2.4 and 3.2.5, p. 137] for the whole variety $|\mathrm{Mth}_J|$; for a proof in a more general case of fractal categories (see [Pt. 1, Representation Theorem 4.7, p. 165] and the remarks following the proof of Corollary 13.7 in [Pt. 3]). But the proof(s) given there can easily seen to be valid for any subvariety of $|\mathrm{Mth}_J|$ as well. Thus, (iv) is true. The reason to give a proof of (iii) separately is to verify the difference between (iii) and (iv), emphasizing the fact that the proof of (iii) is nearly trivial and can be given independently of [50], [Pt. 1], and [Pt. 3]. Consider any identity, say

$$\Phi_1(\underline{X}) = \Phi_2(\underline{X});\ A \to B; \underline{X} = (X_1, \ldots, X_n),\ X_i :\ U_i \to V_i, \qquad (1)$$

where $\Phi_i(\underline{X})$ are terms written in the signature of $\underline{M}$, with variables only from $\underline{X}$ (in connection with $\underline{M}$-algebras over Par, such an identity is also called a *hyperidentity*). Now, in $\underline{M}/\mathrm{Ker}\, \underline{F}$ the validity of (1) is equivalent to:

$$\text{for all evaluations of } \underline{X} \text{ in } \underline{M} : (\Phi_1(\underline{X}), \Phi_2(\underline{X})) \in \mathrm{Ker}\, \underline{F}.$$

Since

$$\mathrm{Ker}\, \underline{F} = \bigcap_{\mathrm{Ker}\underline{F} \subseteq \mathrm{Ker}\, \underline{G}} \mathrm{Ker}\, \underline{G}$$

this is equivalent to $(\Phi_1(\underline{X}), \Phi_2(\underline{X})) \in \mathrm{Ker}\, \underline{G}$ for all evaluations of $\underline{X}$ and all $\underline{G} \in \mathrm{MON}_{\mathrm{Ker}\, \underline{F}}(\underline{M},\ \mathrm{Par})$. Hence, to validity of (1) in all $\underline{G} \in \mathrm{MON}_{\mathrm{Ker}\, \underline{F}}(\underline{M}, \mathrm{Par})$.

26.4. Corollary. *Each subvariety of* $|\mathrm{Mth}_J|$ *is generated by its concrete Mal'cev theories.*

26.5. Remark. Proposition 26.3, (iii) is an analogue of the statement of D. Schweigert [79, 80] on the identities valid in the clone of a total algebra. In the terms of Lawvere theories, Schweigert's result would appear also as a rather simple factorization property.

26.6. Remark. Proposition 26.3 can be generalized, replacing Par by any dht-symmetric category $\underline{K}$ and taking $\underline{F} \in \mathrm{MON}\,(\underline{M}, \underline{K})$. Then one has

$$\mathrm{Ary}^0\, \underline{M}F \cong \underline{M}/\mathrm{Ker}\, \underline{F} \quad (\mathrm{Ker}\, \underline{F} = \underline{F}\, \mathrm{Eq}),$$

and (the symbol $\tilde{\leftrightarrow}$ stands for one-to-one correspondence)

$$\mathrm{MON}_{\mathrm{Ker}\, \underline{F}}(\underline{M}, \underline{K}) \,\tilde{\leftrightarrow}\, \mathrm{MON}\,(\underline{M}/\mathrm{Ker}\, \underline{F}, \underline{K}) \,\tilde{\leftrightarrow}\, \mathrm{MON}\,(\mathrm{Ary}^0 \underline{M}F, \underline{K}),$$

which imply statements as in 26.3, (i)–(iii). To generalize 26.3, (iv) one has to suppose that $|\mathrm{Mth}_J|$ is logically complete over $\underline{K}$, i.e. each $\underline{M} \in |\mathrm{Mth}_J|$ is a subdirect product of Mal'cev theories faithfully embeddable into $\underline{K}$ (say: $|\mathrm{Mth}_J|$ is *subgenerated by* $\underline{K}$). Indeed, an identity $(\Phi_1(\underline{X}), \Phi_2(\underline{X}))$ to be valid in the class $\{\underline{M}/\mathrm{Ker}\, \underline{F};\ \underline{M} \in \underline{\underline{M}},\ \underline{F} \in \mathrm{MON}\,(\underline{M}, \underline{K})\}$ is equivalent to

$$(\Phi_1(\underline{X}), \Phi_2(\underline{X})) \in \bigcap \{\mathrm{Ker}\, \underline{F};\ \underline{F} \in \mathrm{MON}\,(\underline{M}, \underline{K})\} = \underline{O}_{\underline{M}}$$

for all $\underline{M} \in \underline{\underline{M}}$ and evaluations of $\underline{X}$ in $\underline{M}$.

26.7. Definition. A Mal'cev theory $\underline{M} \in |\mathrm{Mth}_J|$ is said to be *equationally complete* (or *equationally maximal*) if $\underline{M}$ is congruence-simple. An $\underline{M}$-algebra $\underline{F} : \underline{M} \to \underline{K}$ is *equationally complete* if $\underline{M}/\mathrm{Ker}\, \underline{F} \cong \mathrm{Ary}^0 \underline{M}F$ is equationally complete. An $\underline{M}$-algebra $\underline{F} : \underline{M} \to \mathrm{Par}$ is said to be *primal* if

$$\begin{aligned} &\mathrm{Ker}\, \underline{F} = \underline{O}_{\underline{M}}, \quad \mathrm{Ker}\, |F| = \underline{O}_{|M|}, \\ &\underline{M} \cong \mathrm{Ary}^0 \underline{M}F = \underline{M}F = \mathrm{Par}\, K_J^0, \\ &\forall A \in J : 2 \leqq \mathrm{Card}\, A|F| < \infty. \end{aligned}$$

By Proposition 26.3, (ii), a primal $\underline{M}$-algebra $\underline{F} : \underline{M} \to \mathrm{Par}$ obviously generates the variety $\mathrm{MON}\,(\underline{M}, \mathrm{Par})$. The main statement on primal algebras is

26.8. Theorem. *Let be* $\underline{M} \in |\mathrm{Mth}_J|$ *a Mal'cev theory. Then one has:*

(a) *there exists a primal* $\underline{M}$*-algebra* $\underline{M} \to \mathrm{Par}$ *iff* $\underline{M}$ *satisfies* 24.1, (i)–(v) *and* 24.4, (vii);

(b) *every primal* $\underline{M}$*-algebra is equationally complete.*

Proof. The item (a) follows from Theorem 24.4, (b) from Theorem 25.2.

27. Mal'cev theories and Mal'cev clones

For any dht-symmetric category $\underline{C}$ and any subset $G \subseteqq \underline{C}$, by a stepwise definition we introduce the notion of a *$\underline{C}$-clone generated by G*: firstly, choose a set J of objects of $\underline{C}$ such that any $g \in G$ is of the form $g : A \to B$, where A and B belong to the submonoid K_J^0 with zero of $(|\underline{C}|, \otimes, I, O)$, which is generated by J (this might be ambiguous); secondly, form the full subcategory $\underline{C}(K_J^0)$ of $\underline{C}$ such that $|\underline{C}(K_J^0)| = K_J^0$ (it is a dht-symmetric category); thirdly, consider the dht-symmetric subcategory $G_{\underline{C}(K_J^0)}$ of $\underline{C}(K_J^0)$, which is generated by G. The *$\underline{C}$-clone generated by G* is the set $\langle G\rangle_J^{\underline{G}}$ of all $f : A \to B \in G_{\underline{C}(K_J^0)}$ such that $B \in J \cup \{I, O\}$. For $\langle G\rangle_J^{\underline{G}} = G$, the set G is called a *$\underline{C}$-clone*. The only essential case is that of the monoid K_J^0 with zero being freely generated by J. This is assumed henceforth.

These notions apply to $\underline{C} = \underline{M} \in |\mathrm{Mth}_J|$, and then one also speaks of an abstract Mal'cev clone G in place of an $\underline{M}$-clone G. Further, the notions apply to $\underline{C} = \mathrm{Par}$ and in this case $\underline{C}$ as an upper index is omitted. Thus, suppose a family J of pairwise different sets each of cardinality at least two to be given. Then K_J^0 is the free submonoid with zero of $(|\mathrm{Par}|, \otimes, I, O)$, which is freely generated by J. One speaks of a concrete Mal'cev clone G instead of a Par-clone G.

Abstract and concrete Mal'cev clones are commonly called *Mal'cev clones.* (Of course, each concrete Mal'cev clone should be an abstract one.) Each $\underline{C}$-clone G, by assumption, generates a Mal'cev theory $G_{\underline{C}(K_J^0)} = \underline{M} \in |\mathrm{Mth}_J|$ and then G is also called the *(Mal'cev) clone-part of $\underline{M}$*. One may consider the problem to characterize the clone-part of a Mal'cev theory $\underline{M}$ independently of its connection to $\underline{M}$. This is elucidated below.

If $\underline{M} \in |\mathrm{Mth}_J|$ and $\underline{F} : \underline{M} \to \underline{K}$ is an $\underline{M}$-algebra over $\underline{K}$ then $\mathrm{Mcl}\,\underline{F} = \{(A, fB, B); f : A \to B \in \underline{M}, B \in J \cup \{I, O\}\}$ is the clone-part of the Mal'cev theory $\mathrm{Ary}^0\underline{M}F$. It is also called the *Mal'cev clone of the $\underline{M}$-algebra $\underline{F}$*. In particular, this notion applies to the case $\underline{K} = \mathrm{Par}$. In this case $\mathrm{Mcl}\,\underline{F}$ is obviously a concrete Mal'cev clone. Observe that $\underline{F}$ maps the clone-part $\mathrm{Mcl}\,Id_{\underline{M}}^-$ of $\underline{M}$ onto $\mathrm{Mcl}\,\underline{F}$.

27.1. Definition. For any Mal'cev theory $\underline{M} \in |\mathrm{Mth}_J|$ we set $\mathrm{Mcl}\,\underline{M} := \mathrm{Mcl}\,\mathrm{Id}_{\underline{M}}$, and in $\mathrm{Mcl}\,\underline{M}$ we introduce the following structure, considering it as a heterogeneous total algebra $\mathrm{Mcl}\,\underline{M} = (\mathrm{Mcl}\,\underline{M}, \theta, p, o)$, according to the decomposition

$$\mathrm{Mcl}\,\underline{M} = \bigcup_{A \in K_J^0, B \in J_{IO}} \mathrm{Mcl}_{AB}\,\underline{M}$$

into carrier sets

$$\mathrm{Mcl}_{AB}\,\underline{M} = \underline{M}(A, B); A = A_1 \otimes \cdots \otimes A_N,$$

where $\underline{A} = (A_1, \dots, A_n) \in J_{IO}^n = J_{IO} \times \dots \times J_{IO}$ (n-times) and $J_{IO} = J \cup \{I, O\}$. The operations are:

$$\theta = \theta_{AC}^{\underline{B}} = [-](-) : (\mathrm{Mcl}_{AB_1} \underline{M} \times \dots \times \mathrm{Mcl}_{AB_m} \underline{M}) \times \mathrm{Mcl}_{\underline{B}C} \underline{M} \to \mathrm{Mcl}_{AC} \underline{M},$$

where A, $\underline{B} = (B_1, \dots, B_m)$, C run through $A \in K_J^0$, $\underline{B} \in J_{IO}^m$, $C \in J_{IO}$, and for

$$\underline{f} = (f_1, \dots, f_m) \in \mathrm{Mcl}_{AB_1} \underline{M} \times \dots \times \mathrm{Mcl}_{AB_m} \underline{M}, \; g \in \mathrm{Mcl}_{\underline{B}C} \underline{M},$$

their θ-composition is given by

$$[\underline{f}]g = d_A^m (f_1 \otimes \dots \otimes f_m) g \in \mathrm{Mcl}_{AC} \underline{M}.$$

Here, $d_A^m : A \to A \otimes \dots \otimes A$ designate the m-fold diagonal, inductively defined by

$$d_A^1 = 1_A, \quad d_A^{n+1} = d_A(1_A \otimes d_A^n).$$

The symbols p and o indicate zero-placed operations, $o : \; I \to O \in \underline{M}$ is already available, and p consists of families of canonical projections, formed by the lax terminal morphism $t_B : \; B \to I$ $(B \in J_{IO})$ as

$$p_{\underline{A}}^{(i)} = t_{A_1} \otimes \dots \otimes t_{A_{i-1}} \otimes 1_{A_i} \otimes t_{A_{i+1}} \otimes \dots \otimes t_{A_n} : \; A \to A_i$$

for $\underline{A} = (A_1, \dots, A_n) \in J_{IO}^n$, $A = A_1 \otimes \dots \otimes A_n$, and $i = 1, \dots, n$. For short, we set $(\underline{A} \in J_{IO}^n; \; B, C \in J_{IO})$

$$p_{\underline{A}} = p_{\underline{A}}^{(1)}, \quad p_{BC} = p_{(B,C)}^{(1)}, \quad p'_{BC} = p_{(B,C)}^{(2)}. \tag{1}$$

For $\underline{A} = (A_1, \dots, A_n) \in J_{IO}^n$, $A = A_1 \otimes \dots \otimes A_n$, we get the derived operations

$$t_A = p_{I\underline{A}} \in \mathrm{Mcl}_{AI} \underline{M} \quad (\text{i.e. } t_A : \; A \to I \in \underline{M}), \tag{2}$$

$$\begin{cases} o_{AO} = [t_A]o \in \mathrm{Mcl}_{AO} \underline{M}, \\ o_{AI} = [o_{AO}]t_O \in \mathrm{Mcl}_{AI} \underline{M}, \\ o_{AB} = [o_{AO}]p_{BO} \in \mathrm{Mcl}_{AB} \underline{M}. \end{cases} \tag{3}$$

Observe that in $\underline{M}$ one has $o_{AB} = t_A \cdot o \cdot (1_B \otimes t_O)$ and $1_B \otimes t_O = p_{BO}$.

The problem, posed by the author on the 2nd Conference of young algebraists at Potsdam, 1987, to characterize the Mal'cev clone structure of Mcl $\underline{M}$ axiomatically by identities was attacked one year later by F. Börner [74] in the homogeneous case (Card $J = 1$) without taking into account the objects I and O and without establishing the relation to $\underline{M}$. The present state is analyzed below. In the general case, there are two different ways, one via normal forms, which is not carried out here, and the other via an analysis of Schreckenberger's representation theorem for Mal'cev theories [50, 3.2.3, p. 134]. The situation is similar to that of fractal categories (cf. [Pt. 1, the proof of Theorem 4.7, p. 165]). The functor of interest is $\underline{H}_e : \; \underline{M} \to \mathrm{Par}$ $(e \leqq A; \; A \in |\underline{M}|)$, which maps $B \in |\underline{M}|$ to $L_B = \{f : \; A \to B; \alpha f = e\}$. The abstraction is applied to describe the sets of the form

$$L = \bigcup_{B \in J_{IO}} L_B$$

without use of the operation α. But this can be done just by the observation that L is one of the equivalence classes of the clone-part of $\underline{M}$,

$$\dot{\bigcup_{A\in K_J^0,\ B\in J_{IO}}} \underline{M}(A,B), \tag{4}$$

with respect to the equivalence ($\sim$) given by $f \sim g \Leftrightarrow \alpha f = \alpha g$ (for $f:\ A \to B$, $g:\ A \to C$) $\Leftrightarrow (\alpha g)f = f \& (\alpha f)g = g$, which (e.g. by [49, 3.7, p. 19]) can be expressed as:

$$f \sim g \Leftrightarrow [f,g]p_{BC} = f \& [g,f]p_{CB} = g. \tag{5}$$

Then there remains only to ensure by suitable axioms that the family of homomorphisms, induced by the $\underline{H}_e$ (in correspondence with the equivalence classes of the clone-part Mcl $\underline{M}$ of $\underline{M}$) yields a subdirect representation of Mcl $\underline{M}$. This leads to the following definition, which slightly extends and sharpnens that of Börner, loc. cit.

27.2. Definition. Let $K_J^0 = (K_J^0, \otimes, I, O)$ be a free object monoid with zero. An *axiomatic Mal'cev clone* $G = (G, \theta, p, o)$ is a heterogeneous total algebra, which decomposes according to

$$G = \dot{\bigcup_{A\in K_J^0,\ B\in J_{IO}}} G_{AB}$$

into carrier sets G_{AB}. The operations are of the following type:

$$\theta = \theta_{AC}^{\underline{B}} = [-](-) : (G_{AB_1} \times \cdots \times G_{AB_m}) \times G_{\underline{B}C} \to G_{AC},$$

where A, $\underline{B}$, C run as in Definition 27.1. The symbol p is for families of zero-placed operations,

$$p_{\underline{A}}^{(i)} \in G_{\underline{A}A_i} \quad (\underline{A} = (A_1, \ldots, A_n) \in J_{IO}^n), \quad i = 1, \ldots, n.$$

o is a zero-placed operations, $o \in G_{IO}$. We adopt the conventions, which correspond to (1,2,3). These data are connected by the following axioms [(SA) stands for "superassociativity"]:

(SA) $$[\underline{f}]([\underline{g}]h) = [[\underline{f}]g_1, \ldots, [\underline{f}]g_1]h \quad (\underline{g} = (g_1, \ldots, g_l)).$$

(L) $$[p_{\underline{A}}^{(1)}, g]p_{A_1C}, \ldots, [p_{\underline{A}}^{(n)}, g]p_{A_nC}]h = [h, g]p_{BC}$$

The class of all axiomatic Mal'cev clones of heterogeneous type J (with I and O adjoined) together with their homomorphisms forms a category AxMcl_J.

27.3. Theorem. *The clone-part* Mcl $\underline{M}$ *of any Mal'cev theory* $\underline{M} \in |\mathrm{Mth}_J|$ *(and hence any abstract Mal'cev clone of type J) is an axiomatic Mal'cev clone,* Mcl $\underline{M} \in |\mathrm{AxMcl}_J|$. *Conversely, each axiomatic Mal'cev clone $G \in |\mathrm{AxMcl}_J|$ is equal to the clone-part $G =$* Mcl $\underline{M}$ *of a suitable Mal'cev theory* $\underline{M} \in |\mathrm{Mth}_J|$.

In other words, abstract Mal'cev clones and axiomatic Mal'cev clones are identical and form a variety (of heterogeneous total algebras).

Proof. The first half of this theorem is a rather easy consequence of Definitions 27.1 and 27.2. The verifications of the second half is done in two steps, (i) and (ii).

(i) We show each axiomatic Mal'cev clone $G \in |\mathrm{AxMcl}_J|$ to be a subdirect product (we use the symbol $\coprod$) $G = \coprod_{L \in \underline{L}} G_L$ of subalgebras $G_L \subsetneqq \mathrm{Mcl}\,\mathrm{Par}\,K_J^0(L)$, following the approach in [74].

Thus, fixing $A \in K_J^0$, in $G_A = \bigcup_{B \in J_{IO}} G_{AB}$ we introduce a quasiorder ($\lesssim$) by

$$f \lesssim g \Leftrightarrow [f,g]p_{BC} = f \text{ for } f \in G_{AB} \text{ and } G_{AC}, \tag{6}$$

leading to an equivalence relation ($\sim$) on G_A by (5):

Reflexivity of ($\lesssim$) is ensured by

(R6) $$[f,f]p_{BB} \overset{(R1)}{=} [[f]p_B,[f]_B]p_{BB} \overset{(\mathrm{SA})}{=} [f]([p_B,p_B]p_{BB}) \overset{R2}{=} [f]p_B \overset{(R1)}{=} f.$$

Transitivity of ($\lesssim$). Suppose that $f \lesssim g(+)$ and $g \lesssim h(++)$ ($f \in G_{AB}$, $g \in G_{AC}$, $h \in G_{AD}$). Then

$$[f,h]p_{BD} \overset{(+)}{=} [[f,g]p_{AC},h]p_{AD} \overset{(\mathrm{R3})}{=} [f,[g,h]p_{CD}]p_{AC}$$

$$(\underline{A} = (A_1,\dots,A_n) \in J_{IO}^n, \quad h \in G_{AB}, \quad g \in G_{AC});$$

(R1) $$[f]p_A = f \in G_{AA} \quad (A \in J_{IO});$$

(R2) $$[f,p_{\underline{A}}^{(i)}]p_{BA_i} = f \in G_{AB} \quad (A = A_1 \otimes \cdots \otimes A_n,$$

$$\underline{A} = (A_1,\dots,A_n) \in J_{IO}^n,\ B \in J_{IO});$$

(R3) $$[[f,g]p_{BC},h]p_{BD} = [f,[g,h]p_{CD}]p_{BC}$$

$$(f \in G_{AB},\ g \in G_{AC},\ h \in G_{AD});$$

(R4) $$[\underline{g}']p_{\underline{C}'}^{(k+1)} = [\underline{g}]p_{\underline{C}}^{(k)} \quad (1 \leqq k < l)$$

$$(\underline{g} = (g_1,\dots,g_l),\ \underline{g}' = (g_l,g_1,\dots,g_{l-1}),\ g_k \in G_{BC_k},$$

$$C = (C_1,\dots,C_l),\ \underline{C}' = (C_l,C_1,\dots,C_{l-1}) \in J_{IO}^l);$$

(R5) $$[g_o,\underline{g}]p_{C_o\underline{C}} = [g_o,[\underline{g}]p_{\underline{C}}]p_{C_oC_1}$$

$$(\underline{g} = (g_1,\dots,g_l);\ g_k \in G_{BC_k}\ (k = 0,1,\dots,l),$$

$$B \in K_J^0,\ \underline{C} = (C_1,\dots,C_l) \in J_{IO}^n,\ C_o \in J_{IO});$$

(I1) $$[f,g]p_{II} = [g,f]p_{II} \quad (f,g \in G_{AI});$$

(I2) $$p^{(i)}_{\underline{A}} = p^{(i')}_{\underline{A}'};\ \underline{A} = (A_1, \dots, A_n) \in J^n_{IO},\ A'_{j'} = A_j,\ A'_k = I,$$

$$\text{(for some } k \leqq n'), j' = j\ (j < k),\ j' = j+1\ (k \leqq j \leqq n);$$

(I3) $$[\underline{g}]h = [\underline{g}']h,\ \underline{g} = (g_1, \dots, g_l),\ g_j \in G_{AC_j},\ C_i = I$$

$$(\text{ for some } i),\ \underline{g}' = (g'_1, \dots, g'_{i-1}, g'_{i+1}, \dots, g'_l),$$

$$g'_k = [g_k, g_i]p_{C_k I} \text{ for some } k \neq i \text{ and } g'_j = g_j \text{ else;}$$

(O1) $$o_{AB} = f \in G_{AB},\ A = A_1 \otimes \dots \otimes A_n \in K^0_J,\ B \in J_{IO},$$

$$(\underline{A} = (A_1, \dots, A_n) \in J^n_{IO},\ O \in \{A_1, \dots A_n, B\});$$

(O2) $$[\underline{f}]o_{BC} = o_{AC},\ \underline{f} = (f_1, \dots, f_m),\ f_i \in G_{AB_i},\ A \in K^0_J,$$

$$(B = B_1 \otimes \dots \otimes B_m,\ \underline{B} = (B_1, \dots, B_m) \in J^m_{IO},\ C \in J_{IO});$$

(O3) $$[\underline{g}]h = o_{AC} \text{if in } \underline{g} = (g_1, \dots, g_l) \text{ one has } g_i = o_{AC_i},$$

$$\text{for some i } (h \in G_{CD}\ C = C_1 \otimes \dots \otimes C_l);$$

By $\overset{(++)}{=} If,gIp_{BC} \overset{(+)}{=} f$ it is true that $f \lesssim h$; we get

(O4) (a) $L^0 = L^0(A) = \{o_{AC};\ C \in J_{IO}\}$ is an $(\sim)$-class of G_A $(A \in K^0_J)$;
(b) if $L = L(A)$ is any $(\sim)$-class of G_A such that $L \neq L^0$ (write $L_B = L \cap G_{AB}$ for $B \in J_{IO}$), then $L_O = \emptyset$.

Proof. (a) For $f \in G_{AB}$ one has

$$f \sim o_{AC} \overset{(5)}{\Leftrightarrow} [f, o_{AC}]p_{BC} = f \ \&\ [o_{AC}, f]p_{CB} = o_{AC} \overset{(O3)}{\Leftrightarrow} f = o_{AB}.$$

By (O1) we get $G_{AO} \subseteqq L^0$, and hence (a) implies $L_O = \emptyset$. For any $(\sim)$-class $L \neq L^0$ of G_A, $K^0_J[L]$ be the submonoid with zero O of $(|\mathrm{Par}|, \otimes, I, O)$ which is generated by the set $J^L = \{L_B;\ B \in J\}$. Identifying L_I and $I = \{\emptyset\}$, the mapping $J \to J^L : B \mapsto L_B$ (because of freeness of K^0_J) uniquely extends according to

$$K^0_J \to K^0_J[L];\ B = B_1 \otimes \dots \otimes B_m \mapsto L_B = L_{B_1} \times \dots \times L_{B_m}$$
$$(\underline{B} = (B_1, \dots, B_m) \in J^m_{IO})$$

to a homomorphism of monoids with zero, which in general is not an isomorphism, since also for $B \in J$ one may have $L_B = \emptyset$. One observes also that $\mathrm{Card}\, L_B = 1$ not necessarily implies $B = I$, but that nevertheless $K^0_J[L]$ as a monoid with zero O is freely generated by the set $J^L \setminus \{O\}$. This suggests, that one could (but we do not) go one step further and in $K_J^0[L]$ identify each L_B for which $\operatorname{Card} L_B = 1$ with L_I (which would give rise to a congruence on $K_J^0[L]$). For technical reasons we proceed in the opposite direction. We replace $K_J^0[L]$ by the free monoid with zero $K_J^0(L) = \{\{B\} \times L_B; B \in K_J^0\}$ with multiplication $(\otimes)$ given by $(\{B\} \times L_B) \otimes (\{C\} \times L_C) = \{B \otimes C\} \times L_{B\otimes C}$, which is isomorphic to K_J^0. The identity element and the zero of $K_J^0(L)$ are $\{I\} \times \{\emptyset\}$ and $\{O\} \otimes \emptyset = \emptyset$ respectively. (Observe that in our language $\{\emptyset\} = I = L_I$ and $\emptyset = 0$.) Accordingly, by $\operatorname{Par} K_J^0(L)$ we understand the Mal'cev theory which is given by $|\operatorname{Par} K_J^0(L)| = K_J^0(L)$ as the object monoid with zero, and by

$$(A, f, B) : \{A\} \times L_A \to \{B\} \times L_B \quad (\text{ for } f : \ L_A \to L_B \in \operatorname{Par} K_J^0[L])$$

as the morphisms; the signature of $\operatorname{Par} K_J^0(L)$ being induced in a natural way by that of the full subcategory $\operatorname{Par} K_J^0[L]$ (which again is a Mal'cev theory) of Par from the object class $K_J^0[L]$.

In accordance with the convention in Definition 26.2, $\operatorname{Par} K_J^0(L)$ can also be designated by $\operatorname{Par} K_J^0$, for short where the dependence from L is clear. There the Mal'cev subtheories $\underline{M}$ of $\operatorname{Par} K_J^0$ were also denoted as the concrete Mal'cev theories. In accordance with this concept we have introduced the notion of a concrete Mal'cev clone as a heterogeneous algebra being of the form $\operatorname{Mcl} \underline{M}$ for $\underline{M} \subseteqq \operatorname{Par} K_J^0$. While by definition a Mal'cev subtheory of a concrete Mal'cev theory again is concrete, there arises the question whether the analogous statement remains true also for any subalgebra G of a concrete Mal'cev clone $\operatorname{Mcl} \underline{M}$ ($\underline{M} \subseteqq \operatorname{Par} K_J^0$)? The answer is "yes," which can be verified as follows.

(C1) *Let* $G_{\operatorname{Par} K_J^0}$ *designate the Mal'cev subtheory of* $\underline{M}$ *(or, equally, of* $\operatorname{Par} K_J^0$*) which is generated by* G.

(a) *Then* $G_{\operatorname{Par} K_J^0}(A, B)$ *is identical with the set*

$$\begin{aligned} H(A,B) = \{f;\ & f = d_A^m(f_1 \otimes \cdots \otimes f_m) :\ A \to B \in \operatorname{Par};\\ & f_i :\ A \to B_i \in G;\ B_1 \otimes \cdots \otimes B_m = B,\\ & \underline{B} = (B_1, \dots, B_m) \in J_{IO}^m\};\ A, B \in K_J^0. \end{aligned}$$

(b) *One has* $G = \operatorname{Mcl} G_{\operatorname{Par} K_J^0}$.

Proof. (a) Form the subset

$$H = \bigcup_{A,B\in K_J^0} H(A,B) \subseteqq G_{\operatorname{Par} K_J^0}.$$

We show that $H = G_{\operatorname{Par} K_J^0}$ by proving that H is a Mal'cev subtheory of $G_{\operatorname{Par} K_J^0}$. For short, as far as the context excludes contradictions, we identify $K_J^0(L)$ and $K_J^0[L]$. The set H is closed with respect to composition of morphisms. Suppose that

$$f = d_A^m(f_1 \otimes \cdots \otimes f_m) \in H(A,B),$$
$$g = d_B^l(g_1 \otimes \cdots \otimes g_l) \in H(B,C).$$

Then we have $d_A^m(f_1 \otimes \cdots \otimes f_m)g_j = [\underline{f}]g_j \in G$ $(j = 1, \dots, l)$ and hence

$$\begin{aligned} fg &= d_A^m(f_1 \otimes \cdots \otimes f_m)d_B^l(g_1 \otimes \cdots \otimes d_l) \\ &= d_A^m d_{A^m}^l((f_1 \otimes \cdots \otimes f_m) \otimes \cdots \otimes (f_1 \otimes \cdots \otimes f_m))(g_1 \otimes \cdots \otimes g_l) \\ &= d_A^{ml}((f_1 \otimes \cdots \otimes f_m)g_1 \otimes \cdots \otimes (f_1 \otimes \cdots \otimes f_m)g_l) \\ &= d_A^l(d_A^m(f_1 \otimes \cdots \otimes f_m)g_1 \otimes \cdots \otimes d_A^m(f_1 \otimes \cdots \otimes f_m)g_l) \in H(A,C). \end{aligned}$$

The set H is closed with respect to monoidal multiplication ($\otimes$). Suppose that $f \in H(A,B)$ be as above and

$$h = d_C^k(h_1 \otimes \cdots \otimes h_k) \in H(C,D).$$

Then we get

$$\begin{aligned} f \otimes g &= d_A^m(f_1 \otimes \cdots \otimes f_m) \otimes d_C^k(h_1 \otimes \cdots \otimes h_k) \\ &= (d_A^m \otimes d_C^k)(f_1 \otimes \cdots \otimes f_m \otimes h_1 \otimes \cdots \otimes h_k) \\ &= d_{A\otimes C}^{m+k}((f_1 \otimes t_C) \otimes \cdots \otimes (f_m \otimes t_C) \otimes (t_A \otimes h_1) \otimes \cdots \otimes (t_A \otimes h_k)) \\ &\in H(A \otimes C, B \otimes D), \end{aligned}$$

since for $\underline{A} = (A_1, \dots, A_n) \in J_{IO}^n$ one has

$$t_A = p_{I\underline{A}}^{(1)} \in H(A,I) \quad (A = A_1 \otimes \cdots \otimes A_n);$$

further,

$$1_A = d_A^n(p_{\underline{A}}^{(1)} \otimes \cdots \otimes p_{\underline{A}}^{(n)}) \in H(A,A);$$

and for $\underline{B} = (B_1, \dots, B_m) \in J_{IO}^m$, $B = B_1 \otimes \cdots \otimes B_m$,

$$\begin{aligned} s_{AB} =& d_{A\otimes B}^{n+m}(p_{\underline{A}\,\underline{B}}^{(n+1)} \otimes \cdots \otimes p_{\underline{A}\,\underline{B}}^{(n+m)} \otimes p_{\underline{A}\,\underline{B}}^{(1)} \otimes \cdots \otimes p_{\underline{A}\,\underline{B}}^{(n)}) \\ &\in H(A \otimes B, B \otimes A); \end{aligned}$$

$$\begin{aligned} d_A =& d_A^{2n}((p_{\underline{A}}^{(1)} \otimes \cdots \otimes p_{\underline{A}}^{(n)}) \otimes (p_{\underline{A}}^{(1)} \otimes \cdots \otimes p_{\underline{A}}^{(n)})) \\ &\in H(A, A \otimes A). \end{aligned}$$

(b) Let $f = d_A(f_1 \otimes \cdots \otimes f_m) \in H(A,B)$, where $A \in K_J^0$, $\underline{f} = (f_1, \dots, f_m)$, $f_i : A \to B_i \in G$, $B = B_1 \otimes \cdots \otimes B_m$, $\underline{B} = (B_1, \dots, B_m) \in J_{IO}^m$. We have to show that $B \in J_{IO}$ implies $f \in G$. Indeed, one has $f = [\underline{f}]1_B \in G$, since $1_B \in G$ $(B \in J_{IO})$ and by assumption $f_i \in G$ $(i = 1, \dots, m)$. We recall the notation which we agreed upon at the beginning of the proof of (i), i.e. we take any $G \in |\mathrm{AxMcl}_J|$, $A \in K_J^0$ (fixed), and $G_A = \bigcup_{B \in J_{IO}} G_{AB}$. The symbol $L \neq L^0$ designates any $(\sim)$-class of G_A.

A mapping $(^{(L)}) : G \to \mathrm{Mcl\,Par}\, K_J^0$ $(K_J^0 = K_J^0(L))$ is defined as follows. Each $h \in G_{CD}$ $(C = C_1 \otimes \cdots \otimes C_l;\ \underline{C} = (C_1, \dots, C_l) \in J_{IO}^l)$ is mapped to some

$$h^{(L)} \in \mathrm{Mcl\,Par}\, K_J^0(C, D) = \mathrm{Mcl\,Par}\, K_J^0(L)(\{C\} \times L_C, \{D\} \times L_D),$$

i.e. to

$$h^{(L)} = (\{C\} \times L_C, h^L, \{D\} \times L_D),$$

where the partial mapping $h^L : L_C \to L_D \in \mathrm{Mcl\,Par}\, K_J^0[L]$ is given by

$$\begin{aligned} \mathrm{Dex}\, h^L &= \{\underline{g} = (g_1, \dots, g_l) \in L_C; \quad [\underline{g}]h \in L\}, \\ \underline{g}h^L &= [\underline{g}]h \in L_D. \end{aligned}$$

One observes that this mapping is proved to remain single-valued if, according to our arrangement each L_I in $L_C = L_{C_1} \times \cdots \times L_{C_l}$, i.e. each L_{C_i}, for which $C_i = I$, is identified with $I = \{\emptyset\}$ (that means, is cancelled, as long as $l > 1$). Correspondingly, in $\underline{g} = (g_1, \dots, g_l)$ (for $l > 1$) the component g_i can be omitted without changing the value of $[\underline{g}]h$, since all components of $\underline{g}$ belong to L, i.e. have the same "abstract domain of existence," i.e. $g_k \sim g_i$ holds which by Axiom $(I3)$ implies that $g_k' = [g_k, g_i]p_{C_k I} = g_k$.

We claim $(^{(L)})$ to be a homomorphism:

(C2) *For* $\underline{C} = (C_1, \dots, C_l) \in J_{IO}^l$,

$$p_{\underline{C}}^{(k)L} : L_{C_1} \times \cdots \times L_{C_l} \to L_{C_k}$$

is the kth canonical projection in $\mathrm{Par}\, K_J^0$. *If some* $C_k = 0$, *then we get the empty projection.*

Proof. We remark that projections, which are identical in G by Axiom (I2), by (C2) are mapped to concrete projections in $\mathrm{Par}\, K_J^0$, which are identified because of the identification of L_I and $I = \{\emptyset\}$. Now, for $g \in L_C$ and $g_i \in L_{C_i}$ $(i = 1, 2)$ one has

$$g\, p_C^L \overset{(\mathrm{R1})}{=} g$$

and, because of $g_1 \lesssim g_2$, $[g_1, g_2]p_{C_1C_2} = g_1 \in L$, i.e.

$$(g_1, g_2)p_{C_1C_2}^L = g_1.$$

(C2) is proved by induction on k. We get

$$\begin{aligned} &[g_1, \dots, g_k, g_o, g_{k+1}, \dots, g_l]p_{\underline{C}}^{(k+1)} \overset{(\mathrm{R4})}{=} [g_o, \underline{g}]p_{C_o\underline{C}} \\ &\overset{(\mathrm{R5})}{=} [g_o, [\underline{g}]p_{\underline{C}}]p_{C_oC_1} = [g_o, g_1]p_{C_oC_1} = g_o. \end{aligned}$$

Since for all $g:\ A \to I \in G:\ [g]o = o_{AO} \notin L$, we have:

(C3) $o^L = o:\ I \to O = \emptyset$ (*the empty mapping of* $\operatorname{Par} K_J^0$).

Next we show:

(C4) $(^{(L)})$ *preserves composition* $[-](-)$.

Let we have

$$\begin{aligned} &\underline{f} = (f_1, \dots, f_m) \in L_{B_1} \times \cdots \times L_{B_m}; \underline{B} = (B_1, \dots, B_m) \in J_{IO}^m, \\ &\underline{g} = (g_1, \dots, g_l) \in G_{BC_1} \times \cdots \times G_{BC_l}; B = B_1 \otimes \cdots \otimes B_m, \\ &h \in G_{CD};\ C = C_1 \otimes \cdots \otimes C_l,\ \underline{C} = (C_1, \dots, C_l) \in J_{IO}^l. \end{aligned}$$

Writing for short

$$a = [g_1^L, \dots, g_l^L]h^L, \quad b = ([\underline{g}]h)^L,$$

we have to verify that $a = b$. Respecting Definition 27.1 for $\underline{M} = \operatorname{Par} K_J^0$, the assumption $\underline{f} \in \operatorname{Dex} a$ implies

$$\underline{f}a = [[\underline{f}]g_1, \dots, [\underline{f}]g_l h \overset{(SA)}{=} [\underline{f}]([\underline{g}]h) = I\underline{f}]b.$$

Hence, $\operatorname{Dex} a \subseteqq \operatorname{Dex} b$, and on $\operatorname{Dex} a$, the mappings a and b are identical. Conversely, let $\underline{f} \in \operatorname{Dex} b$, i.e. $[\underline{f}]([\underline{g}]h) \in L$. We have to show that $[\underline{f}]g_j \in L$ $(j = 1, \dots, l)$. For its proof we need:

(R7) $$[f, g]p_{BC} \lesssim f \text{ for } (f, g) \in G_{AB} \times G_{AC},$$

which follows from

$$\begin{aligned} [[f, g]p_{BC}, f]p_{BB} &\overset{(R3)}{=} [f, [g, f]p_{CB}]p_{BC} \\ &\overset{(R6)}{=} [f, [[g, g]p_{CC}, f]p_{CB}]p_{BC} \\ &\overset{(R3)}{=} [[f, g]p_{BC}, [g, f]p_{CB}]p_{BC} \\ &\overset{(R4)}{=} [[f, g]p_{BC}, [f, g]p'_{BC}]p_{BC} \\ &\overset{(SA)}{=} [f, g]([p_{BC}, p'_{BC}]p_{BC}) \overset{(R2)}{=} [f, g]p_{BC}. \end{aligned}$$

By induction for $\underline{u} = (u_1, \dots, u_l) \in G_{AC_1} \times \cdots \times G_{AC_l}$ and $h \in G_{CD}$ $(C = C_1 \otimes \cdots \otimes C_l)$ regarding (R4), (R5), this implies

$$[\underline{u}]p_{\underline{C}}^{(i)} \lesssim u_i \quad (\underline{C} = (C_1, \dots, C_l) \in J_{IO}^l).$$

Since

$$[[\underline{u}]h, [\underline{u}]p_{\underline{C}}^{(i)}]p_{DC_i} \overset{(SA),(R2)}{=} [\underline{u}]h,$$

we get

$$\text{(R8)} \qquad [\underline{u}]h \lesssim [\underline{u}]p^{(i)}_{\underline{C}} \lesssim u_i \quad (i = 1, \ldots, l).$$

For $\underline{f} \in \operatorname{Dex} b$ this implies

$$\underline{f}b \overset{(\mathrm{SA})}{=} [[\underline{f}]g_1, \ldots, [\underline{f}]g_l]h \lesssim [\underline{f}]g_i \lesssim f_1$$

Since f_1 and $[\underline{f}]b$ belong to L, we have $f_1 \sim [\underline{f}]b$, and by transitivity of $(\lesssim)$ we get $[\underline{f}]g_i \in L$. Hence $a = b$.

Consider the set

$\underline{L} = \{L;$ there exists $A \in K^0_J$ such that $L = L(A)$ is a $(\sim)$-class of G_A which is different from $L^0\}$.

(C5) *The mapping*

$$(\overset{L}{-}) : \ G \to \prod_{L \in \underline{L}} \operatorname{Mcl} \operatorname{Par} K^0_J(L),$$

such that the image of h *is* $(h^{(L)})_{L \in \underline{L}}$, *is an injective homomorphism.*

Proof. Let $f \in G_{AB}$, $g \in G_{AC}$ and for all $L \in \underline{L}$:

$$f^{(L)} = g^{(L)}; \ A = A_1 \otimes \cdots \otimes A_n, \ \underline{A} = (A_1, \ldots, A_n) \in J^n_{IO}.$$

We have

$$\begin{aligned} g &\overset{(\mathrm{R2}),(\mathrm{R6})}{=} [[g, p^{(i)}_{\underline{A}}]p_{CA_i}, g]p_{CC} \\ &\overset{R3}{=} [g, [p^{(i)}_{\underline{A}}, g]p_{A_iC}]p_{CA_i}. \end{aligned}$$

Hence, by (R8) and the transitivity of $(\lesssim)$ we get

$$g \sim [p^{(i)}_{\underline{A}}, g]p_{A_iC}, \ i = 1, \ldots, n.$$

Choosing $L \in \underline{L}$ such that $g \in L$, one has

$$\begin{aligned} g &\overset{(\mathrm{R6})}{=} [g, g]p_{CC} \\ &\overset{(\mathrm{L})}{=} ([p^{(1)}_{\underline{A}}, g]p_{A_1C}, \ldots, [p^{(n)}_{\underline{A}}, g]p_{A_nC})g^{L} \\ &= ([p^{(1)}_{\underline{A}}, g]p_{A_1C}, \ldots, [p^{(n)}_{\underline{A}}, g]p_{A_nC})f^{L} \\ &\overset{(\mathrm{L})}{=} [f, g]p_{CC}. \end{aligned}$$

Hence, $g \lesssim f$. Quite similarly, one obtains $f = [g, f]p_{CB}$. But $g \lesssim f$ implies $[g, f]p_{CB} = g$, and thus one has $f = g$. The property (C5) implies that G is isomorphic to a subdirect product $G \cong \coprod\!\!\coprod_{L \in \underline{L}} G_L$ of subalgebras $G_L \subseteqq \operatorname{Mcl} \operatorname{Par} K^0_J(L)$ and thus of concrete Mal'cev clones

$$G_L = \operatorname{Mcl}(G_L)_{\operatorname{Par} K^0_J(L)}; \ (G_L)_{\operatorname{Par} K^0_J(L)} \subseteqq \operatorname{Par} K^0_J(L). \tag{7}$$

(ii) We have

$$G \cong G' \subseteqq \underset{L\in\underline{L}}{\times} \operatorname{Mcl}\operatorname{Par} K_J^0(L) = \operatorname{Mcl} \underset{L\in\underline{L}}{\times} \operatorname{Par} K_J^0(L), \tag{8}$$

thus

$$G' = \operatorname{Mcl}(G')_{\underset{L\in\underline{L}}{\times} \operatorname{Par} K_J^0(L)},$$

therefore, $G \cong G'$ is an abstract Mal'cev clone.

27.4. Corollary.

(a) *Each axiomatic Mal'cev clone is a subdirect product of concrete Mal'cev clones* [74].

(b) *Let* Mcl_J *designate the category of all axiomatic Mal'cev clones with object monoid* K_J^0 *with zero. Then each subvariety of* $|\mathrm{Mcl}_J|$ *is generated by its concrete Mal'cev clones.*

(c) *If the Mal'cev clone* G *is given as* $G = \operatorname{Mcl}\underline{M}$, *then each* G_L *in* (7) *is of the form*

$$G_L = G^{(L)} = \operatorname{Mcl}\operatorname{Ary}^0 \underline{M}H_e \text{ for } e(\leqq A) \in L = L(A) \in \underline{L}.$$

Proof. The item (a) is essentially the topic of the proof (i) of 27.3, (b) rests upon the fact that each G_L in (7) is the homomorphic image $G^{(L)}$ of G by the homomorphism $(^{(L)})$. The item (c) already mentioned in Definition 27.1.

28. The clone-part functor

The connection between the structures of Mal'cev theories $\underline{M}$ and their clone-parts $\operatorname{Mcl}\underline{M}$ can be derived from the clone-part functor $\underline{Cp} = (|Cp|, Cp) : \mathrm{Mth}_J \to \mathrm{Mcl}_J$, which is defined by:

$$\underline{M}|Cp| = \operatorname{Mcl}\underline{M} \ (\underline{M} \in |\mathrm{Mth}_J|),$$
$$\underline{F}\, Cp = F|_{\operatorname{Mcl}\underline{M}} \ (\underline{F} = (|F|, F) : \ \underline{M} \to \underline{M'} \in \mathrm{Mth}_J).$$

The study of the clone-part functor $\underline{Cp}$ rests upon the following basic equations which can be easily seen to hold as in the proof of (C1),(a) of 27.3, for any Mal'cev theory $\underline{M}$.

28.1. Lemma. *Let be* $f : A \to B, g : B \to C, h : C \to D \in \underline{M} \in |\mathrm{Mth}_J|$; $A, B, C, D \in K_J^0$, *where*

$$A = A_1 \otimes \cdots \otimes A_n,\ \underline{A} = (A_1, \dots, A_n) \in J_{IO}^n,$$
$$B = B_1 \otimes \cdots \otimes B_m,\ \underline{B} = (B_1, \dots, B_m) \in J_{IO}^m,$$
$$C = C_1 \otimes \cdots \otimes C_l,\ \underline{C} = (C_1, \dots, C_l) \in J_{IO}^l,$$
$$D = D_1 \otimes \cdots \otimes D_k,\ \underline{D} = (D_1, \dots, D_k) \in J_{IO}^k.$$

Then

$$f = d_A^m(fp_{\underline{B}}^{(1)} \otimes \cdots \otimes fp_{\underline{B}}^{(m)}), \tag{1}$$

$$fp_{\underline{B}}^{(j)} = [fp_{\underline{B}}^{(1)}, \dots, fp_{\underline{B}}^{(m)}]p_{\underline{B}}^{(j)}, \quad j = 1, \dots, m, \tag{2}$$

$$fg = d_A^l(fgp_{\underline{C}}^{(1)} \otimes \cdots \otimes fgp_{\underline{C}}^{(l)}), \tag{3}$$

$$fgp_{\underline{C}}^{(j)} = [fp_{\underline{B}}^{(1)}, \dots, fp_{\underline{B}}^{(m)}](gp_{\underline{C}}^{(j)}), \quad j = 1, \dots, l, \tag{4}$$

$$f \otimes h = d_{A\otimes C}^{(m+k)}(p_{AC}f_1 \otimes \cdots \otimes p_{AC}f_m \otimes p'_{AC}h_1 \otimes \cdots \otimes p'_{\underline{A}C}h_k),$$

$$p_{AC}f_j = f_j \otimes t_C,\ f_j = fp_{\underline{B}}^{(j)},\ j = 1, \dots, m,$$

$$p'_{AC}h_j = t_A \otimes h_j,\ h_j = hp_{\underline{D}}^{(j)}, \quad j = 1, \dots, k, \tag{5}$$

$$f \otimes h = d_{A\otimes C}^{(m+k)}((f \otimes h)p_{\underline{B}\,\underline{D}}^{(1)} \otimes \cdots \otimes (f \otimes h)p_{\underline{B}\,\underline{D}}^{(m+k)}), \tag{6}$$

$$(f \otimes h)p_{\underline{B}\,\underline{D}}^{(i)} = [p_{AC}f_1, \dots, p_{AC}f_m, p'_{AC}h_1, \dots, p'_{AC}h_k]p_{\underline{B}\,\underline{D}}^{(i)}, \quad i = 1, \dots, m + k, \tag{7}$$

$$1_A = d_A^n(p_{\underline{A}}^{(1)} \otimes \cdots \otimes p_{\underline{A}}^{(n)}), \tag{8}$$

$$s_{AB} = d_{A\otimes B}^{(n+m)}(s_{AB}p_{\underline{B}\,\underline{A}}^{(1)} \otimes \cdots \otimes s_{AB}p_{\underline{B}\,\underline{A}}^{(n+m)}), \tag{9}$$

$$s_{AB}p_{\underline{A}\,\underline{B}}^{(i)} = p_{\underline{A}\,\underline{B}}^{(n+i)}, \quad i = 1, \dots, m,$$

$$s_{AB}p_{\underline{B}\,\underline{A}}^{(i+m)} = p_{\underline{A}\,\underline{B}}^{(i)}, \quad i = 1, \dots, n, \tag{10}$$

$$d_A = d_A^{2n}(d_Ap_{\underline{A}\,\underline{A}}^{(1)} \otimes \cdots \otimes d_Ap_{\underline{A}\,\underline{A}}^{(2n)}), \tag{11}$$

$$d_Ap_{\underline{A}\,\underline{A}}^{(i)} = d_Ap_{\underline{A}\,\underline{A}}^{(n+i)} = p_{\underline{A}}^{(i)}, \quad i = 1, \dots, n, \tag{12}$$

$$t_A = p_{I\underline{A}}^{(1)}. \tag{13}$$

28.2. Theorem. *The clone-part functor* $\underline{Cp}$: $\mathrm{Mth}_J \to \mathrm{Mcl}_J$ *is full and faithful.*

Proof. The functor $\underline{Cp}$ is faithful. Indeed, let $\underline{F}$, $\underline{F}' : \underline{M} \to \underline{M}' \in \mathrm{Mth}_J$ be such that $\underline{F}\ Cp = \underline{F}' Cp$. Then (1) implies

$$\begin{aligned} fF &= d_A^m((fp_{\underline{B}}^{(1)})F \otimes \cdots \otimes (fp_{\underline{B}}^{(m)})F) \\ &= d_A^m((fp_{\underline{B}}^{(1)})F' \otimes \cdots \otimes (fp_{\underline{B}}^{(m)})F') = fF', \end{aligned}$$

i.e. $\underline{F} = \underline{F}'$. The functor $\underline{Cp}$ is full. In fact, suppose that

$$H : \ \mathrm{Mcl}\,\underline{M} \to \mathrm{Mcl}\,\underline{M}' \in \mathrm{Mcl}_J.$$

We have to construct a functor $\underline{F} = (|F|, F) : \underline{M} \to \underline{M}' \in \mathrm{Mth}_J$ such that $\underline{F}\ Cp = H$. Applying (1) we define

$$A|F| = A \quad \text{for all } A \in K_J^0,$$

$$fF = d_A^m((fp_{\underline{B}}^{(1)})H \otimes \cdots \otimes (fp_{\underline{B}}^{(m)})H). \tag{14}$$

By Axioms 27.2 (I2), (O1), this definition is unique. Especially, we get

$$fF = fH \quad \text{for all } f \in \mathrm{Mcl}\,\underline{M}, \tag{15}$$

such that $\underline{F}$ extends H. We have to show $\underline{F} \in \mathrm{Mth}_J$: $\underline{F}$ preserves the identities 1_A of $\underline{M}$. In fact, since H preserves p, we get [cf. (8)]

$$\begin{aligned} 1_A F &= d_A^n((1_A p_{\underline{A}}^{(1)})H \otimes \cdots \otimes (1_A p_{\underline{A}}^{(m)})H) \\ &= d_A^n(p_{\underline{A}}^{(1)} \otimes \cdots \otimes p_{\underline{A}}^{(m)}) = 1_A \in \underline{M}'. \end{aligned}$$

Similarly, (9)–(13) imply the preservations

$$s_{AB}F = s_{AB}, \quad d_A F = d_A, \quad t_A F = t_A,$$

and for $o = d_I(op_O)$ (where $op_O = o : \ I \to O$)

$$oF = d_I(oH) = d_I o = o.$$

The functor $\underline{F}$ preserves composition. Indeed, (3) implies

$$\begin{aligned} fg\ F =& d^l_A([(fp^{(1)}_{\underline{B}})H, \dots, (fp^{(m)}_{\underline{B}})H](gp^{(1)}_{\underline{C}})H \otimes \cdots \\ & \otimes [(fp^{(1)}_{\underline{B}})H, \dots, (fp^{(m)}_{\underline{B}})H](gp^{(l)}_{\underline{C}})H) \\ =& d^l_A(d^m_A((fp^{(1)}_{\underline{B}})H \otimes \cdots \otimes (fp^{(m)}_{\underline{B}})H)(gp^{(1)}_{\underline{C}})H \otimes \cdots \\ & \otimes d^m_A((fp^{(1)}_{\underline{B}})H \otimes \cdots \otimes (fp^{(m)}_{\underline{B}})H)(gp^{(l)}_{\underline{C}})H) \\ =& d^l_A((fH)(gp^{(1)}_{\underline{C}})H \otimes \cdots \otimes (fH)(gp^{(l)}_{\underline{C}})H) \\ =& (fF)d^l_B((gp^{(1)}_{\underline{C}})H \otimes \cdots \otimes (gp^{(l)}_{\underline{C}})H) \\ =& (fF)(gF). \end{aligned}$$

The functor $\underline{F}$ preserves the monoidal product ($\otimes$). Respecting (1), (5), one has

$$fF = d^m_A(f_1H \otimes \cdots \otimes f_mH), \quad hF = d^k_C(h_1H \otimes \cdots \otimes h_kH).$$

Since we already know that $\underline{F}$ preserves composition, we get

$$(p_{AC}f_j)H = (p_{AC}f_j)F = (p_{AC}F)(f_jF) = p_{AC}(f_jH),$$

similarly

$$(p'_{AC}h_j)H = p'_{AC}(h_jH).$$

Hence,

$$\begin{aligned} ((f\otimes h)p^{(i)}_{\underline{B}\,\underline{D}})H &= [p_{AC}(f_1H), \dots, p_{AC}(f_mH), \\ & \qquad p'_{AC}(h_1H), \dots, p'_{AC}(h_kH)]p^{(i)}_{\underline{B}\,\underline{D}} \\ &= (fF \otimes hF)p^{(i)}_{\underline{B}\,\underline{D}} \end{aligned}$$

and (6), (7) imply

$$\begin{aligned} (f\otimes h)F &= d^{(m+k)}_{A\otimes C}(((f\otimes h)p^{(1)}_{\underline{B}\,\underline{D}})H \otimes \cdots \otimes ((f\otimes h)p^{(m+k)}_{\underline{B}\,\underline{D}})H) \\ &= d^{(m+k)}_{A\otimes C}((fF\otimes hF)p^{(1)}_{\underline{B}\,\underline{D}} \otimes \cdots \otimes (fF\otimes hF)p^{(m+k)}_{\underline{B}\,\underline{D}}) \\ &= fF \otimes hF. \end{aligned}$$

28.3. Corollary.

(a) *Let* $\underline{F} := \ \underline{M} \to \underline{M}' \in \mathrm{Mth}_J$. *Then* $\underline{F}$ *is an isomorphism (the identity) iff* $\underline{F}\ \underline{Cp}\colon\ \mathrm{Mcl}\,\underline{M} \to \mathrm{Mcl}\,\underline{M}'$ *is an isomorphism (the identity).*

(b) *Every Mal'cev theory* $\underline{M} \in |\mathrm{Mth}_J|$ *is uniquely determined by its clone-part* $\mathrm{Mcl}\,\underline{M}$.

(c) *The word problem for* $\underline{M} \in |\mathrm{Mth}_J|$ *is solvable iff it is solvable for* $\mathrm{Mcl}\,\underline{M}$.

(d) *The subvarieties of* $|\mathrm{Mth}_J|$ *(i.e. of Mal'cev theories) and of* $|\mathrm{Mcl}_J|$ *(i.e. of Mal'cev clones) correspond biuniquely to one another.*

Proof. The assertions (a) and (b) are immediate conclusions of 28.2.

(d) Let $\underline{\underline{V}} \subseteqq |\mathrm{Mcl}_J|$ and $\underline{\underline{W}} \subseteqq |\mathrm{Mth}_J|$ be subvarieties. We prove that again

$$\mathrm{Mcl}^{-1}\,\underline{\underline{V}} = \{\underline{M} \in |\mathrm{Mth}_J|; \quad \mathrm{Mcl}\,\underline{M} \in \underline{\underline{V}}\} \subseteqq |\mathrm{Mth}_J|, \tag{16}$$

$$\mathrm{Mcl}\,\underline{\underline{W}} = \{\mathrm{Mcl}\,\underline{M};\ \underline{M} \in \underline{\underline{W}}\} \subseteqq |\mathrm{Mcl}_J|. \tag{17}$$

are subvarieties.

Prove (16). Let

$$\Phi_1(\underline{X}) = \Phi_2(\underline{X}), \quad \underline{X} = (X_1, \dots, X_n), \tag{18}$$

be any equation in the signature of Mal'cev clones with respect to the free variables X_i of type $(U_i \in K_J^0, V_i \in J_{IO})$. We express the signature θ, p, o of Mal'cev clones formally according to Definition 27.1 by that of Mal'cev theories, thus we get from (18) an equation

$$\Psi_1(\underline{X}) = \Psi_2(\underline{X}), \quad \underline{X} = (X_1, \dots, X_n) \tag{19}$$

with the same set of variables $\underline{X}$, now considering each X_i as an arrow $X_i : U_i \to V_i$. Then, applying the usual universal validity concept ($\models$) to these equations, evaluating each variable X_i by all morphisms of $\mathrm{Mcl}\,\underline{M}$ of suitable size, one obtains

$$\underline{M} \models \Psi_1(\underline{X}) = \Psi_2(\underline{X}) \Leftrightarrow \mathrm{Mcl}\,\underline{M} \models \Phi_1(\underline{X}) = \Phi_2(\underline{X})$$

and thus

$$\mathrm{Mcl}^{-1}\,\underline{\underline{V}} \models \Psi_1(\underline{X}) = \Psi_2(\underline{X}) \Leftrightarrow \underline{\underline{V}} \models \Phi_1(\underline{X}) = \Phi_2(\underline{X}).$$

Prove (17). By 28.2, $\mathrm{Mcl}\,\underline{\underline{W}}$ is closed with respect to subalgebras and to homomorphic images. As $\times_{k\in K} \mathrm{Mcl}\,\underline{M}_k = \mathrm{Mcl} \times_{k\in K} \underline{M}_k$ shows, $\mathrm{Mcl}\,\underline{\underline{W}}$ is also closed with respect to cartesian product. This implies that $\mathrm{Mcl}\,\underline{\underline{W}}$ is a variety. Finally, one has $\mathrm{Mcl}\,\mathrm{Mcl}^{-1}\,\underline{\underline{V}} = \underline{\underline{V}}$ and $\mathrm{Mcl}^{-1}\,\mathrm{Mcl}\,\underline{\underline{W}} = \underline{\underline{W}}$.

(c) We prove more: if $\underline{\underline{W}}$ is a subvariety of $|\mathrm{Mth}_J|$, then the word problem of $\underline{\underline{W}}$ is solvable iff it is solvable for $\mathrm{Mcl}\,\underline{\underline{W}}$ (17).

Let $\underline{X} = (X_1, \dots, X_n)$ be a set of free arrows $X_i : U_i \to V_i$, where as in the proof of (d) we assume that $V_i \in J_{IO}$. By $F(\underline{X})$ we designate the free algebra of $\underline{\underline{W}}$ which is freely generated by $\underline{X}$. Then $\mathrm{Mcl}\,F(\underline{X}) \in \mathrm{Mcl}\,\underline{\underline{W}}$ is the free algebra of the variety $\mathrm{Mcl}\,\underline{\underline{W}}$ which is freely generated by $\underline{X}$. This follows from the commutativity of the following diagram ($\underline{M} \in \underline{\underline{W}}$)

$$\begin{array}{ccccc} & & \mathrm{Mcl}\, F(\underline{X}) & \longrightarrow & F(\underline{X}) \\ & \nearrow & | & & | \\ \underline{X} & & | & & | \\ & \searrow & \downarrow & & \downarrow \\ & & \mathrm{Mcl}\, \underline{M} & \longrightarrow & \underline{M} \end{array}$$

in which the existence and uniqueness of the arrow $(- \rightarrow)$ is implied by 28.2.

Now, for $f, g: A \to B \in F(\underline{X})$ by (1) one has

$$f = g \text{ in } F(\underline{X}) \Leftrightarrow f p_{\underline{B}}^{(j)} = g p_{\underline{B}}^{(j)} \text{ in } \mathrm{Mcl}\, F(\underline{X}) \ (j = 1, \ldots, m),$$

which proves the assertion.

Let $\underline{M} \in |\mathrm{Mth}_J|$. By $\underline{\underline{W}}_{\underline{M}}$ and $\underline{\underline{V}}_{\underline{M}}$ we designate the subvarieties of $|\mathrm{Mth}_J|$ and of $|\mathrm{Mcl}_J|$ which are generated by $\underline{M}$ and $\mathrm{Mcl}\, \underline{M}$ respectively. As the proof of (d) shows, $\mathrm{Mcl}^{-1} \underline{\underline{V}}_{\mathrm{Mcl}\, \underline{M}} = \underline{\underline{W}}_{\underline{M}}$. Hence $\mathrm{Mcl}\, \underline{\underline{W}}_{\underline{M}} = \underline{\underline{V}}_{\mathrm{Mcl}\, \underline{M}}$. Let $\underline{M} \in |\mathrm{Mth}_J|$. By $\mathrm{End}\, \underline{M}$, $\mathrm{Sal}\, \underline{M}$, and $\mathrm{Con}\, \underline{M}$ respectively we designate the monoid of endomorphisms of $\underline{M}$, the lattice of subalgebras of $\underline{M}$, and the lattice of congruences on $\underline{M}$; similarly for $\mathrm{Mcl}\, \underline{M}$. These data are connected by:

28.4. Corollary. *For each* $\underline{M} \in |\mathrm{Mth}_J|$ *one gets the isomorphisms*

$$\mathrm{End}\, \underline{M} \cong \mathrm{End}\, \mathrm{Mcl}\, \underline{M}, \tag{20}$$

$$\mathrm{Sal}\, \underline{M} \cong \mathrm{Sal}\, \mathrm{Mcl}\, \underline{M}, \tag{21}$$

$$\mathrm{Con}\, \underline{M} \cong \mathrm{End}\, \mathrm{Mcl}\, \underline{M}. \tag{22}$$

Proof. The assertion (20) is a straightforward consequence of 28.2.

Prove (21). Obviously the chain of inclusions of Mal'cev subtheories

$$\underline{M}'' \subseteqq \underline{M}' \subseteqq \underline{M} \tag{23}$$

implies

$$\mathrm{Mcl}\, \underline{M}'' \subseteqq \mathrm{Mcl}\, \underline{M}' \subseteqq \mathrm{Mcl}\, \underline{M}. \tag{24}$$

Conversely, (24) and (14) imply (23).

Prove (22). The restriction of a congruence on $\underline{M}$ induces a congruence on $\mathrm{Mcl}\, \underline{M}$. Conversely, let $(\equiv)$ be a congruence on $\mathrm{Mcl}\, \underline{M}$. We extend $(\equiv)$ to $\underline{M}$ according to

$$f \equiv f' : A \to B \Leftrightarrow f p_{\underline{B}}^{(j)} \equiv f' p_{\underline{B}}^{(j)}, \ j = 1, \ldots, m.$$

($B = B_1 \otimes \cdots \otimes B_m; \underline{B} = (B_1, \dots, B_m) \in J^m_{IO}$). We claim that ($\equiv$) is a congruence on $\underline{M}$. Let $f \equiv f' : A \to B$, $g \equiv g' : B \to C$, $h \equiv h' : C \to D \in \underline{M}$. We have to show that $fg \equiv f'g'$ and $f \otimes h \equiv f' \otimes h'$. The first congruence is obtained as follows. The equality (4) implies

$$\begin{aligned} fgp^{(j)}_{\underline{C}} &= [fp^{(1)}_{\underline{B}}, \dots, fp^{(m)}_{\underline{B}}](gp^{(j)}_{\underline{C}}) \\ &\equiv [f'p^{(1)}_{\underline{B}}, \dots, f'p^{(m)}_{\underline{B}}](g'p^{(j)}_{\underline{C}}) = f'g'p^{(j)}_{\underline{C}} \end{aligned}$$

($j = 1, \dots l$). Thus $fg \equiv f'g'$. For the second congruence we prove at first:

$$f \equiv f' : A \to B \in \operatorname{Mcl} \underline{M} \Rightarrow$$

$$\forall C \in K^0_J : \; t_C \otimes f \equiv t_C \otimes f', \; f \otimes t_C \equiv f' \otimes t_C. \tag{25}$$

By symmetry, it is sufficient to prove the second congruence in (25). Observing

$$1_A \otimes t_C = d^n_{A \otimes C}(p^{(1)}_{\underline{A}\,\underline{C}} \otimes \cdots \otimes p^{(n)}_{\underline{A}\,\underline{C}}), \tag{26}$$

one gets

$$\begin{aligned} f \otimes t_C &= (1_A \otimes t_C)f = d^n_{A \otimes C}(p^{(1)}_{\underline{A}\,\underline{C}} \otimes \cdots \otimes p^{(n)}_{\underline{A}\,\underline{C}})f \\ &= [p^{(1)}_{\underline{A}\,\underline{C}}, \dots, p^{(n)}_{\underline{A}\,\underline{C}}]f \equiv [p^{(1)}_{\underline{A}\,\underline{C}}, \dots, p^{(n)}_{\underline{A}\,\underline{C}}]f' = f' \otimes 1_C. \end{aligned}$$

The assertion (25) implies $p_{AC} f_j = fp^{(j)}_{\underline{B}} \otimes 1_C \equiv f'p^{(j)}_{\underline{B}} \otimes 1_C = p_{AC} f'_j$; similarly $p'_{AC} h_j \equiv p'_{AC} h'_j$. Thus by (7) we get $(f \otimes h)p^{(i)}_{\underline{B}\,\underline{D}} \equiv (f' \otimes h')p^{(i)}_{\underline{B}\,\underline{D}}$, $i = 1, \dots, m+k$, i.e. $f \otimes h \equiv f' \otimes h'$.

28.5. Corollary.

(a) *Let P be congruence on a Mal'cev theory $\underline{M}$. Then there is a canonical isomorphism* $\operatorname{Mcl}(\underline{M}/P) \cong (\operatorname{Mcl} \underline{M})/\underline{O}$, $\underline{O} = P \cap ((\operatorname{Mcl} \underline{M}) \times (\operatorname{Mcl} \underline{M}))$.

(b) *The Mal'cev clone* $\operatorname{Mcl} \operatorname{Par} K^0_J$ *is a congruence-simple* [75].

Proof. The item (a) follows immediately from the definition; $\underline{M}/P$ designates the (Mal'cev) factor theory of $\underline{M}$ with respect to the congruence P (analogously, $(\operatorname{Mcl} \underline{M})/Q$). The item (b) is implied by Theorem 25.2 and (22).

28.6. Remark. There is an essential difference between the word problem for $\underline{M} \in |\mathrm{Mth}_J|$ in the sense of Corollary 28.3, (c) and the word problem for a representation of $\underline{M}$ as a factor algebra $\underline{M} \cong \langle \underline{X} \rangle / \langle P \rangle$ of the free Mal'cev theory $\langle \underline{X} \rangle$ freely generated by $\underline{X} = (X_1, \dots, X_n)$ with respect to a set of relations P ($\langle P \rangle$ designates the congruence on $\langle \underline{X} \rangle$ generated by P). Indeed, the free algebra $F(\underline{X})$ occurring in the proof of 28.3, (c) is of the form $\langle \underline{X} \rangle / \rangle P \langle$, where $\rangle P \langle$ designates the fully invariant congruence on $\langle \underline{X} \rangle$ which is generated by P.

References

[Pt. 1] Hoehnke, H.-J., On certain classes of categories and monoids constructed from abstract Mal'cev clones. I. In: Universal and Applied Algebra: Proc. of the Vth Universal Algebra Symposium (Turawa, Poland, 1988), 149–176. Singapore 1989.

[Pt. 2] Hoehnke, H.-J., On certain classes of categories and monoids constructed from abstract Mal'cev clones. II. In: Proc. of the 18th Conf. on Universal Allgebra and Related Topics (Nahmitz, GDR, 1988), 30–64. Potsdamer Forschungen der Päd. Hochschule "Karl Liebknecht" Potsdam, Reihe B, Heft 62, Potsdam 1989.

[Pt. 3] Hoehnke, H.-J., On certain classes of categories and monoids constructed from abstract Mal'cev clones. III. Report (in preparation).

[17][1] Haddad, L.; Rosenberg, I. G.; Schweigert, D., A maximal partial clone and a Slupecki-type criterion. Preprint, Univ. de Montréal, 1986.

[23] Hoehnke, H.-J., On partial algebras. Universal Algebra Esztergom (Hungary) 1977, Colloq. Math. Soc. J. Bolyai Vol. 29, Amsterdam 1981, 373–412.

[26] Hoehnke, H.-J., Monoidal structure of Mal'cev clones, their theories and completions. Preprint P-Math-34/87, Karl-Weierstrass-Inst. f. Math., Akad. d. Wiss. d. DDR, Berlin 1987.

[30] Mac Lane, S., Kategorien. Begriffssprache und mathematische Theorie. Berlin 1972.

[49] Schreckenberger, J., Über die Einbettung von dht-symmetrischen Kategorien in die Kategorie der partiellen Abbidungen zwischen Mengen. Preprint P-12/80, Zentralinst. f. Math., Akad. d. Wiss. d. DDR, Berlin 1980.

[50] Schreckenberger, J., Zur Theorie der dht-symmetrischen Kategorien. Diss. (B), Päd. Hochschule Potsdam, Math.-Naturwiss. Fak., Potsdam (DDR) 1984.

[64] Mal'cev, A. I., Iterative Algebras and Post Varieties, Algebra i Logika 5 (1966), 5–24 (Russian).

[74] Börner, F., Abstrakte Partielle Klone, § 4.2 (pp. 80–86) in his Diss. (A) "Operationen auf Relationen", Karl-Marx-Universität Leipzig 1988.

[75] Denecke, K., On Primal Partial Algebras. Preprint, Potsdam 1989.

[76] Denecke, K.; Lau, D., Congruences on Maximal Partial Clones and Strong Regular Varieties Generated by Preprimal Partial Algebras, I. Preprint, Potsdam; Rostock 1989.

[77] Foster, A. L., Generalizad "Boolean" Theories of Universal Algebras, I; II. Math. Z. 58 (1953), 306–336; 59 (1953), 191–199.

[78] Mal'cev, A. I., Iterative Post Algebras. Novosibirsk 1976 (Russian)

[79] Schweigert, D., On Varieties of Clones. Semigroup Forum 26 (1983), 275–285.

[80] Schweigert, D., On Algebras and Clones. Semigroup Forum 35 (1987), 85–99.

[1] The following seven references are selected from [Pt. 1], [Pt. 2].

Decomposition matrices of finite quasisimple groups

Anatoliĭ S. Kondrat'ev *

Abstract. The aim of this paper is to describe some results and methods of the calculation of decomposition matrices of finite quasisimple groups (i.e., covering groups of simple non-Abelian groups). Mainly, we concentrate on covering groups of sporadic simple groups and simple groups of Lie type with exceptional Schur multiplier.

1991 Mathematics Subject Classification: 20C33, 20C34.

An important problem in representation theory is to calculate the decomposition matrix of a finite group with given character table. The *decomposition matrix* of a finite group for a prime p consists of the multiplicities of irreducible constituents in reduction modulo p of the irreducible complex representations of this group. A general method of solving the mentioned problem is unknown even in the relatively simple case of p-blocks with cyclic defect group. Neither is known whether the decomposition matrix (up to permutation of rows and columns) is uniquely determined by the character table of the group (a problem of W. Feit [1]). A typical situation in concrete calculations is that the decomposition matrix is found up to several parameters. The decomposition matrices of quasisimple groups (i.e., covering groups of simple non-abelian groups) are of special interest in connection with the problem of determination of the subgroup structure of finite simple groups (see, for example, [2]).

The aim of this paper is to describe some results and methods of the calculation of the decomposition matrices of finite quasisimple groups. We use the notations of [3, 4]. The universal covering group for a finite simple non-abelian group G is denoted by $\widehat{G}$. We write p for a prime number.

At first, we discuss briefly the situation in calculating the decomposition matrices of quasisimple groups.

*This work was supported in part by the Russian Fund for Fundamental Research (grant 93-011-1529) and the Fund of Russian State Committee on High Education (grant 2-11-29).

Offprint from: Algebra, Eds.: Ershov/Khukhro/Levchuk/Podufalov

For groups $\widehat{A_n}$, the representation theory [5] and the projective representation theory [6] of the symmetric groups are used. The final results for all p are obtained only for $n \leqslant 13$. Let $G = G(q)$ be a simple group of Lie type over the field $GF(q)$ of characteristic r. In the investigations of p-modular representation of $\widehat{G}$, the cases $p = r$ and $p \neq r$ are crucially different. The first case is studied in the frame of the representation theory of linear algebraic groups (see [7]). An information about the decomposition matrices of G in this case can be found in the survey [8]. The results on the decomposition numbers of finite groups of Lie type in the second case are discussed in the survey [9]. In the general case, the results on the decomposition matrices for all p and for all p-blocks are known only for the following infinite series of finite groups of Lie type: $SL_2(q)$, $SL_3(2^m)$, $SU_3(2^m)$, $Sp_4(q)$, $GL_n(q)$ ($n \leqslant 10$), $Sz(q)$, ${}^2G_2(q)$, and $G_2(q)$. Furthermore, Parker's collection [10] contains some Brauer characters tables of individual groups of Lie type.

We consider more explicitly the remaining case of the covering groups of sporadic simple groups and simple groups G of Lie type with exceptional Schur multiplier (i.e., $\widehat{G}$ is not a group of Lie type over the field of the same characteristic r). These groups appear as exceptions in various investigations, therefore their decomposition matrices are necessary in the first place.

The all p-modular decomposition matrices are known for the following groups of sporadic type:

Group	p	References
M_{11}	any	James [11]
$\widehat{M_{12}}$	any	James [11], Humphreys [12]
$\widehat{M_{22}}$	any	James [11], Humphreys [13]
M_{23}	any	James [11]
M_{24}	any	James [11], Parker [10]
J_1	any	Fong [14], Feit [15]
$\widehat{J_2}$	any	Feit [15], Hiss–Lux [16], Kondrat'ev [17, 18]
$\widehat{J_3}$	any	Feit [15], Wilson [19, 20]
$\widehat{\text{HiS}}$	any	Thackray [21], Humphreys [22], Feit [15], Parker [10], Suleiman–Wilson [23]
McL	2	Thackray [21]
	3	Suleiman–Wilson [24]
$\widehat{\text{McL}}$	$p > 3$	Feit [15], Woldar [25], Hiss–Lux–Parker [26]
He	7	Ryba [27]
Co_2	5	Lux (unpublished)

In addition, G. Hiss and K. Lux [28] determined all Brauer trees of the blocks with cyclic defect group for all covering groups of sporadic groups, excluding 10 concrete situations for the groups F_3, $\widehat{F}_2$, and F_1. The decomposition matrices for the p-blocks with non-cyclic defect groups are almost determined in a unpublished work of Lux for He ($p = 3, 5$), Suz ($p = 5$), Fi_{22} ($p = 5$), $\widehat{Co}_1$ ($p = 7$).

For a simple group G of Lie type with exceptional Schur multiplier $M(G)$, complete information about the decomposition matrices of G can be found in [29–31], excluding the following groups:

G	$\Omega_7(3)$	$U_6(2)$	$G_2(3)$	$G_2(4)$	$F_4(2)$	${}^2E_6(2)$
$M(G)$	6	6×2	3	2	2	6×2

Moreover the p-modular decomposition matrices are known (or almost known) for the following cases:

Group	p	References
$\Omega_7(3)$	$p > 3$	Kondrat'ev [32]
$G_2(3)$	2	Kondrat'ev [32]*, Hiss–Shamash [33]*, Ryba [27]
	3	Jantzen [34], Parker [10]
$\widehat{G_2(3)}$	2	Kondrat'ev [32]*
	$p > 3$	Kondrat'ev [32]*, Parker [10]
$G_2(4)$	2	Jantzen [34]
	3	Kondrat'ev [32]*, Hiss–Shamash [35]*
	5	Kondrat'ev [32]*, Hiss (see [36])
	$p > 5$	Kondrat'ev [32]
$\widehat{G_2(4)}$	3	Kondrat'ev [32]*
	$p > 3$	Kondrat'ev [32]*
$\widehat{U_6(2)}$	$p > 3$	Kondrat'ev [32]
$U_6(2)$	3	Kondrat'ev [31]
$2{}^{\cdot}U_6(2)$	3	Kondrat'ev [31]*
$\widehat{F_4(2)}$	$p > 7$	White [37]

Here the symbol * means that the appropriate decomposition matrix is determined up to several parameters.

In the following two theorems, we state our recent result [31] on the 3-modular decomposition matrix of $\widehat{U_6(2)}$.

Table 1

Decomposition matrix for the principal 3-block B_0 of $U_6(2)$

	1	21	210	229	364	560	925	925	925	1365	1540	3465	3465	3465	3850
	1	2	3	4	5	6	7	8	9	10	11	12	13	14	15
1	1														
22	1	1													
231		1		1											
252	2	1		1											
385		1			1										
440	1		1	1											
560						1									
616	2	1		1	1										
$770_{\frac{1}{2}}$			1			1									
1155_1	1			1			1								
1155_2	1			1				1							
1155_3	1			1					1						
1386		1								1					
1540											1				
$3080_{\frac{1}{2}}$		1	1		1	2				1					
3520	1	1	2	1	1	2				1					

Continued on next page

Table 1. — continued

	1	2	3	4	5	6	7	8	9	10	11	12	13	14	15
4620		1	1		1	2				1	1				
4928	1		1	2	1	2	1	1	1						
5544			1		1	2									1
$6160_{\frac{1}{2}}$			1			1					1				1
6930	3	1	2	3	1	2	1	1	1		1				
8064	1	1	2	2	1	2	1	1	1	1	1				
9240_1	1	1	1	1	1	2	1			1	1	1			
9240_2	1	1	1	1	1	2		1		1	1		1		
9240_3	1	1	1	1	1	2			1	1	1			1	
$10395_{\frac{1}{2}}$			2			3				1	2				1
10395_3											2	1			1
10395_4											2		1		1
10395_5											2			1	1
11264			1	1		2	1	1	1		2				1
13860	3	2	3	3	2	4	1	1	1	1	1				1
14784					1	2				1	1	1	1	1	
$20790_{\frac{1}{2}}$			2			3				1	2	1	1	1	1
24640		1	2		1	4				2	3	1	1	1	1
32768			2	1	1	6	1	1	1	1	4	1	1	1	2

Theorem 1. *The group $U_6(2)$ has three 3-blocks of defect* 0: $\{25515_1\}$, $\{25515_2\}$, $\{40095\}$ *and one* 3*-block of defect* 1 *with Brauer tree:*

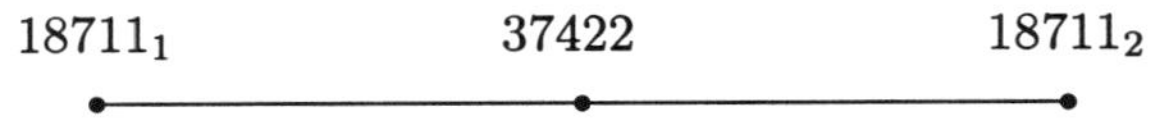

The remaining irreducible characters of $U_6(2)$ are all in the principal 3*-block B_0. The decomposition matrix of B_0 is given in Tabl.* 1.

Theorem 2. *The group $2{\cdot}U_6(2)$ has one* 3*-block of faithful characters of defect* 1 *with Brauer tree:*

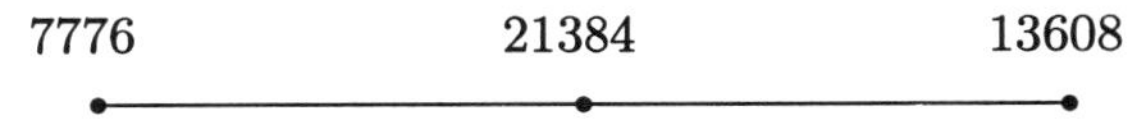

The remaining faithful characters of $2{\cdot}U_6(2)$ are all in the 3*-block B_1. The decomposition matrix of B_1 is given in Tabl.* 2.

In Tabl. 1 and 2, missing entries are 0. The ith irreducible character (ordinary or Brauer) of degree n of a group is denoted by n_i or by n if such character is unique.

In [31, 32], we used systematically some approach for finding the decomposition matrices. The following is a formalized account of this approach.

Let G be a finite group with given character table, p be a prime, and G_0 be the set of all p'-elements of G. For simplicity, we shall assume that the irreducible characters of G are distributed into p-blocks. Let B be a p-block of G which consists of the irreducible ordinary characters $\chi_1, \dots, \chi_n$, the irreducible Brauer characters $\varphi_1, \dots, \varphi_m$, and the principal indecomposable characters $\Phi_1, \dots, \Phi_m$. Then we may assume that the following relations hold on G_0:

$$\chi_i = \sum_{j=1}^{m} d_{ij}\varphi_j, \qquad 1 \leqslant i \leqslant n, \tag{1}$$

$$\Phi_i = \sum_{k=1}^{n} d_{ki}\chi_k, \qquad 1 \leqslant i \leqslant m, \tag{2}$$

$$\Phi_i = \sum_{k=1}^{m} c_{ik}\varphi_k, \qquad 1 \leqslant i \leqslant m, \tag{3}$$

where $0 \leqslant d_{ij}, c_{ij} \in Z$. Write $D = (d_{ij})_{n,m}$ and $C = (c_{ij})_{n,n}$. Then the matrices D and C ($C = D^tD$) are called the (p-*modular*) *decomposition matrix* and the *Cartan*

Table 2

Decomposition matrix for the 3-block B_1 of $2^{\cdot}U_6(2)$; here $\alpha \in \{0,1\}$

	56	120	560	1056	$1848\text{-}\alpha\cdot 56$	2184	2184	7272	8568
56	1								
176	1	1							
616	1		1						
792	2	1	1						
1232	1	1		1					
2464	$1+\alpha$		1		1				
3080	$1+\alpha$	1		1	1				
3360_1		1		1		1			
3360_2		1		1			1		
5544	$2+2\alpha$	1	1	1	2				
6160_1	1	1	1	1		1	1		
6160_2	1	1	1	1		1	1		
6160_3	1	1	1	1		1	1		
9856	$3+\alpha$	2	2	2	1	1	1		
11264	$3+\alpha$	1	1		1				1
12320_1	$3+\alpha$	1	1	1	1				1
12320_2		1	1			1	1	1	
12320_3		1	1			1	1	1	
15840								1	1
18480	$4+\alpha$	2	2	2	1	1	1		1
22176	$4+2\alpha$	2	2	1	2			1	1
24640_1	$3+\alpha$	2	2	1	1	1	1	1	1
24640_2	$3+\alpha$	2	2	1	1	1	1	1	1
24640_3	$3+\alpha$	2	2	1	1	1	1	1	1
27720	$3+\alpha$	4	3	3	1	3	3	1	
30240	$8+2\alpha$	2	4	2	2	1	1		2
32768	$4+2\alpha$	4	3	2	2	2	2	1	1
44352	$5+\alpha$	5	4	3	1	3	3	2	1

matrix of block B, respectively. Introduce the vector-functions $\chi = (\chi_1, \dots, \chi_n)^t$, $\varphi = (\varphi_1, \dots, \varphi_m)^t$, and $\Phi = (\Phi_1, \dots, \Phi_m)^t$ on G_0. Define the scalar product (,) on the vector space of the complex-valued class function on G_0 by

$$(f, g) = \frac{1}{|G|} \sum_{x \in G_0} f(x)\overline{g(x)}.$$

Since the rank of D is equal to m and χ is known, the calculations of D, φ, Φ are equivalent.

To calculate D we apply the following approach. In the set of Brauer characters of the blocks B we find a subset φ' which is a Z-basis for the ring of generalized Brauer characters of B, i.e. φ' is linearly independent over Z and $\varphi_1, \dots, \varphi_m$ are in the Z-span of φ'. In particular, φ' is a basis for the space of complex-valued class functions on G_0, therefore $|\varphi'| = m$. Set $\varphi' = (\varphi'_1, \dots, \varphi'_m)^t$. By [38, Theorem 61.6], we can take as φ' a maximal linearly independent over Z subset of the set $\{\chi_i|_{G_0} : 1 \leqslant i \leqslant n\}$. The Z-basis φ' is an initial approximation for the desired Z-basis φ.

Such a Z-basis was considered independently by G. Hiss [36]. He called it a basis set (of Brauer characters) of B and applied for the calculation of the decomposition matrices.

By the definition of φ', we have

$$\chi_i = \sum_{j=1}^{m} d'_{ij}\varphi'_j \quad \text{on } G_0, \qquad 1 \leqslant i \leqslant n, \tag{1'}$$

where $d'_{ij} \in Z$. The relations (1′) are analogous to the relations (1). By analogy with (2), we define the following generalized characters Φ'_i of G:

$$\Phi'_i = \sum_{k=1}^{n} d'_{ki}\chi_k, \qquad 1 \leqslant i \leqslant m. \tag{2'}$$

Since φ' consists of the Brauer characters of B, we have

$$\varphi'_i = \sum_{j=1}^{m} a_{ij}\varphi_j, \qquad 1 \leqslant i \leqslant m, \tag{4}$$

where $0 \leqslant a_{ij} \in Z$.

Let $D' = (d'_{ij})_{n,m}$, $\Phi' = (\Phi'_1, \dots, \Phi'_m)^t$, $A = (a_{ij})_{m,m}$, and $C' = (c'_{ij}) = (D')^t D'$. Then we obtain

Theorem 3. *The following statements hold:*

(a) $D = D'A$;

(b) $\Phi = A^t\Phi'$;

(c) $\Phi' = C'\varphi'$;

(d) *the vector-function* Φ' *vanishes on* $G - G_0$, *and* $|G|_p$ *divides any coordinate of the vector* $\Phi'(1)$;

(e) *the matrix* A^{-1} *is integral*;

(f) $(\Phi'_i, \varphi'_j) = \delta_{ij}$, $1 \leqslant i, j \leqslant m$;

(g) $(\Phi'_i, \Phi'_j) = c'_{ij}$, $1 \leqslant i, j \leqslant m$;

(h) $((\varphi'_i, \varphi'_j)) = (C')^{-1}$.

Proof. (a) By the relations (1), (1′), (2′), and (4), we have $\chi = D\varphi = D'\varphi'$ and $\varphi' = A\varphi$, hence $D = D'A$.

(b) By (2) and (a), we have $\Phi = D^t\chi = (D'A)^t\chi = A^t((D')^t\chi)$, hence $\Phi = A^t\Phi'$ in view of (2′).

(c) By (2′) and (1′), we have $\Phi' = (D')^t\chi$ and $\chi = D'\varphi'$, hence $\Phi' = C'\varphi'$.

(d) By (b), we have $\Phi' = (A^t)^{-1}\Phi$. Since Φ vanishes on $G - G_0$, the same holds for Φ'. Since $\Phi'_1, \dots, \Phi'_m$ are generalized characters, Corollary 61.7 of [38] implies that they are integral linear combinations of $\Phi_1, \dots, \Phi_m$. But $|G|_p$ divides Φ_i [see (1)], $1 \leqslant i \leqslant m$, hence (d) holds.

(e) Since $\Phi'_1, \dots, \Phi'_m$ are integral linear combinations of $\Phi_1, \dots, \Phi_m$ [see the proof of (d)], the assertion (e) follows from (2).

(f) Let $(A^t)^{-1} = (\gamma_{ij})$. By (b), (e), and [38, Theorem 60.3], we have

$$(\Phi'_i, \varphi'_j) = \left(\sum_{k=1}^{m} \gamma_{ik}\Phi_k, \sum_{l=1}^{m} a_{jl}\varphi_l\right) = \sum_{k=1}^{m}\sum_{l=1}^{m} \gamma_{ik}(\Phi_k, \varphi_l)$$

$$= \sum_{l=1}^{m}\sum_{k=1}^{m} \gamma_{ik}a_{jl}\delta_{kl} = \sum_{k=1}^{m} \gamma_{ik}a_{jk} = \delta_{ij}.$$

(g) By (c), (f), and [38, Theorem 60.3], we have

$$(\Phi'_i, \Phi'_j) = \left(\sum_{k=1}^{m} c'_{ik}\varphi'_k, \Phi'_j\right) = \sum_{k=1}^{m} c'_{ik}(\varphi'_k, \Phi'_j) = \sum_{k=1}^{m} c'_{ik}\delta_{kj} = c'_{ij}.$$

(h) By (4) and [38, Theorem 60.5], we have

$$(\varphi_i', \varphi_j') = \left(\sum_{k=1}^{m} a_{jk}\varphi_k, \sum_{l=1}^{m} a_{il}\varphi_l\right) = \sum_{k=1}^{m}\sum_{l=1}^{m} a_{ik}a_{jl}(\varphi_k, \varphi_l)$$

$$= \sum_{k=1}^{m}\sum_{l=1}^{m} a_{ik}a_{jl}\widetilde{c}_{kl} = \sum_{l=1}^{m}\left(\sum_{k=1}^{m} a_{ik}\widetilde{c}_{kl}\right)a_{jl},$$

where $C^{-1} = (\widetilde{c}_{kl})$. Hence $((\varphi_i', \varphi_j')) = AC^{-1}A^t$. But, by (a),

$$C = D^tD = (D'A)^tD'A = A^t((D')^tD')A = A^tC'A,$$

hence $AC^{-1}A^t = (C')^{-1}$. Theorem 3 is proved.

Theorem 3 implies that φ', Φ', D', C' have the main properties of φ, Φ, D, C, respectively. In addition, for the calculation of D, it is sufficient to find the matrix A, i.e. the decomposition into irreducible constituents of all Brauer characters from φ'. For this, we construct the set $\mathcal{P}$ of projective characters (i.e., sums of principal indecomposable characters) of G and consider their scalar products with the Brauer characters of φ'. By Theorem 3, this gives us upper bounds for elements of the matrix A. The set $\mathcal{P}$ is obtained from some known projective characters (for example, from characters of defect 0) of the group G, or their products with some Brauer characters of G, or by inducing projective characters from some subgroups of G. It is well known that the operation of product and inducing preserve the projectivity of characters.

If the decomposition matrices of some large subgroups of G are known, then the decomposition into irreducible constituents of the restrictions of the Brauer characters from φ' to these subgroups allows to refine the Z-basis φ', for example to make all elements of the matrix D' nonnegative, and the matrix A unitriangular. The decompositions of some products $\varphi_i'\varphi_j'$ into elements of φ' is used to refine φ' too. If G is a quasisimple group and p divides the order of its Schur multiplier, then sometimes it is useful to choose some elements of φ' from the reductions modulo p of characters of G.

References

[1] Feit, W., The Representation Theory of Finite Groups (North-Holland Math. Libr. 25). North–Holland, Amsterdam–New York–Oxford 1982.

[2] Aschbacher, M., On the maximal subgroups of the finite classical groups. Invent. Math. 76 (1984), 469–514.

[3] Conway, J. H., Curtis, R. T., Norton, S. P., Parker, R. A., Wilson, R. A., Atlas of Finite Groups. Maximal Subgroups and Ordinary Characters for Simple Groups. Clarendon Press, Oxford 1985.

[4] Gorenstein, D., Finite Simple Groups. An Introduction to Their Classification (Univ. Ser. Math. 10). Plenum Press, New York–London 1982.

[5] James, G. D., Kerber, A., The Representation Theory of the Symmetric Group (Encycl. Math. Appl. 16). Addison-Wesley, Reading, Mass. 1981.

[6] Hoffman, P. N., Humphreys, J. F., Projective Representations of the Symmetric Groups. Clarendon Press, Oxford 1992.

[7] Steinberg, R., Lectures on Chevalley Groups. Yale University, New Haven 1967.

[8] Humphreys, J. E., Cartan invariants. Bull. Lond. Math. Soc. 17:1 (1985), 1–14.

[9] Hiss, G., Decomposition numbers of finite groups of Lie type in nondefining characteristic. In: Representation Theory of Finite Groups and Finite-dimensional Algebras. Bielefeld, 1991 (Progr. Math. 95), 405–418. Birkhäuser, Basel 1991.

[10] Parker, R. A., A collection of modular characters. Prepr. Cambridge University Press, Cambridge 1989.

[11] James, G. D., The modular characters of the Mathieu groups. J. Algebra 27:1 (1973), 57–111.

[12] Humphreys, J. F., The projective characters of the Mathieu group M_{12} and of its automorphism group. Math. Proc. Camb. Philos. Soc. 87:3 (1980), 401–412.

[13] Humphreys, J. F., The projective characters of the Mathieu group M_{22}. J. Algebra 76:1 (1982), 1–24.

[14] Fong, F., On decomposition numbers of J_1 and $R(q)$. In: Gruppi Abeliani (Symp. Math. 13), 415–422. Academic Press, London–New York 1974.

[15] Feit, W., Blocks with cyclic defect groups for some sporadic groups. In: Representation Theory II. Groups and Orders. Proc. of the Forth Int. Conf. on Representations of Algebras (Ottawa, August 16–25, 1984) (Lect. Notes Math. 1178), 25–63. Springer-Verlag, Berlin–Heidelberg–New York–Tokyo 1986.

[16] Hiss, G., Lux, K., The Brauer characters of the Hall–Janko group. Commun. Algebra 16:2 (1988), 357–398.

[17] Kondrat'ev, A. S., Decomposition numbers of the group J_2. Algebra Logika 27:5 (1988), 535–561. English translation: Algebra Logic 27:5 (1988), 333–349.

[18] Kondrat'ev, A. S., Decomposition numbers of the groups $\hat{J}_2$ and $\mathrm{Aut}(J_2)$. Algebra Logika 27:6 (1988), 690–710. English translation: Algebra Logic 27:6 (1988), 429–444.

[19] Wilson, R. A., The 2- and 3-modular characters of J_3, its covering group and automorphism group. J. Symb. Comput. 10:6 (1990), 647–656.

[20] Wilson, R. A., The Brauer tree for J_3 in characteristic 17. Prepr. 1990.

[21] Thackray, J. G., Modular representations of finite groups. Ph. D. Thesis. University of Cambridge, Cambridge 1981.

[22] Humphreys, J. F., The modular characters of the Higman–Sims simple group. Proc. R. Soc. Edinb. Sect. A 92:3–4 (1982), 319–335.

[23] Suleiman, I. A., Wilson, R. A., The 3- and 5-modular characters of the covering and the automorphism groups of the Higman–Sims group. J. Algebra 148:1 (1992), 225–242.

[24] Suleiman, I. A., Wilson, R. A., The 3-modular characters of the McLaughlin group McL and its automorphism groups McL. 2. In Groups, Combinatorics

and Geometry (Durham, 1990) (Lond. Math. Soc. Lect. Note Ser. 165), 422–437. Cambridge University Press, Cambridge 1992.

[25] Woldar, A. J., On the 5-decomposition matrix for McLaughlin's sporadic simple group. Commun. Algebra 14:2 (1986), 277–291.

[26] Hiss, G., Lux, K., Parker, R., The 5-modular characters of the McLaughlin group and its covering group. Manuscr. Math. 73:1 (1991), 91–114.

[27] Ryba, A. J. E., Calculation of the 7-modular characters of the Held group. J. Algebra 117:1 (1988), 240–255.

[28] Hiss, G., Lux, K., Brauer Trees of Sporadic Groups (Oxf. Sci. Publ.). Clarendon Press, Oxford 1989.

[29] Khosraviyani, F., Decomposition numbers of exceptional Weyl groups. I. J. Algebra 91:2 (1984), 390–409.

[30] Khosraviyani, F., Morris, A. O., Decomposition numbers of exceptional Weyl groups. II. J. Algebra 92:2 (1985), 525–531.

[31] Kondrat'ev, A. S., The decomposition matrices of the group $\widehat{U_6(2)}$ (to appear).

[32] Kondrat'ev, A. S., Modular representations of small degrees of groups of Lie type. Prepr. Ural'skiĭ Nauchnyĭ Tsentr 1989.

[33] Hiss, G., Shamash, J., 2-blocks and 2-modular characters of the Chevalley groups $G_2(q)$. Math. Comput. 59:200 (1992), 645–672.

[34] Jantzen, J. C., Darstellungen halbeinfacher Gruppen und kontravariante Formen. J. Reine Angew. Math. 290 (1977), 117–141.

[35] Hiss, G., Shamash, J., 3-blocks and 3-modular characters of $G_2(q)$. J. Algebra 131:2 (1990), 371–387.

[36] Hiss, G., On the decomposition numbers of $G_2(q)$. J. Algebra 120:2 (1989), 339–360.

[37] White, D. L., Brauer trees of $2.F_4(2)$. Commun. Algebra 20:11 (1992), 3353–3368.

[38] Dornhoff, L., Group representation theory. Pt. B. Modular representation theory. In: Pure Appl. Math., Marcel Dekker 7, 255–511. Marcel Dekker, New York 1972.

Characterization of dilatations expressible as a product of independent simple transformations

Alekseĭ A. Korobov *

Abstract. This paper is concerned with the presentation of dilatations of $SL_n(K)$ as a product of minimal numbers of transvections. For commutative K, there are no proper dilatations in $SL_n(K)$. If K is not commutative, the dilatation s with ratio Λ belongs to $SL_n(K)$ whenever Λ is a conjugacy class composed by elements of the commutator subgroup $K^{*\prime}$ of K^* in K^* (see [1, Theorem 4.2]). Thus, if an element of $K^{*\prime}$ is a product of at most $n-1$ commutators, Corollary (see below) and the author's result [2, Theorem 1] imply that any dilatation of $SL_n(K)$ is a product of at most $n+1$ transvections. Moreover, there is a dilatation of $SL_n(K)$ which is not a product of n transvections.

1991 Mathematics Subject Classification: 20H25.

Let V be a right vector space of dimension at least 2 over a division ring K, and let V^* be its dual space. The tensor product $V \otimes V^*$ is an algebra over the center F of K whose multiplication is characterized by

$$(u \otimes f)(v \otimes g) = uf(v) \otimes g, \quad u, v \in V, \quad f, g \in V^*. \tag{1}$$

The canonical map $\varphi : V \otimes V^* \to \mathrm{End}_K(V)$, $\varphi(u \otimes f)(v) = uf(v)$, $v \in V$, is an injective morphism of F-algebras. Using this map we identify $V \otimes V^*$ with its image, i.e. with F-algebra of K-linear operators of V of finite rank.

With each $a \in GL(V)$ we associate two subspaces of V:

$$\mathrm{Fix}\,(a) = \mathrm{Ker}\,(a-1), \quad \mathrm{Res}\,(a) = \mathrm{Im}\,(a-1).$$

An element $s \in GL(V)$ is called *simple* if $s = 1 + u \otimes f$ and $s \neq 1$. The conjugacy class $\mathrm{Con}\,(1+f(u))$ of $1+f(u)$ in the multiplicative group K^* of K is called the *parameter* of s and is denoted by $\mathrm{Par}\,(s)$. A simple element is called *dilatation* if $\mathrm{Par}\,(s) \neq 1$ and *transvection* if $\mathrm{Par}\,(s) = 1$. We say that simple transformations $a_i = 1 + v_i \otimes f_i$, $i = 1, \dots, n$, are *independent* if vectors $v_{i_1}, \dots, v_{i_{n-1}}$ are linear independent for any $1 \leqslant i_1 < \dots < i_{n-1} \leqslant n$.

---

* Supported by the Russian Fund for Fundamental Research (grant 93-011-1513).

Offprint from: Algebra, Eds.: Ershov/Khukhro/Levchuk/Podufalov

Theorem. *Let n be a positive integer, and Λ_i, $i = 1, \dots, n$, given conjugacy classes in K^*. Assume that* $\dim V \geqslant n-1$. *A dilatation s is a product $a_1 \dots a_n$ of independent simple transformations a_i such that* $\operatorname{Par}(a_i) = \Lambda_i$, $i = 1, \dots, n$, *if and only if* $\operatorname{Par}(s) = \operatorname{Con}([x_1, y_1] \dots [x_{n-2}, y_{n-2}]\lambda_1 \dots \lambda_n)$ *for some $x_i, y_i \in K^*$, $\lambda_i \in \Lambda_i$, where $[x,y] = x^{-1}y^{-1}xy$.*

Corollary. *Assume that* $\dim V \geqslant n-1$. *A dilatation s is a product of n independent transvections if and only if* $\operatorname{Par}(s) = \operatorname{Con}([x_1, y_1] \dots [x_{n-2}, y_{n-2}])$.

We start with the following lemma whose proof is straightforward.

Lemma 1. *If $a = 1 + u \otimes f$ is a simple transformation then*

$$a^{-1} = 1 - u(1 + f(u))^{-1} \otimes f_i.$$

We also need the following lemma.

Lemma 2. *Let $a_i = 1 + v_i \otimes f_i$, $i = 1, \dots, n$, be simple. Assume that $a = a_1 \dots a_n$ is a simple transformation.*

(i) *If $f_1, \dots, f_n$ are linearly independent then $v_iK = v_jK$ for any i, j.*

(ii) *If $v_iK \neq v_jK$ for $i \neq j$ and vectors $f_{i_1}, \dots, f_{i_{n-1}}$ are linearly independent for any $1 \leqslant i_1 < \dots < i_{n-1} \leqslant n$ then* $\dim(v_iK + v_jK + v_kK) \leqslant 2$ *for any i, j, k.*

Proof. It follows from (1) that

$$a_1 \dots a_n = a_1 \dots a_{n-1} + a_1 \dots a_{n-1}(v_n) \otimes f_n. \tag{2}$$

Assume that there exists the smallest number k such that $v_kK \neq v_1K$. Since $f_1, \dots, f_n$ are linearly independent, there are $u_1, u_2 \in V$ such that $f_1(u_1) = f_k(u_2) = 1$, $f_i(u_j) = 0$ for another i, j. Consider vectors $a(u_j)$. By the definition of simple transformation and the relation (2), the vectors $v_1, a_1 \dots a_{k-1}(v_k)$ are linearly independent and belong to $\operatorname{Res}(a)$. But it is impossible because a is simple.

To prove (ii) it suffices to show that $v_i \in v_1K + v_2K$ for all i. Assume that there exists the smallest number k such that v_1, v_2, v_k are linearly independent. Since $f_2, \dots, f_n$ are linearly independent, there are $u_1, u_2 \in V$ such that $f_2(u_1) = 1$, $f_i(u_1) = 0$, $i \neq 1, 2$, and $f_k(u_2) = 1$, $f_i(u_2) = 0$, $i \neq 1, k$. Consider vectors $a(u_i)$. By the definition of simple transformation and the relation (2) the vectors $v_1 f_1(u_1) + a_1(v_2)$, $v_1 f_1(u_2) + a_1 \dots a_{k-1}(v_k)$ are linearly independent and belong to $\operatorname{Res}(a)$. Since a is simple it is impossible. Lemma 2 is proved.

Lemma 3. *Let $a_i = 1 + v_i \otimes f_i$, $i = 1, \dots, n$, be simple. Assume that $a = a_1 \dots a_n$ is a simple transformation.*

(i) *If* $v_1, \dots, v_n$ *are linearly independent then* $Kf_i = Kf_j$ *for any* i, j.

(ii) *If* $Kf_i \neq Kf_j$ *for* $i \neq j$ *and vectors* $v_{i_1}, \dots, v_{i_{n-1}}$ *are linearly independent for any* $1 \leqslant i_1 < \dots < i_{n-1} \leqslant n$ *then* $\dim(Kf_i + Kf_j + Kf_k) \leqslant 2$ *for any* i, j, k.

Proof. Fix a basis of V. Applying Lemma 2 to the transposed matrix of a, we obtain the desired results.

Proof of the theorem. We first prove that the assumption of the theorem is necessary. We now proceed by induction on n. For $n = 1$ the assertion is trivial. Let $n \geqslant 2$, and let $s = a_1 \dots a_n$ be a dilatation. Assume that there are l, k such that $l < k$ and Fix (a_l)=Fix (a_k). Since $a_1, \dots, a_n$ are independent, it follows from Lemma 1 that $a'_k = a_l a_k$ is a simple transformation with parameter Con $(\lambda_l \lambda_k)$, $\lambda_i \in \Lambda_i$. Consequently,

$$s = a_1 \dots a_{l-1} a'_{l+1} \dots a'_k a_{k+1} \dots a_n, \tag{3}$$

where $a'_j = a_l a_j a_l^{-1}$ is a simple transformation with parameter Λ_j. It is easy to check that $a_1 \dots a_{l-1}, a'_{l+1}, a'_k, a_{k+1}, \dots, a_n$, are independent. To complete the proof of the induction step, it suffices to apply the assumption of induction to (3) and use the relations between parameters of a_i and a'_i.

We may assume that Fix $(a_i) \neq$ Fix (a_j) for $i \neq j$. For $n = 2$, the easy comutation leads to the desired result. Assume that $n \geqslant 3$ Since $a_1, \dots, a_n$ are independent, without loss of generality we can assume, by Lemma 3, that

$$a_i = 1 + v_i \otimes f_i, \quad i = 1, \dots, n, \tag{4}$$

$$v_1 + \dots + v_n = 0 \tag{5}$$

and, for some $\alpha_i, \beta_i \in K^*$, $i = 3, \dots, n$,

$$f_i = \alpha_i f_1 + \beta_i f_2, \quad i = 1, \dots, n, \tag{6}$$

where $\alpha_1 = \beta_2 = 1$, $\alpha_2 = \beta_1 = 0$.

Let $\xi_i \in K, i = 1, \dots, n$, be given elements. Consider

$$v_k^\xi = v_k \xi_k + a_k(v_{k+1}^\xi)\xi_{k+1}, \quad v_{n+1}^\xi = 0. \tag{7}$$

Put $v^\xi = v_1^\xi$. In view of (2) the following assertion holds.

Lemma 4. *If* $a_i = 1 + v_i \otimes f_i$, $i = 1, \dots, n$, *and*

$$f_i = \alpha_i f_1 + \beta_i f_2, \quad \alpha_i, \beta_i \in K, \quad i = 1, \dots, n, \tag{8}$$

then $a_1 \dots a_n = 1 + v^\alpha \otimes f_1 + v^\beta \otimes f_2$.

Since f_1, f_2 are linearly independent, there are $u_1, u_2 \in V$ such that $f_1(u_1) = f_2(u_2) = 1$, $f_1(u_2) = f_2(u_1) = 0$. Consider vectors $s(u_i)$. By Lemma 4 we obtain $v^\alpha, v^\beta \in \mathrm{Res}\,(s)$. Note that $v^\beta \neq 0$. Indeed, the condition $v^\beta = 0$ implies $\dim(\mathrm{Res}\,(a_1^{-1}s)) \leqslant 1$. But from Lemma 3 it follows that $\dim(\mathrm{Res}\,(a_2 \dots a_n)) \geqslant 2$.

Since s is simple and $v^\alpha, v^\beta \in \mathrm{Res}\,(s)$, we have

$$v^\alpha = v^\beta \sigma \tag{9}$$

for some $\sigma \in K$. Multiplying both sides of (9) by $(a_1 \dots a_{n-1})^{-1}$, from (2) and Lemma 1 we find that

$$v_n \gamma_n + \sum_{i=1}^{n-1} a_n^{-1} \dots a_{i+1}^{-1}(v_i)\lambda_i^{-1}\gamma_i = 0 \tag{10}$$

where $\gamma_i = \alpha_i - \beta_i \sigma$, $\lambda_i = 1 + f_i(v_i)$, $i = 1, \dots, n$. Put

$$u_0 = 0, \quad u_k = v_k \lambda_k^{-1} \gamma_k + a_k^{-1}(u_{k-1}), \quad k = 1, \dots n-1, \tag{11}$$

and

$$r_k = \lambda_k^{-1}\gamma_k - \lambda_k^{-1} f_k(u_{k-1}).$$

By the relations (4), (11) and Lemma 1 we obtain

$$u_k = u_{k-1} + v_k r_k, \quad k = 1, \dots, n-1, \tag{12}$$

consequently,

$$u_k = \sum_{i=1}^{k} v_i r_i. \tag{13}$$

By (12), the expression for r_k takes the form

$$r_k = \lambda_k^{-1}\gamma_k - \lambda_k^{-1} \sum_{i=1}^{k-1} f_k(v_i) r_i, \quad r_1 = \lambda_1^{-1}. \tag{14}$$

Since $v_1, \dots, v_{n-1}$ are linearly independent and $v_n\gamma_n + u_{n-1} = 0$, it follows from (5) and (13) that

$$r_k = \gamma_n, \quad k = 1, \dots, n-1. \tag{15}$$

Since $f_n(v_n) = \lambda_n - 1$, from (14), (15) and (6) we find

$$\gamma_n \lambda_1 - 1 = 0, \tag{16.1}$$

$$\eta_{k,1} + \dots + \eta_{k,k-1} + \lambda_k - \gamma_k \lambda_1 = 0, \quad k = 2, \dots, n, \tag{16.k}$$

where $\eta_{i,j} = f_i(v_j)$. By Lemma 3, there are $\widetilde{\alpha}_i, \widetilde{\beta}_i \in K$, $k = 3, \dots, n$, such that

$$f_{k+1} = \widetilde{\alpha}_{k+1} f_{k-1} + \widetilde{\beta}_{k+1} f_k, \quad k = 2, \dots, n-1. \tag{17}$$

Since f_1, f_2 are linearly independent, there is $u \in V$ such that $f_1(u) = 1$, $f_2(u) = -\sigma$. Multiplying the equation (17) by u, we arrive at the formulas

$$\gamma_{k+1} = \widetilde{\alpha}_{k+1}\gamma_{k-1} + \widetilde{\beta}_{k+1}\gamma_k, \quad k = 2, \dots, n-1. \tag{18}$$

Choose $u \in V$ such that $f_1(u) = 0$ and $f_2(u) = 1$. Then

$$\beta_{k+1} = \widetilde{\alpha}_{k+1}\beta_{k-1} + \widetilde{\beta}_{k+1}\beta_k, \quad k = 2, \dots, n-1. \tag{19}$$

Finally, choose $u \in V$ such that $f_1(u) = 1$ and $f_2(u) = -\beta_k^{-1}\alpha_k$. Then

$$\widetilde{\alpha}_{k+1}c_k = -\beta_{k+1}\beta_k^{-1}c_{k+1}, \quad k = 2, \dots, n-1, \tag{20}$$

where

$$c_k = \alpha_{k-1} - \beta_{k-1}\beta_k^{-1}c_{k+1}, \quad k = 2, \dots, n. \tag{21}$$

By (17) we get $\eta_{k,i} = \widetilde{\alpha}_k\eta_{k-2,i} + \widetilde{\beta}_k\eta_{k-1,i}$. Substituting the expression for $\eta_{k,i}$ into (16) and taking into account (18) we can assert that

$$\widetilde{\alpha}_{k+1}(\eta_{k-1,k} - 1) - \widetilde{\beta}_{k+1} + \lambda_{k+1} = 0, \quad k = 2, \dots, n-1. \tag{22}$$

Put

$$q_k = \beta_k - \beta_{k-1} - f_{k-1}(v_k^\beta) + f_k(v_{k+1}^\beta), \quad k = 2, \dots, n. \tag{23}$$

Employing (7), (17), (19), and (22), it is not difficult to deduce the recursion relation

$$q_k = -\widetilde{\alpha}_{k+1}^{-1}\lambda_{k+1}q_{k+1}, \quad k = 2, \dots, n-1. \tag{24}$$

Moreover, from (16.$n-1$) and (16.1) we get

$$q_n = c_n\lambda_1\beta_n. \tag{25}$$

From the relation (2) and Lemma 4 it follows that

$$\lambda = 1 + (\sigma f_1 + f_2)(v^\beta) \in \mathrm{Par}\,(s).$$

By (7) we get

$$\lambda = (\eta_{2,1} + \sigma)f_1(v_2^\beta) + \lambda_2(f_2(v_3^\beta) + 1). \tag{26}$$

By (16.2) we obtain $\lambda = \lambda_2 q_2$. Put

$$p_k^{(s)} = -\widetilde{\alpha}_{k+1}^{-1}\lambda_{k+1}p_{k+1}^{(s)}, \quad k = 2, \dots, s-1, \quad p_s^{(s)} = c_s\lambda_1\beta_s.$$

Substituting the expression for $\widetilde{\alpha}_k$ [see (20)] into $p_k = \lambda_2 p_2^{(k)}$, we can verify that

$$p_k = p_{k-1}[(c_k\lambda_1)^{\beta_{k-1}}, \beta_{k-1}^{-1}\beta_k(\lambda_k^{-1})^{\beta_k}]\lambda_k{}^{\beta_k}, \quad k = 3, \dots, n. \tag{27}$$

From the last relations and $\lambda = \lambda_2 q_2 = p_n$ we conclude that the assumption of the theorem is necessary if s is expressible as a product of n independent transformations. The proof of the induction step is completed.

It remains to prove that the assumption of the theorem is also sufficient. Assume that $\lambda_1 \dots \lambda_n \notin \mathrm{Par}\,(s)$ for any $\lambda_i \in \Lambda_i$, $i = 1, \dots, n$.

Lemma 5. *Let $\omega_k = [x_3, y_3] \dots [x_n, y_n]z_1 \dots z_n$, $k = 3, \dots, n$, $\omega_2 = z_1z_2$; x_i, y_i, $z_i \in K^*$. Then there are $\lambda_i \in K^*$, $i = 1, \dots, k$, $c_i, \beta_i \in K^*$, $i = 3, \dots, k$ such that $\omega_k = p_k$ where p_k is defined by (27).*

Proof. We now proceed by induction on k. For $k = 2$ the assertion of Lemma 5 is trivial. Assume that there are $\lambda_i \in K^*$, $i = 1, \dots, k-1$, and $c_i, \beta_i \in K^*$, $i = 3, \dots, k-1$, such that $\omega_{k-1} = p_{k-1}$. Put $c_k = x_k^{z_1 \dots z_{k-1} \beta_{k-1}^{-1}} \lambda_1^{-1}$, $\beta_k = \beta_{k-1} y_k^{z_1 \dots z_{k-1}} z_k$, $\lambda_k = z_k^{\beta_k}$. It is easy to check that $\omega_k = \omega_{k-1}[x_k, y_k]^{z_1 \dots z_{k-1}} = p_k$. From this Lemma 5 follows.

By the theorem, there is $\lambda \in \operatorname{Par}(s)$ such that $\lambda = [x_3, y_3] \dots [x_n, y_n] z_1 \dots z_n$. By Lemma 5, there are $\lambda_i \in K^*$, $i = 1, \dots, n$, and $c_i, \beta_i \in K^*$, $i = 3, \dots, n$, such that $\lambda = p_n$. Define $\widetilde{\alpha}_i \in K^*$, $\alpha_i \in K$, and $\widetilde{\beta}_i \in K$, $i = 3, \dots, n$, by the formulas (20), (21), and (19) respectively. Put

$$\sigma = \beta_n^{-1}(\alpha_n - \lambda_1^{-1}), \quad \gamma_i = \alpha_i - \beta_i \sigma, \quad i = 3, \dots, n. \tag{28}$$

Choose linearly independent vectors $v_1, \dots, v_{n-1} \in V$ and put

$$\eta_{i,i} = \lambda_i - 1, \quad i = 1, 2, \tag{29}$$

$$\eta_{2,1} = -\lambda_2 - \sigma. \tag{30}$$

Let the elements $\eta_{i,k}$, $i \neq k$, $i = 1, 2$; $k = 2, \dots, n-1$, be defined by the equations

$$\widetilde{\alpha}_{k+1}(\alpha_{k-1}\eta_{1,k} + \beta_{k-1}\eta_{2,k} - 1) - \widetilde{\beta}_{k+1} + \lambda_{k+1} = 0, \quad k = 2, \dots, n-1,$$

$$\alpha_k \eta_{1,k} + \beta_k \eta_{2,k} = \lambda_k - 1, \quad k = 2, \dots, n-1. \tag{31}$$

There are $f_1, f_2 \in V^*$ such that $f_i(v_k) = \eta_{i,k}$, $i = 1, 2$; $k = 1, \dots, n-1$. Define f_i, $k = 3, \dots, n$, by the formulas (6).

Lemma 6. *Define r_i, $i = 1, \dots, n-1$, by (14). For any $k = 1, \dots, n-1$, the relations (16.$k+1$) and $r_k = \lambda_1^{-1}$ are valid, where $f_i(v_j) = \eta_{i,j}$.*

Proof. We proceed by induction on k. For $k = 1, 2$ we obtain the assertion from (29) and (30). The relations (19), (20), (21), (6), and (28) yield (17) and (18). Moreover, (31) and (6) imply (22). By the induction assumption we have (16.k), (16.$k-1$) and $r_1 = \dots = r_{k-1} = \lambda_1^{-1}$. With the help of (17) we can calculate the left side of (16.$k+1$). Then from (16.k), (16.$k-1$), (22), and (18) we obtain (16.$k+1$). Finally, the following chain of the equalities

$$r_k = \lambda_k^{-1}\gamma_k - \lambda_k^{-1} \sum_{i=1}^{k-1} f_k(v_i) r_i = \lambda_k^{-1}\gamma_k - \lambda_k^{-1} \sum_{i=1}^{k-1} f_k(v_i) \lambda_1^{-1} = \lambda_1^{-1}$$

completes the proof of Lemma 6.

Define nonsingular transformations a_i, $i = 1, \dots, n$, by the formulas (4) and (5). Since (28) is equivalent to (16.1), Lemma 6 implies (15). We obtain (10) from (15) because (14) and (15) imply (11). Since (10) and (9) are equivalent, we have

$$a = 1 + v^\beta \otimes (\sigma f_1 + f_2). \tag{32}$$

Put $\mu = 1 + (\sigma f_1 + f_2)(v_\beta)$. Then

$$a_1 \dots a_n(v^\beta) = v^\beta \mu. \tag{33}$$

By the above, we obtain $\mu = p_n$. Hence $\mu = \lambda$.

Assume that the forms f_1 and f_2 are linearly dependent. There is $l \in \{1, 2\}$ such that

$$f_k = \delta_k f_l, \quad \delta_k \in K, \quad k = 1, \dots, n. \tag{34}$$

Applying Lemma 4 to the transformations a_i, $i = 1, \dots, n$, and taking into account (34), we obtain $a = a_1 \dots a_n = 1 + v^\delta \otimes f_l$ and, consequently,

$$a(v^\delta) = v^\delta(1 + f_l(v^\delta)). \tag{35}$$

Set $\delta'_k = \delta_k$ for $\delta_k \neq 0$ and $\delta'_k = 1$ for $\delta_k = 0$. By (7), (31) and (6) it is easy to verify that

$$1 + f_l(v^\delta) = \lambda_1^{\delta'_1} \dots \lambda_n^{\delta'_n}. \tag{36}$$

Since s is a dilatation and $\mu \in \operatorname{Par}(s)$, we have $\mu \neq 1$. From (35) it follows that $f_l \neq 0$. Hence a is a dilatation too. Moreover, from (35), (36) and (33) we find that $\operatorname{Con}(\lambda_1^{\delta'_1} \dots \lambda_n^{\delta'_n}) = \operatorname{Con}(\lambda)$. The last equality contradicts our assumption. Therefore, from (32) it follows that $\lambda \in \operatorname{Par}(a)$. Then s is a conjugate in $GL(V)$ to the product $a_1 \dots a_n$. Hence s is also a product of n independent simple transformations with given parameters.

Thus, we may assume that there are $\lambda_i \in \Lambda_i$, $i = 1, \dots, n$ such that $\lambda_1 \dots \lambda_n \in \operatorname{Par}(s)$. Let $s = 1 + u \otimes f$. Since s is a dilatation, it follows that $\lambda_k \neq 1$ for some k. Choose linearly independent vectors $v_1, \dots, v_{n-1} \in V$ such that $f(v_i) = \lambda_i - 1$, $i = 1, \dots, n$, where $v_n = \sum_{i=1}^{n-1} v_i \tau_i$ and $\tau_i \neq 0$ for all $i = 1, \dots, n-1$. Put $a_i = 1 + u_i \otimes f$, $i = 1, \dots, n$, $a = a_1 \dots a_n$. Apply Lemma 4 to a_i, $i = 1, \dots, n$, to find that $a = 1 + v^\varepsilon \otimes f$ where $\varepsilon_i = 1$ for all $i = 1, \dots, n$. From (7) we get $1 + f(v^\varepsilon) = \lambda_1 \dots \lambda_n$. Hence a is a dilatation. Moreover, $\operatorname{Par}(s) = \operatorname{Par}(a)$. The above reasoning completes the proof of the theorem.

References

[1] Artin, E., Geometric Algebra. Interscience Publishers, New York 1957.

[2] Korobov, A. A., Parameter of the simple transformation which is decomposed into product of simple transformation. In: Abstracts of the IIIrd Int. Conf. on Algebra, 165–166. Krasnoyarsk 1993.

Monoidal and distributive intervals in clone lattices

Andreĭ A. Krokhin

Abstract. Monoidal and distributive intervals in clone lattices are studied in the paper. Wide classes of examples with finite or continual monoidal intervals are given. For some class of monoids on a finite set A, containing at least two elements, a countable direct product of finite chains is proved to be isomorphic to a suitable interval in the monoidal interval of the given monoid, which gives a class of distributive lattices isomorphic to intervals in the complete lattice of all clones on A. Sketches of the proofs are given.

1991 Mathematics Subject Classification: 06A23, 06B05, 06D05.

Introduction

For the most part, the notation is taken from the book [1]. Let A be a finite set, containing at least two elements. A *clone* on A is a set of operations on A that is closed under all compositions and contains all projections, i.e., operations of the form $e_{n,i}(x_1, \dots, x_n) = x_i$ $(n \in \mathbb{N}, 1 \le i \le n)$. For more complete definitions we refer to [1, 2]. It is clear that the set O_A of all operations on A is a clone. The intersection of a set of clones on A is again a clone. Thus, all clones on A form a complete lattice $\operatorname{Lat} A$ and for a set F of operations on A there exists a least clone containing F which is called the *clone generated by* F and denoted by $\langle F \rangle$. For simplicity of notation, we write $\langle f_1, \dots, f_n \rangle$ instead of $\langle \{f_1, \dots, f_n\} \rangle$.

Let C be a clone on A. The set of all n-ary operations from C is called the nth *layer* of C and is denoted by $C^{(n)}$. It is evident that, for each clone C on A, its first layer $C^{(1)}$ is a transformation monoid on A with identity $id = e_{1,1}(x)$, the identity transformation. Let M be an arbitrary transformation monoid on A. Consider the set

$$\operatorname{St} M = \{f(x_1, \dots, x_n) \in O_A \mid \forall m_1, \dots, m_n \in M \;\; f(m_1(x), \dots, m_n(x)) \in M\}.$$

It is easy to see that $\operatorname{St} M$ is a clone, which we call the *stabilizer* of M. The inclusion $M \subseteq \operatorname{St} M$ is obvious. The interval $[\langle M \rangle; \operatorname{St} M]$ in $\operatorname{Lat} A$ is called a *monoidal interval* of M and is denoted by $\operatorname{Int} M$. The interval $\operatorname{Int} M$ may be one-element. This means that only a unique clone on A, namely $\langle M \rangle$, has M as its first layer. Such monoids M are called *collapsing*. As is known, $C \in \operatorname{Int} M$ iff $C^{(1)} = M$ (see [1]). So, we have the partition of the lattice $\operatorname{Lat} A$ into intervals $\operatorname{Int} M$.

In the case of $|A| = 2$, the lattice $\operatorname{Lat} A$ is known in its entirely [3]. Hence we know all monoidal intervals. Three of them are finite and the other three are countable. Further, we assume that $|A| \geq 3$. For $|A| \geq 3$, the equality $|\operatorname{Lat} A| = 2^{\aleph_0}$ holds [4] and there are only finitely many transformation monoids on A. Hence there is a monoid M such that $|\operatorname{Int} M| = 2^{\aleph_0}$.

A. Szendrei [1] posed the following problem: for which monoids M on A

(a) $\operatorname{Int} M$ is finite,

(b) $|\operatorname{Int} M| = 2^{\aleph_0}$?

She wrote that a solution of the problem would contribute to a better understanding of the structure of $\operatorname{Lat} A$.

Some examples of monoids M for which $\operatorname{Int} M$ is a finite chain are given in [1]. List some of them:

1) every transformation monoid containing all non-permutations,

2) the monoid of affine (linear) transformations of a finite vector space over a finite field,

3) nonregular transitive permutation group in which all nontrivial normal subgroups are transitive.

I. A. Mal'cev [5] proved that if $M = \{f(x) \mid |f(A)| \leq l\} \cup \{\mathrm{id}\}$ $(2 \leq l \leq |A|)$ then $\operatorname{Int} M$ is a $(l+1)$-element chain. For transformation monoids and permutation groups, Lemma 1.3.1 of [2] gives criterions for the property $|\operatorname{Int} M| = 1$. For permutation groups, another criterion is described in [6]. Unfortunately, these criterions are hard to apply.

One result should be mentioned specially because it has important applications in universal algebra. P. P. Palfy [7] proved that if M consists of all constant operations and some permutations then $|\operatorname{Int} M| \leq 2$ where the equality holds iff M is the monoid of unary polynomial operations of a vector space.

As for continual intervals, it follows from results of [8, 9] that $|\operatorname{Int}\{\mathrm{id}\}| = 2^{\aleph_0}$.

It is easy to verify that if M consists of all constants and the identity operation then $|\operatorname{Int} M| = 1$. Here, we consider all the cases when M consists of some constants and the identity operation. Let l be the number of constants in M. As was mentioned, we have $|\operatorname{Int} M| = 1$ if $l = |A|$. The case $l = |A| - 1$ is handled in Section 1 (see Theorems 1 and 2) and intervals $\operatorname{Int} M$ turn out to be finite. If $l \leq |A| - 2$ then M satisfies the assumptions of Theorem 3 (see Section 2) and $|\operatorname{Int} M| = 2^{\aleph_0}$. Theorem 3 and Corollary 2 to Theorem 4 yield a wide class of monoids M such that $|\operatorname{Int} M| = 2^{\aleph_0}$.

Another interesting question: what is the local structure of the lattice $\operatorname{Lat} A$, i.e. which lattices are isomorphic to some interval in $\operatorname{Lat} A$? It was proved [10] that the lattice of all subsets of a countable set (i.e., the countable direct power of the two-element lattice) is isomorphic to some interval in $\operatorname{Lat} A$. In [11], it is proved that each lattice, which is a countable direct product of finite lattices, is isomorphic to some sublattice of Lat A for $|A| \geq 4$. Can we change here the word "sublattice" for "interval"? Corollary 1 to Theorem 4 shows that if these finite lattices are chains then, even if $|A| \geq 3$, we can. Consequently, there is a wide class of distributive lattices isomorphic to suitable intervals in Lat A.

We give only sketches of proofs. It will cause no confusion if we use the same letter a to designate the constant operation equal to a.

1. Finite intervals

Let $A = \{0, 1, 2\}$. Denote by min (max) the binary function on A of taking minimum (maximum) with respect to the order $0 < 2 < 1$.

Theorem 1. *Let* $M = \{0, 1, \mathrm{id}\}$. *Then* Int M *is as follows:*

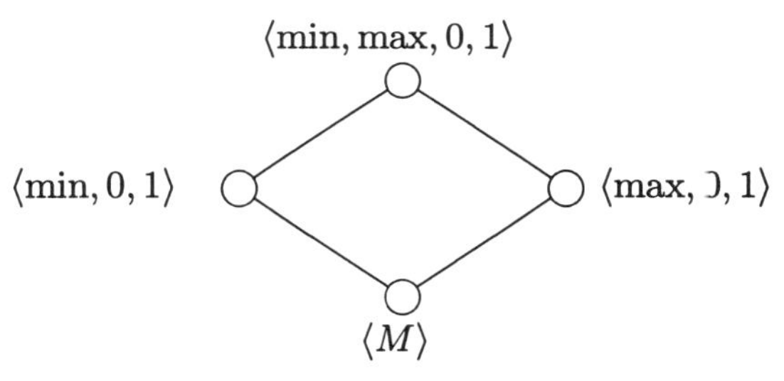

Proof. We notice the following property of functions from St M. For any $f_1, f_2 \in$ St M we have

$$\text{if} \quad \begin{array}{l} f_1(x_1, \ldots, x_n, 0) = f_2(x_1, \ldots, x_n, 0) \\ f_1(x_1, \ldots, x_n, 1) = f_2(x_1, \ldots, x_n, 1) \end{array} \quad \text{then} \quad f_1 = f_2. \tag{1}$$

We denote $\langle \min, \max, 0, 1\rangle$ briefly by K. We need to prove $K = \mathrm{St}\, M$. It is easy to see that $\{\min, \max, 0, 1\} \subseteq \mathrm{St}\, M$ whence $K \subseteq \mathrm{St}\, M$. We prove the inverse inclusion by induction on arity of operations. Clearly, $K^{(1)} = (\mathrm{St}\, M)^{(1)} = M$. Let $(\mathrm{St}\, M)^{(n)} \subseteq K^{(n)}$. Assume that an operation f belongs to $(\mathrm{St}\, M)^{(n+1)}$ and has no fictitious variables. By (1), we are to prove that f belongs to K. Consider a mapping $\theta : L \to K$, where L is the bounded free distributive lattice with free generators $x_1, \ldots, x_n, \ldots$, the least element 0, and the greatest element 1 $(0, 1 \in A)$. It is constructed by induction as follows: $\theta(0) = 0$, $\theta(1) = 1$, $\theta(x_i) = x_i$.

If t_1, t_2 are terms in L then

$$\theta(t_1 \wedge t_2) = \min(\theta(t_1), \theta(t_2)), \theta(t_1 \vee t_2) = \max(\theta(t_1), \theta(t_2)).$$

In view of the mutual distributivity of min and max, the mapping θ is well defined and injective. Consider operations g_0, g_1 from K, where

$$g_i(x_1, \dots, x_n) = f(x_1, \dots, x_n, i) \ (i = 0, 1).$$

Choose terms $t_0, t_1 \in L$ such that g_i can be obtained from $\theta(t_i)$ by adding fictitious variables. Starting from t_0 and t_1 we construct a new term $t \in L$. For operations $\theta(t)$ and f, all conditions of (1) hold. Therefore $\theta(t) = f$ and $f \in K$. Thus, we have the equality $K = \operatorname{St} M$. It is obvious that $\min \in \langle f, 0, 1\rangle$ for every $f \in \langle \min, 0, 1\rangle \backslash \langle M\rangle$, and $\langle \min, 0, 1\rangle = \langle f, 0, 1\rangle$. The same is true for the operation max. To finish the proof we show that $f \in K \backslash (\langle \min, 0, 1\rangle \cup \langle \max, 0, 1\rangle)$ implies $\{\min, \max\} \subseteq \langle f, 0, 1\rangle$ and $\langle f, 0, 1\rangle = K$. □

Theorem 2. *Let $|A| \geq 4$, and let M consist of $|A| - 1$ constant operations and the identity operation. Then M is collapsing.*

Proof. We now proceed by induction on arity and show that every operation from $\operatorname{St} M$ has not more than one essential variable. The base of induction is obvious. Let $(\operatorname{St} M)^{(n)} \subseteq \langle M\rangle$ and $f \in (\operatorname{St} M)^{(n+1)}$. For all $a \in M$, $f(a, x_2, \dots, x_{n+1})$ is either a constant operation from $\langle M\rangle$ or a projection. The proof is reduced to a combinatorial analysis of some cases; one of them uses essentially that M contains more than two constant operations. Each case lead either to contradiction or to the conclusion that f belongs to $\langle M\rangle$. □

2. Intervals Int M of cardinality $2^{\aleph_0}$ and distributive intervals in Lat A

Denote $N_k = \{n \in N \cup \{0\} \mid n \geq k\}$ where N is the set of all natural numbers. Let B be a nonempty subset of A such that $|B| \leq |A| - 2$. Denote by B^A the set of transformations m of A such that $m(A) \subseteq B$.

Theorem 3. *For every monoid M on A containing more than one element and such that $M \subseteq B^A \cup \{\mathrm{id}\}$, the interval $\operatorname{Int} M$ is of cardinality $2^{\aleph_0}$.*

Proof. In view of the equality $|\operatorname{Lat} A| = 2^{\aleph_0}$, we only have to find a system of clones of cardinality $2^{\aleph_0}$ in $\operatorname{Int} M$.

In the case $|B| = 1$, this system can be derived from the proofs exposed in [4]. Suppose that $|B| \geq 2$. Fix a transformation $m_0 \in M \backslash \{\mathrm{id}\}$ and elements $a, b \in A \backslash B$ $(a \neq b)$. Choose an element $c \in B$ such that $c \neq m_0(a)$. Consider the following system of operations on A:

$$f_n(x_1,\dots,x_n)=\begin{cases} c & \text{if } x_1=\dots=x_{i-1}=x_{i+1}=\dots=x_n=a,\\ & x_i=b \quad\text{or}\\ & x_1=\dots=x_{i-1}=x_{i+1}=\dots=x_n=b, x_i=a,\\ & i=1,\dots,n;\\ m_0(x_1) & \text{otherwise}\end{cases}$$

where $n\in N_3$. Define the clone $C_I=\langle\{f_i\mid i\in I\}\cup M\rangle$ where $I\subseteq N_3$. We show that $\{C_I\mid I\subseteq N_3\}$ is the desired system. Indeed, we have $f_n(m_1(x),\dots,m_n(x))=m_0m_1(x)\in M$ for every $n\in N_3$ and $m_1,\dots,m_n\in M$. Hence every f_n belongs to $\operatorname{St}M$ and every C_I belongs to $\operatorname{Int}M$. Then $I\neq J$ implies $C_I\neq C_J$ $(I,J\subseteq N_3)$. Note that every operation from C_I, with all variables essential, belongs to $D_I=\{m(f_i(x_{j_1},\dots,x_{j_i}))\mid i\in I, m\in M\}\cup M$. We take operation f_j such that $j\in J\backslash I$ and prove that $f_j\in D_I$ implies $c=m_o(a)$, which contradicts the choice of c. Thus, $|\{C_I\mid I\subseteq N_3\}|=2^{\aleph_0}$ and the theorem is proved. □

Note that for every finite abstract monoid M' one can choose a suitable finite set A and a transformation monoid M on A such that M satisfies the conditions of Theorem 3 and M' is isomorphic to M.

Now, consider a partition π of A on two subsets A_0 and A_1 such that $|A_0|\geq 2$. Denote by M_π the monoid of all operations m satisfying one of the following conditions:

(m1) $$m(A_0)\subseteq A_0, m(A_1)\subseteq A_1;$$

(m2) $$m(A)\subseteq A_0;$$

(m3) $$m(A)\subseteq A_1.$$

An operation $^-$ satisfying conditions $\overline{a}=a_0$, $a\in A_0$, and $\overline{a}=a_1$, $a\in A_1$, where a_0 and a_1 are two distinct elements of A_0, is called a *characteristic operation of the partition* π.

Theorem 4. *Let $M\subseteq M_\pi$, and let a characteristic operation $^-$ of π belong to M. There is an interval isomorphic to the countable direct power of a chain of the type $(\omega+1)$ in the lattice* $\operatorname{Int}M$.

Proof. Denote by L_2 the lattice $\langle\{a_0,a_1\};+,\cdot\rangle$; here $a_0<a_1$. Consider the following system of operations on A:

$$
\begin{aligned}
f_{n,0}(x_1, \dots, x_{n+1}) &= \overline{x}_1 \dots \overline{x}_n + \overline{x}_1\overline{x}_2\overline{x}_3\overline{x}_{n+1} + \Big(\sum_{1 \le i < j \le n, j \ge 4} \overline{x}_i\overline{x}_j\overline{x}_{n+1}\Big) \\
f_{n,l}(x_1, \dots, x_{n+l+1}) &= f_{n,0}(x_1, \dots, x_{n+1}) + \overline{x}_{n+2} + \dots + \overline{x}_{n+l+1}, \\
& \quad n \in N_5, l \in N.
\end{aligned}
$$

It is easy to see that $f_{n,l} = f_{p,q}$ implies $n = p$ and $l = q$. These operations act on A in the following way: given a vector of arguments, every argument is transferred by $^-$ into $\{a_0, a_1\}$ and there we calculate by means of the lattice operations of L_2. We point out the following identities:

$$\overline{f_{n,l}(x_1, \dots, x_{n+l+1})} = a_0 \quad \text{for every} \quad n, l,$$

$$\overline{m(x)} = \begin{cases} \overline{x} & \text{if} \quad m \quad \text{satisfies (m1)}, \\ a_0 & \text{if} \quad m \quad \text{satisfies (m2)}, \\ a_1 & \text{if} \quad m \quad \text{satisfies (m3)}. \end{cases} \tag{2}$$

Denote by C_∞ the clone $\langle \{f_{n,l} \mid n \in N_5, l \in N_0\} \cup M\}\rangle$ and by C_0 the set of operations from C_∞ which are not (up to fictitious variables and permutations of variables) $f_{n,l}$ for any n, l. By (2), C_0 is a clone, $C_0, C_\infty \in \operatorname{Int} M$, and

$$C_1 \vee C_2 = C_1 \cup C_2 \tag{3}$$

for every $C_1, C_2 \in [C_0, C_\infty]$. It is easy to see that $C \in [C_0, C_\infty]$ and $f_{n,l} \in C$ imply $f_{n,k} \in C$ for every $k \le l$. Hence, we can characterize every clone $C \in [C_0, C_\infty]$ by means of the sequence $(l_5(C), \dots, l_s(C), \dots)$ where $l_s(C) = \sup\{l \mid f_{s,l} \in C\}$. For every such sequence, there is a suitable clone. Thus, this correspondence φ is a bijection. It is clear that φ is a meet homomorphism. By (3), we can prove that φ is a join homomorphism. Consequently, φ is an isomorphism between the interval $[C_0, C_\infty]$ and the lattice L of all countable sequences $(l_5, l_6, \dots)$ where every l_i belongs to N_0 or is the infinity. The lattice L is the countable direct power of a chain of the type $(\omega + 1)$. Hence $[C_0, C_\infty]$ is the required interval. □

Corollary 1. *Let M satisfy the conditions of Theorem 4. Then each lattice, which is at most countable direct product of finite chains, is isomorphic to a suitable interval of* Int M *and* Lat A.

Proof. The assertion follows from the facts that the above-mentioned lattices are isomorphic to suitable intervals of the countable direct power of a chain of the type $(\omega + 1)$ and that $\operatorname{Int} M$ is an interval of $\operatorname{Lat} A$. □

Applying Corollary 1 to the direct product of countably many two-element chains, we arrive at the result of [10].

Corollary 2. *Let M satisfy the conditions of Theorem 4. Then* $|\operatorname{Int} M| = 2^{\aleph_0}$.

Corollary 3 [12]. *Let A be a finite set, $|A| \geq 3$. Then the number of clones containing all constants is $2^{\aleph_0}$.*

Proof. The monoid M_π contains all constants and satisfies the conditions of Theorem 4. Corollary 3 follows from Corollary 2 with $M = M_\pi$. □

To conclude, we mention that many questions related to monoidal intervals are still open and the most interesting among them is the following:

Does there exist a countable monoidal interval (if $|A| \geq 3$)?

References

[1] Szendrei, A., Clones in universal algebra. Montreal 1986.

[2] Pöschel, R., Kaluznin, L. A., Funktionen- und Relationenalgebren. Berlin, VEB DVW 1979.

[3] Post, E., Two-valued iterative systems of mathematical logic. Princeton University Press, Princeton 1941.

[4] Yanov, Yu. I., Muchnik, A. A., On the existence of k-valued closed classes without finite basis. Dokl. Akad. Nauk SSSR 127 (1959), 44–46.

[5] Mal'cev, I. A., Some propeties of cellular subalgebras of Post algebras and their basic cells. Algebra Logika 11:5 (1972), 571–587.

[6] Ihringer, Th., A remark on clones and permutation groups. In: Contributions to General Algebra 7: Proc. Conf. Wien, Wien, Verlag Hölder–Pichler–Tempsky, 197–200.

[7] Palfy, P. P., Unary polinomials in algebras, I. Algebra Universalis 18 (1984), 262–273.

[8] Szendrei, A., Algebras of prime cardinality with cyclic automorphism. Arch. Math. (Basel) 39 (1982), 417–427.

[9] Marchenkov, S. S., On closed classes of self-dual functions in a many-valued logic. Probl. Cibern. 36 (1976), 5-22.

[10] Haddad, L., Rosenberg, I., Un intervalle Boolean de clones sur un univers fini. Ann. Sci. Math., Quebec 12:2 (1988), 211–231.

[11] Bulatov, A. A., Finite sublattices of clone lattices. In: Abstracts of the IIIrd Int. Conf. on Algebra, 55. Krasnoyarsk 1993.

[12] Agoston, I., Demetrovic, J., Hannak, L., On the number of clones containing all constants. Lect. in Universal Algebra, Coll. Math. Soc. J. Bolyai 43 (1983).

On a class of Moufang loops

E. K. Loginov and David I. Moldavanskiĭ

Abstract. A loop G is said to be *R-regular* if its loop ring $F[G]$ is alternative but not associative. Regular Moufang loops are described in the paper.

1991 Mathematics Subject Classification: 20N05.

Throughout the paper, R will denote a commutative associative ring with identity in which the equation $2x = 0$ holds. A loop G is called *R-regular* if its loop ring $F[G]$ is alternative but not associative. Loops of this type known as RA loops were studied in [1–3]. It is well known [4] that if a loop has an alternative loop ring over some coefficient ring of characteristic $\neq 2$, then it has an alternative loop ring over any coefficient ring of characteristic $\neq 2$. Therefore we may speak about regular Moufang loops without mention on the ring R.

A *Moufang loop* L is a loop satisfying the identity $(xy)(zx) = (x(yz))x$ for $x, y, z \in L$. By [5] the equivalent identities $((xy)z)y = x(y(zy))$ and $x(y(xz)) = ((xy)x)z$ also hold. Any Moufang loop L is diassociative, i.e. any two elements of L generate a subgroup. The subloop L' of a Moufang loop L generated by all commutators $[x, y] = (yx)(xy)^{-1}$ and associators $(x, y, z) = (x(yz))^{-1}((xy)z)$ is called the *commutator-associator subloop of the loop* L. It is known [5] that L' is a normal subloop of L and the mapping $x \to xL'$ is a homomorphism of L onto an abelian group L/L' in which multiplication is defined by $(xL')(yL') = (xy)L'$. The subloop $Z(L) = \{a \in L \mid [a, b] = (a, b, c) = 1;\ b, c \in L\}$, is called the *center of the Moufang loop* G.

We say that a *diassociative loop* L *has property* LC if it is not abelian and, for any pair of elements x, y of L,

$$xy = yx \quad \text{if and only if} \quad x \in Z(L),\ y \in Z(L) \text{ or } xy \in Z(L).$$

Assume that the group G has property LC and $G' \simeq A_2$, where A_2 is a cyclic group of order 2. Then the set $L = G \cup Gh$, h an indeterminate, together with multiplication defined by the rule

$$f \cdot gh = (gf)h,$$

$$fh \cdot g = (fg^*)h,$$

$$fh \cdot gh = c(g^* f),$$

where $f, g \in G$, c is a fixed element of the center $Z(C)$ of G and

$$g^* = \begin{cases} g & \text{if } g \in Z(G), \\ ge & \text{if } g \notin Z(G), \end{cases} \quad e \in G',\ e \neq 1.$$

Moreover,

$$L' = G', \quad Z(L) = Z(G), \quad L/Z(L) \simeq A_2 \times A_2 \times A_2.$$

Any regular Moufang loop can be constructed in such a way (see [1, 2]). Furthermore any periodic regular Moufang loop is represented in the form

$$L = \widetilde{L} \times A,$$

where the loop $\widetilde{L}$ is indecomposable (i.e. it is not the direct product of two proper subloops), and A is an abelian group. Any periodic indecomposable regular Moufang loop is a 2-loop. Thus we describe all regular Moufang loop if we describe all groups having property LC and a unique non-identity commutator. Such groups are described in Theorems 1–3.

Theorem 1. *The group G has property* LC *and an unique non-identity commutator if and only if the factor-group of group G by center $Z(G)$ is the Klein four-group.*

Theorem 2. *Let A be an abelian group and e an element of A of order* 2. *There exists a group G such that*

(i) $Z(G) = A$;

(ii) *the factor-group G/A is the Klein four-group*;

(iii) *the commutant G' of G is a cyclic group generated by the element e.*

The group G above depends on the group A and its elements e, a, b. Therefore, we shall denote it by $G = G(A, e, a, b)$.

Theorem 3. *Let G be a group such that $G/Z(G)$ is the Klein four-group. Then the commutant G' of G is the cyclic group of order* 2 *and the group G is isomorphic to a group $G(A, e, a, b)$, where $A = Z(G)$, $e, a, b \in A$ and e is the generator of G'.*

The arguments above show that, for any abelian group A with an element e of order 2, there exists the regular Moufang loop $L = L(A, e, a, b, c)$ where $a, b, c \in A$, which can be defined by generators and relations of the form

$$L = \langle A, f, g, h \mid \Omega, \Theta, \Sigma \rangle;$$

here Ω is the set of defining relators of the group A, Θ is the set of commutators and associators connecting the elements f, g, h with elements of the group A, and

$$\Sigma = \{f^2 = a, g^2 = b, h^2 = c, [f, g] = [g, h] = [h, f] = (f, g, h) = e\}.$$

The problem of isomorphism is solved by the following theorem:

Theorem 4. *Let A_1 and A_2 be abelian groups, e_i, a_i, b_i, c_i elements of A_i, where $|e_i| = 2$ $(i = 1, 2)$. The loops $L(A_1, e_1, a_1, b_1, c_1)$ and $L(A_2, e_2, a_2, b_2, c_2)$ are isomorphic if and only if there exists an isomorphism $\varphi : A_1 \to A_2$ such that $e_1\varphi = e_2$ and the triple $(a_1\varphi, b_1\varphi, c_1\varphi)$ is congruent modulo A_2^2 (where A_2^2 is the subgroup of the group A_2 generated by squares of elements A_2) with one of the triples (a, b, c), where the symbols a, b, c are distinct and denote the elements a_2, b_2, c_2, $a_2b_2e_2$, $b_2c_2e_2$, $c_2a_2e_2$, $a_2b_2c_2$.*

All regular Moufang loops of order < 64 were found in [2]. As an illustration of the theorems above we find all regular Moufang loops of order < 128. We write all abelian groups of order ≤ 8:

$$A_n = \langle x \mid x^n = 1\rangle, \quad n = 2, 3, 4, 5, 6, 7, 8,$$

$$A_2 \times A_2 = \langle x, y \mid x^2 = y^2 = [x, y] = 1\rangle,$$

$$A_4 \times A_2 = \langle x, y \mid x^4 = y^2 = [x, y] = 1\rangle,$$

$$A_2 \times A_2 \times A_2 = \langle x, y, z \mid x^2 = y^2 = z^2 = [x, y] = [y, z] = [z, x] = 1\rangle,$$

where the upper index distinguishes the groups and lower one indicates the order. Further we use the theorem 4 and write all nonisomorphic indecomposable regular Moufang loops of the orders 16, 32, and 64:

$$L_{16}^{(1)} = L(A_2, x, 1, 1, 1), \qquad L_{16}^{(2)} = L(A_2, x, x, x, x),$$

$$L_{32}^{(1)} = L(A_4, x^2, 1, 1, 1), \qquad L_{32}^{(2)} = L(A_4, x^2, x, x, x),$$

$$L_{32}^{(3)} = L(A_2 \times A_2, x, y, y, y), \qquad L_{32}^{(4)} = L(A_2 \times A_2, x, 1, x, y),$$

$$L_{64}^{(1)} = L(A_8, x^4, 1, 1, 1), \qquad L_{64}^{(2)} = L(A_8, x^4, x, x, x),$$

$$L_{64}^{(3)} = L(A_4 \times A_2, y, x, x, x), \qquad L_{64}^{(4)} = L(A_4 \times A_2, y, 1, x, y),$$

$$L_{64}^{(5)} = L(A_4 \times A_2, x^2, 1, x, y), \qquad L_{64}^{(6)} = L(A_2 \times A_2 \times A_2, x, 1, y, z),$$

$$L_{64}^{(7)} = L(A_2 \times A_2 \times A_2, x, y, y, z).$$

As stated above, any finite regular Moufang loop can be represented as the direct product of an abelian group and indecomposable regular Moufang loops of order 2^n, where $n \geq 4$ (if the order of the Moufang loop G is < 12 then G is a group (see [6])). Hence we have all nonisomorphic decomposable regular Moufang loops of the orders < 128:

$$L_{16}^{(i)} \times A_p, \quad L_{16}^{(i)} \times A_2 \times A_s, \quad L_{32}^{(j)} \times A_s,$$

where $i = 1, 2$, $j = 1, 2, 3, 4$, $p = 2, 3, 4, 5, 7$, $s = 2, 3$. Thus, there exist 35 nonisomorphic regular Moufang loops of the order < 128, among them 13 ones are indecomposable.

References

[1] Goodaire, E. G., Alternative loop rings. Publ. Math. 30:1–2 (1983), 31–38.

[2] Chein, O., Goodaire, E. G., Loops whose loop rings are alternative. Commun. Algebra 14:2 (1986), 293–310.

[3] Goodaire, E. G., Parmenter, M. M., Semisimplicity of alternative loop rings. Acta Math. Hung. 50:3–4 (1987), 241–247.

[4] Chein, O., Goodaire, E. G., Is a right alternative loop ring alternative? Algebras Groups Geom. 5:3 (1988), 297–304.

[5] Brack, R. H., A Survey of Binary Systems. 3rd ed. Springer-Verlag, Berlin–Heidelberg–New York 1971.

[6] Chein, O., Pelugfelder, H. O., The smallest Moufang loop. Arch. Math. 22:6 (1971), 573–576.

On extensions of partial geometries

Aleksandr A. Makhnev *

Abstract. In this paper, we investigate strongly φ-uniform extensions $EpG_\alpha(s,t)$ of α-partial geometries of order s,t with $\varphi = s$ or $s-1$, geometries with small number of points on a line, and geometries with minimal μ (for points P,Q at distance 2, we set $\mu(P,Q) = |\mathcal{S}_P \cap \mathcal{S}_Q|$ where $\mathcal{S}_p$ is the residue of $\mathcal{S}$ at point P). There are no proofs in the paper.

1991 Mathematics Subject Classification: 51E14, 51E12, 05B25.

1. Introduction

A *geometry of rank* 2 is a structure with two types of elements, which are said to be *points* and *blocks*, and an incidence relation between points and blocks without repetitions. Thus, each block coincides with the set of points incident to this block. We use informal language, such as *contains, lies on*, etc.

If $\mathcal{S}$ is a geometry of rank 2 then the *point-graph* $\Gamma = \Gamma(\mathcal{S})$ is the graph whose vertices are points of $\mathcal{S}$, with two distinct vertices being *adjacent* if there is a block containing them both. Two distinct points of $\mathcal{S}$ are *collinear* if they are adjacent in Γ. The geometry $\mathcal{S}$ is *connected, regular* and so on if Γ has the corresponding property.

Let $\mathcal{S}$ be a geometry of rank 2. The *residue* $\mathcal{S}_P$ *of* $\mathcal{S}$ *at point* P is the geometry whose point-set is the set of points adjacent to P and whose set of blocks is the set of ones incident with P (with the induced *natural* incidence). Let $\mathcal{F}$ be a family of geometries of rank 2 such that every residue $\mathcal{S}_P$ is in $\mathcal{F}$, then we say that $\mathcal{S}$ is an extension of $\mathcal{F}$. A geometry $\mathcal{S}$ is *triangular* if whenever X, Y, Z are three pairwise collinear points, then there is at least one block of $\mathcal{S}$ which contains all three X, Y, Z. The *point graph* $\Gamma(\mathcal{S}_P)$ is a (possibly, proper) subgraph of the graph induced on Γ_P by Γ and these two graphs are the same iff $\mathcal{S}$ is triangular.

A point-block pair (P, y) is a *flag* if $P \in y$, and an *antiflag* otherwise. If (P, y) is an antiflag, then $\varphi(P, y)$ is the number of points on y which are collinear with P.

* Research partially supported by Russian Fund for Fundamental Research (grant 93–011–1529) and Fund of Russian State Committee on High Education (grant 2–11–6–29).

We say that $\mathcal{S}$ is *(strongly) φ-uniform* if $\varphi(P,y) = 0$ or φ $(= \varphi)$ for every antiflag (P,y) and some constant φ. The blocks of $\mathcal{S}$ are called *lines* if distinct blocks intersect in at most one point.

If $\mathcal{S}$ is a geometry such that every line has $s+1$ points, every point is on $t+1$ lines (with $s>0, t>0$) and $\mathcal{S}$ is strongly α-uniform $(\alpha>0)$, then $\mathcal{S}$ is an *α-partial geometry of order* (s,t), or a $pG_\alpha(s,t)$, or a pG_α. A connected extension of the family of pG_α's is an EpG_α or an EpG.

In an EpG_α $\mathcal{S}$, it is possible to show that for each antiflag (P,y) we have $\varphi(P,y)\geq\alpha+1$ in the case $\varphi(P,y)\neq 0$ and that $\mathcal{S}$ is triangular iff $(\alpha+1)$-uniform.

A *generalized quadrangle* $GQ(s,t)$ of *order* (s,t) is a $pG_1(s,t)$. A geometry $pG_t(s,t)$ is a net and $pG_{s+1}(s,t)$ is a 2-design with $\lambda=1$. A coclique of $1+st/\alpha$ points in $pG_\alpha(s,t)$ is called an *ovoid.*

In this paper, we investigate strongly φ-uniform geometries $EpG_\alpha(s,t)$ with $\varphi = s$ or $s-1$, geometries with a small number of points on a line, and geometries with minimal μ (for points P,Q at distance 2, we set $\mu(P,Q) = |\mathcal{S}_P \cap \mathcal{S}_Q|$).

2. Strongly φ-uniform geometries with a large φ

Let $\mathcal{S}$ be a φ-uniform $EpG_\alpha(s,t)$. If $\varphi = s+2$, then $\mathcal{S}$ is called an *one-point extension* (and $\Gamma(\mathcal{S})$ is complete). *Mathieu design* 3-(22,6,1) is an one-point extension of the projective plane $PG(2,4)$.

If $\varphi = s+1$ then $\mathcal{S}$ is strongly $(s+1)$-uniform and the complement of Γ is a union of $s+2$ complete graphs with $1+st/\alpha$ points each. In this case for each point P, the residue $\mathcal{S}_P$ contains a partition point-set by $s+1$ ovoids. It is a rather strong restriction. Among known generalized quadrangles only $GQ(s,1)$, $GQ(1,t)$, $GQ(2^\alpha,2^\alpha)$, $GQ(q^2,q)$, $GQ(q-1,q+1)$, $GQ(q+1,q-1)$, where q is a prime power, possess partitions into ovoids.

Example 2.1. For any $t\leq 1$, there is the unique $EGQ(1,t)$. Its blocks are all transversals to three disjoint $(t+1)$-sets.

Example 2.2. A strongly $(s+1)$-uniform $EGQ(s,1)$ exists whenever $s=q-1$, where q is a power of 2. For $s=1$ and $s=3$, the examples are unique [1].

Let $\pi = PG(2,q)$, P be a point of π, C a hyperoval containing P, and T the group (of order q^2) of elations with the centre P. As points of the EGQ we take points of π different from P, as blocks lines not containing P and translates of $C-\{P\}$ under T.

Theorem 2.3. *Let $\mathcal{S}$ be a s-uniform $EpG_\alpha(s,t)$, $\overline{\Gamma}$ a complement of $\Gamma(\mathcal{S})$. Then $s=2$ (and the geometry $\mathcal{S}$ is known), or $\mathcal{S}$ is the geometry of points and maximal* *cliques of a strongly regular graph complemented to a graph with* $\lambda = 0, \mu = 2$, *or* $\mathcal{S}$ *is strongly* φ*-uniform and one of the following conditions holds:*

1) $t = \alpha$ *and* $\overline{\Gamma}$ *is a square lattice graph on* $(s+2)^2$ *vertices;*

2) $t = 2\alpha$, $s \le (2\alpha - 1)(\alpha + 1)^2$, $s + \alpha + 1$ *divides* $2s(s+1)(2\alpha + 1)$ *and* $\overline{\Gamma}$ *is a triangular graph on* $(s+2)(2s+3)$ *vertices.*

Example 2.4. A strongly s-uniform $EGQ(s,1)$ exists whenever $s = q - 2$, where q is a power of 2 [1].

Let $\pi = PG(2,q)$, P and Q be points of π, l the line PQ, C a hyperoval containing P and Q, and T the group of all central collineations with centre P and axis containing Q. Then $|T| = q(q-1)$ and T fixes every line on P and is sharply 2-transitive on the lines on Q different from l.

The EGQ is formed by taking as the point-set the set of points of π not on l and as the block-set the set of lines of π not containing P or Q and images of $C - \{P, Q\}$ under T.

Example 2.5. A strongly 3-uniform $EpG_2(3,2)$ exists.

Let the point-set of $\mathcal{S}$ be the set of places $< i,j >$ of a square matrix of order 5 and the block-set be the set of collections $\{1i_1, 2i_2, \ldots, 5i_5\}$ such that the permutation $(i_1 i_2 \ldots i_5)$ is even.

Theorem 2.6. *Let* $\mathcal{S}$ *be a strongly* $(s-1)$*-uniform* $EpG_\alpha(s,t)$. *Then* $\mathcal{S}$ *is an* $EpG_{2x}(3x, 3x-1)$, x *is odd or* $\mathcal{S}$ *is one of the following geometries:*

$EGQ(3,1)$, $EGQ(3,9)$; $EpG_2(6,15)$, $EpG_2(6,55)$, $EpG_2(7,6)$, or $EpG_3(7,9)$.

Theorems 2.3 and 2.6 extend the corresponding results of P. Cameron and P. Fisher on EGQ to the case of partial geometries.

3. Extensions of geometries with short lines

Let $\mathcal{S}$ be a φ-uniform $EpG_\alpha(s,t)$ with a small s. At first assume that $s = 2$. If $\varphi = 4$ then $\alpha = 3$ and by Corollary 3.2 of [2] $\mathcal{S}$ is the unique design 3-(8,4,1), or $\alpha = 2$ and $\mathcal{S}$ is an one-point extension of the dual net $pG_2(2,t)$, or $\alpha = 1$ and $\mathcal{S}$ is an extension of $GQ(2,t)$, $t = 1, 2$ or 4.

Extensions of nets and dual nets were considered in [3], and an one-point extension of $GQ(2,t)$ exists and is unique [1].

Example 3.1. Let G be a $GQ(2,t)$ and ∞ a new point, the points of the extension $\mathcal{S}$ be those of G together with ∞. The blocks of $\mathcal{S}$ are of the two types: $l\cup\{\infty\}$, where l is a line of G, and symmetric differences $x\Delta y$ of intersecting lines x, y of G.

For $\varphi = 3$, $\alpha\neq 1$ because of the parameter φ to be even in EGQ. Hence, $\alpha = 2$ and $\mathcal{S}$ is a triangular extension of the dual net $pG_2(2,t)$. But in this case the graph $\Gamma(\mathcal{S}_P)$ is induced by Γ. Contradiction with $\overline{\Gamma}$ being a union of $s+2$ complete subgraphs on $1+st/\alpha$ vertices.

If $\varphi = 2$ then $\alpha = 1$ and $\mathcal{S}$ is a triangular $EGQ(2,t)$. All such geometries were classified by F. Buekenhout and X. Hubaut (see Example 9.6 [1]).

Let now $s = 3$. If $\varphi = 5$ then $\alpha = 4$ and $\mathcal{S}$ is the unique design 2-(21,6,4) (see Corollary 3.2 of [2]), or $\mathcal{S}$ is an one-point extension of the dual net $pG_3(3,t)$, or $\mathcal{S}$ is an $EpG_2(3,t)$. In the last case $2(t+2)$ divides $12t(t+1)$, so $t = 1, 2, 4$, or 10. It is known that $pG_2(3,4)$ and $pG_2(3,10)$ do not exist. Further, an one-point extension of $pG_\alpha(s,t)$ is a 2-design with $w = 1+(s+1)(1+st/\alpha)$ points, $k = s+2$ points on block, $\lambda = t+1$ blocks on two points and $b = \lambda w(w-1)/k(k-1)$ blocks. In our case $b = (t+1)(5+6t)(4+6t)/20$, so $t\neq 2$. Thus in the case $\alpha = 2$, $\mathcal{S}$ is an $EpG_2(3,1)$.

If $\varphi = 4$ then $\alpha\neq 3$ (as above), so $\mathcal{S}$ is an $EpG_2(3,t)$ or an $EGQ(3,t)$, $t = 1$ or 5. In the case of an $EpG_2(3,t)$, the integrity condition of $1+st/\alpha$ implies $t = 2$.

Lemma 3.2. *If $pG_2(3,2)$ contains an ovoid, then the complement $\overline{\Gamma}$ of the point graph is a square lattice graph on 16 vertices.*

Proof. Let $\mathcal{S}$ be a geometry $pG_2(3,2)$ and P, Q be noncollinear points of $\mathcal{S}$. Then $\mathcal{S}_P$ and $\mathcal{S}_Q$ contain 9 points and $|\mathcal{S}_P\cap\mathcal{S}_Q| = 6$. If P belongs to an ovoid O, then $\mathcal{S} - (\mathcal{S}_P\cup\{P\})$ is a bipartite graph of degree 3, and any point of O lies in two ovoids, and ovoids distinct from O split $\mathcal{S}$. Hence, any point of $\mathcal{S}$ belongs to some ovoid and $\overline{\Gamma}$ is a square lattice graph. □

An example of a 4-uniform $EpG_2(3,2)$ see in [3]. The question on existence of a 4-uniform $EGQ(3,5)$ is still open.

If $\varphi = 3$ then by Theorem 2.3 $\mathcal{S}$ is the geometry from Example 2.5 or the geometry of point and 5-cliques of Clebsh graph. Finally, if $\varphi = 2$ then $\mathcal{S}$ is a triangular extension of $GQ(3,t)$.

Theorem 3.3. *If Γ be a strongly regular locally $GQ(3,t)$-graph, then $t = 1$ and Γ is a quotient of Johnson graph $J(8,4)$, or $t = 3$ and Γ is graph of rank 3 for $U_5(2)$, or $t = 9$ and Γ is McLaghlin graph.*

Theorem 3.4. *If Γ is a connected locally $GQ(3,3)$-graph, then Γ is a graph of rank 3 for $U_5(2)$.*

4. Geometries, containing small μ-subgraphs

Let $\mathcal{S}$ be a φ-uniform $EpG_\alpha(s,t)$. Let $\beta_1 = \varphi(t(\varphi-1)+\alpha)/\alpha$ and $\beta_2 = (s+1)(\varphi(t+1)-s-t-1+\alpha)/\alpha$. It is proved that $\mu(P,Q) \geq \max\{\beta_1, \beta_2\}$ (see [4]), and also $\beta_1 = \beta_2$ iff $\varphi = s+1$ or $\varphi = (s+t+1-\alpha)/t$.

Theorem 4.1. *Let $\mathcal{S}$ be a φ-uniform $EpG_\alpha(s,t)$, $\varphi \leq s$, containing an antiflag (P,y) such that $\varphi(P,y) = \varphi$ and $\mu(P,Q) = \mu(P,R) = \beta_1$ for distinct points Q, R from $y - \mathcal{S}_P$. Then $\mathcal{S}$ is one from the following list:*

1) *a triangular $EGQ(s,t)$, $s \geq t$, or a 6-uniform $EGQ(25,5)$;*

2) *$EpG_t(s,t)$, $t \leq 2$;*

3) *$EpG_2(s,1)$.*

References

[1] Cameron, P. J., Hughes, D. R., Pasini, A., Extended generalized quadrangles. Geom. Dedic. 35 (1990), 193–228.

[2] Hughes, D. R., Extended partial geometries: dual 2-design. Europ. J. Comb. 11 (1990), 459–472.

[3] Hobart, S. A., Hughes, D. R., Extended partial geometries: nets and dual nets. Europ. J. Comb. 11 (1990), 357–372.

[4] Hobart, S. A. , Hughes, D. R., EpGs with minimal μ, II. Geom. Dedic. 42 (1992), 129–138.

Generators of simple groups and their applications

Jakov N. Nuzhin

Abstract. The paper is devoted to a method of choosing generators of finite simple groups. This method is based on the description of subgroups of Lie type groups, whose intersections with all root subgroups are nontrivial. Using only this method, for majority of groups of Lie type we give answers to some problems related to the search of generators of particular order. The method of rigidity and our method are applied to solve the inverse Galois problem for Weyl groups and for $PSL_n(p)$, where $n \geq 5$, p is an odd prime number and $(n, p-1) = 1$.

1991 Mathematics Subject Classification: 20D05, 20D06, 12F12.

Introduction

Many problems in group theory are reduced to the search of generators of finite simple groups. In applications, the generators are usually required to be in an explicit form. Therefore, it is important to have a sufficient number of methods to choose generators. The paper is devoted to one of such methods and its applications. This method is based on the description of subgroups of groups of Lie type, whose intersections with all root subgroups are nontrivial. The description was given by V. M. Levchuk [1] and the author [2]. It is, in turn, based on the well-known Dickson's theorem, its strengthening [3], and its analogs for the other three types of groups of Lie type of rank 1 [4].

Using only this method for majority of groups of Lie type we give answers to some problems related to the search of generators of specific orders. On the other hand, together with the method of rigidity developed by G. V. Belyi, M. Fried, B. Matzat, and J. Thompson, this method is applied to solve inverse Galois problem for Weyl groups and for $PSL_n(p)$, where $n \geq 5$, p is an odd prime number and $(n, p-1) = 1$.

The author would like to thank Professor John Wilson for providing copies of still unpublished papers.

1. Some problems and their state

As far back as the end of the last century, the following well-known problems were in the area of interests of many investigators.

Problem 1. Is a finite nonabelian simple group generated by two elements?

Problem 2. Is a finite nonabelian simple group generated by two elements, one of which is an involution?

Problem 3. Which finite nonabelian simple groups are generated by an involution and an element of order 3?

If a group is generated by an element of order m and an element of order n, then we call it (m, n)-*generated*. The well-known Dickson theorem [5] gives pairs of generators of $PSL_2(q)$ for odd q. G. A. Miller [6] has shown every simple alternating group A_n to be $(2, m)$-generated for any integer $m \geq 4$ such that A_n has an element of order m. R. Steinberg [7] found pairs of generators for each group of Lie type. It follows, from results of M. Aschbacher and R. Guralnick [8], that every sporadic group can be generated by two elements. Thus, Problem 1 received the affirmative solution (modulo the classification of finite simple groups). Of course, pairs of generating elements of finite simple groups were found in many other works, for example, in works concerning free groups to be residually in the class of finite simple groups. We mention only the works covering sufficiently wide classes of groups. Thus, A. A. Albert and J. G. Thompson [9] indicated pairs of generators for $PSL_n(q)$, and T. G. Room and R. J. Smith [10], T. G. Room [11], and P. Stanek [12] for $PSp_{2n}(q)$. The author [13, 14] indicated pairs of generators for each group of Lie type of rank $l \geq 3$, one of which is monomial and the other unipotent (see Section 4).

Problem 2 was positively solved in the above-mentioned papers of G. A. Miller (for A_n), A. A. Albert and J. G. Thompson (for $PSL_n(q)$), T. G. Room and P. Stanek (for $PSp_{2n}(q)$), the author (for each group of Lie type of rank $l \geq 4$). The last point here was posed by G. Malle, J. Saxl, and T. Weigel [15] (see also [16]). Therefore, Problem 2 received the affirmative solution as well.

In contrast to the first two problems, Problem 3 has no uniform solution, it is considered fully only for the cases of alternating and the sporadic groups. G. A. Miller [17] has shown that the alternating group A_n, $n \geq 5$, is $(2, 3)$-generated iff $n \neq 6, 7, 8$ (see also the paper of I. M. S. Dey and J. Wiegold [18]). A. J. Woldar [19] proved that a sporadic group is $(2, 3)$-generated iff it is not M_{11}, M_{22}, M_{23} and McL. Problem 3 received the affirmative answer for some groups of Lie type of sufficiently large rank and some groups of low rank, too. For details, see the review of J. S. Wilson [16], where the conjecture on finite simple groups of Lie type to be $(2, 3)$-generated except for some groups of low rank in characteristic 2 and 3 is stated also.

The following theorem of M. C. Tamburini and J. S. Wilson [20] yields pairs of generators of prescribed orders: let G be $PSL_n(q)$ or A_n and A, B be nontrivial subgroups of G. If $|A||B| \geq 12$, then for each $n \geq 4|A||B| + 12$, the group G has subgroups $A^* \simeq A$ and $B^* \simeq B$ such that $G = \langle A^*, B^* \rangle$ (see also [21]).

The following two problems are closely related to the ones noted above.

Problem 4. Which finite simple groups are generated by three involutions?

Problem 5 (Mazurov; see [22]). Which finite simple groups are generated by three involutions, two of which commute?

Indeed, let G be a finite nonabelian simple group. If G is generated by an involution a and an element b of order 3, then $\langle a, a^b, a^{b^2} \rangle$ is a nontrivial normal subgroup of G. Thus, G is generated by three involutions a, a^b, a^{b^2}. On the other hand, if G is generated by three involutions a, b, c (and $ab = ba$), then $\langle ab, ca \rangle$ is a nontrivial normal subgroup of G. Hence G is generated by two elements ab and ca (and ab is an involution).

If Problem 4 (Problem 5) is solved affirmatively for a group G, then the group G is said to be $(2, 2, 2)$-*generated* ($(2 \times 2, 2)$-*generated*).

Thus, by the result of G. A. Miller [17], A_n, $n \neq 6, 7, 8$, is (2,2,2)-generated. F. Dalla Volta [23] has shown that every sporadic group is (2,2,2)-generated. For special classes of linear groups, Problem 4 was solved by many authors (also by those who were solving Problem 3). Solving a stronger Problem 5, the author [13, 24] has shown that, in particular, every simple group of Lie type over a finite field of characteristic 2 and every simple group of Lie type of rank $l \geqslant 4$ over a field of odd characteristic are (2,2,2)-generated. The last point to Problem 4 was added by G. Malle, J. Saxl and T. Weigel [15] (see also [16]), namely the following theorem holds: except $PSU_3(9)$, every finite nonabelian simple group can be generated by three involutions. It was shown by A. Wagner [25] that $PSU_3(9)$ can not be generated by three involutions.

As for Problem 5, the answer is unknown only for sporadic groups and for groups of Lie type of ranks 2 and 3 over finite fields of odd characteristic, see [13, 14, 24] and Section 5 of the present paper.

From the results of M. C. Tamburini, J. S. Wilson, and N. Gavioli [26] (see also [16]), it follows that the direct powers $(PSL_n(q))^r$ and $(PSp_{2n}(q))^s$ are (2,3)-generated, where $r = q^{[(n-28)/12]}$ and $s = q^{[(n-37)/12]}$. This corollary is interesting in connection with the next problem.

Problem 6 (Syskin; see [22]). For each known finite simple group G find a maximal integer n that the direct product of n copies of G is generated by two elements.

2. Generators of groups of Lie type of rank 1

The following theorem is well-known.

Theorem 1 (Dickson [5]). *Let t be a generator of $GF(q)$, q odd, and*

$$L = \left\langle \begin{pmatrix} 1 & 0 \\ t & 1 \end{pmatrix}, \begin{pmatrix} 1 & 1 \\ 0 & 1 \end{pmatrix} \right\rangle .$$

Then we have either $L = SL_2(q)$ or $q = 9$ and $L/Z(L) \simeq A_5$.

Using the methods of Dickson and the theorem of Schur that a periodic linear group over a field is locally finite, V. M. Levchuk [3] described periodic linear groups of degree 2 over a field. In particular, when the base field is finite, from [3] we have

Proposition 1. *Let t be a generator of $GF(q)$, V an additive subgroup of $GF(q)$, $t \in V$ and*

$$L = \left\langle \begin{pmatrix} 1 & 0 \\ V & 1 \end{pmatrix}, \begin{pmatrix} 1 & 1 \\ 0 & 1 \end{pmatrix} \right\rangle .$$

Then one of the following occurs:

1) $L = SL_2(q)$;

2) $|V| = 3$ *and* $L/Z(L) \simeq A_5$;

3) $|V| = 2$ *and L is a dihedral group.*

Moreover, in all cases for some s, the subgroup L contains the monomial element

$$\begin{pmatrix} 0 & -s^{-1} \\ s & 0 \end{pmatrix} .$$

By Proposition 1, it is easy to choose generating sets (triples) from unitriangular matrices of $SL_2(q)$ for the case of even q, too. The inclusion of the monomial element is a strengthening of the Dickson theorem, in particular, in the case when q is odd. This is useful in applications of Proposition 1 to groups of Lie type.

Taken together, the groups $SL_2(q)$, Ree groups ${}^2G_2(q)$, Suzuki groups ${}^2B_2(q)$, and $SU_3(q)$ form the class of groups of Lie type of rank 1. They play the role of "building blocks" for the groups of Lie type. By virtue of the results of C. Hering, W. M. Kantor, and G. Seitz [27], the simple groups of Lie type of rank 1 form a subclass of the class of doubly transitive permutation groups. The doubly transitive permutation representation of $PSL_2(q)$ was essentially used in the proof of

Dickson's theorem (and Proposition 1) and it is natural to expect its analogs to be valid for the other three classes of groups of Lie type of rank 1. These analogs were obtained in [4], where the author has described the subgroups of the groups ${}^2G_2(q)$, ${}^2B_2(q)$, and $SU_3(q)$ having nontrivial intersections with both the upper and lower unitriangular subgroups of the indicated groups in their natural matrix representations (see also [29]). We formulate only a proposition, which follows from the description.

Proposition 2. *Let t be a generator of $GF(q)$. Then*

1) ${}^2G_2(q) = \langle \gamma'(t), \gamma(1)\rangle$ *if* $q > 3$;

2) ${}^2B_2(q) = \langle \alpha'(t), \beta(1)\rangle$;

3) $SU_3(q) = \langle (o,u)', (1,t) | u \neq 0\rangle$ *if* $q \geq 4$.

Here (as in [29]),

$$\gamma(t) = x_{2a+b}(t^\theta)x_{3a+b}(t),$$
$$\gamma'(t) = x_{-2a-b}(t^\theta)x_{-3a-2b}(t),$$
$$\beta(t) = x_{a+b}(t)x_{2a+b}(t^{2\theta}),$$
$$\alpha'(t) = x_{-a}(t^\theta)x_{-b}(t)x_{-a-b}(t^{\theta+1}),$$

$$(t,u) = \begin{pmatrix} 1 & t & -u^\theta \\ 0 & 1 & t^\theta \\ 0 & 0 & 1 \end{pmatrix}, \qquad u + u^\theta = -tt^\theta,$$

$$(t,u)' = \begin{pmatrix} 1 & 0 & 0 \\ t & 1 & 0 \\ u & t^\theta & 1 \end{pmatrix}, \qquad u + u^\theta = tt^\theta,$$

where θ is the automorphism of $GF(q)$ which is used in the construction of the three indicated groups of Lie type of rank 1.

By Dickson's theorem for odd $q \neq 9$ the group $SL_2(q)$ is generated by the matrices

$$\begin{pmatrix} 1 & 0 \\ t & 1 \end{pmatrix}, \quad \begin{pmatrix} 0 & -1 \\ 1 & 0 \end{pmatrix}$$

if t and t^2 are generators of $GF(q)$. Hence, $SL_2(q)$ is $(2, p)$-generated, where p is the characteristic of the field $GF(q)$. Similarly, it follows from Proposition 2 that ${}^2G_2(q)$ is (2,3)-generated for $q > 3$

3. Generators of Weyl groups

Here and in the sequel, Φ denotes an indecomposable root system with a basis $\Pi = \{r_1, \dots, r_l\}$, where r_1 is a short root and W is the Weyl group of type Φ which is generated by the fundamental reflections w_{r_i}. We also fix the following disposition of the fundamental roots on the Dynkin diagram:

A_l

B_l

C_l

D_l

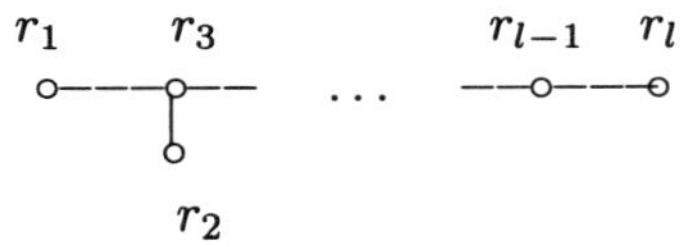

E_l

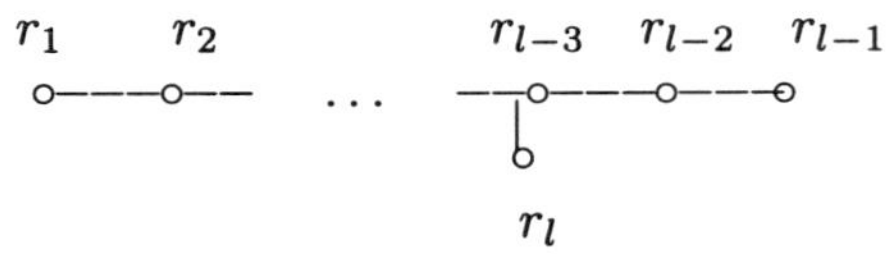

F_4

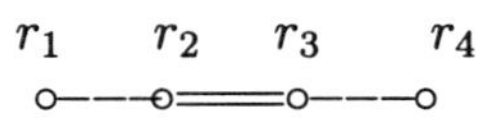

G_2

For $l \geq 2$ the set Π may be divided into two disjoint subsets Π_1 and Π_2, each containing pairwise orthogonal roots. Let $r_1 \in \Pi_1$, and let w_{Π_i} be the product of all w_{r_j}, $r_j \in \Pi_i$. Note that the elements w_{Π_i} are uniquely defined and of order 2, since each of them is a product of reflections and with respect to orthogonal roots.

Further, Coxeter elements of the Weyl group W (i.e. elements of the form $w_{s_1} w_{s_2} \dots w_{s_l}$, where $s_1, \dots, s_l$ is a fundamental system in Φ), play an essential role and we fix two such elements:

$$c = w_{r_1} w_{r_2} \dots w_{r_l}, \quad c_1 = w_{\Pi_1} w_{\Pi_2}.$$

Any two Coxeter elements are conjugate and each Coxeter element can be included in a dihedral subgroup of the Weyl group. In particular, the element c_1 is contained in the dihedral subgroup generated by involutions w_{Π_1} and w_{Π_2}. If $l = 2$ then $c_1 = c$. The following technical but important lemma gives the information about the element w for which $wc_1w^{-1} = c$ in the case $l \geq 3$.

Lemma 1 [24]. *Let $l \geq 3$, $\delta_j = 0$ or 1. Let $k = 2$ if $\Phi = D_l$, $l \geq 4$, and $k = 1$ otherwise. Then there exists an element $w \in \langle w_{r_i} \mid i \geq k+3 \rangle w_{r_{k+2}}$ such that*

$$c_1^w = wc_1w^{-1} = c,$$

$$w\Big(\sum_{j=1}^{k+1} \delta_j r_j\Big) = \sum_{j=1}^{k} \delta_j r_j + \delta_{k+1} \sum_{j=k+1}^{l} r_j,$$

$$w(r_{k+2}) = - \sum_{j=k+2}^{l} r_j.$$

Lemma 2 (Steinberg [7]). *We have $W = \langle c, w_r \mid r \in A \rangle$, where*

$$A = \begin{cases} \{r_1, r_2\} & \text{if } \Phi = B_l,\ l \geq 3, D_l, l \text{ even}, \\ \{r_{l-1}, r_l\} & \text{if } \Phi = C_l,\ l \geq 3, \\ \{r_2, r_3\} & \text{if } \Phi = F_4, \\ \{r_1\} & \text{in all other cases.} \end{cases}$$

If $\Phi = B_l$, $l \geq 3$, or $\Phi = C_l$, $l \geq 3$, then the set A may also be a singleton.

Lemma 3 [24]. *The group W of the types B_l or C_l is generated by two elements c and w_{r_i}, where r_i is any fixed fundamental root, and $i \neq 1$ for $\Phi = B_l$, and $i \neq l$ for $\Phi = C_l$.*

The next follows from Lemmas 1, 2, and 3.

Proposition 3 [24]. *Let Φ be different from A_1, B_l if $l \geq 3$, D_l if l is even, and F_4. Then W is generated by the three involutions w_{r_1}, w_{Π_1}, w_{Π_2}, the first two of which commute.*

Proposition 4 [30]. *Let $s_i = w_{r_l} w_{r_{l-1}} \cdots w_{r_{i+1}}(r_i)$ $i = 1, \ldots, l$, Φ_i be an orbit of Φ under $\langle c \rangle$ and $s_i \in \Phi_i$. Then the sets Φ_i are mutually disjoint and form the whole group of orbits of Φ under $\langle c \rangle$.*

4. Generators of groups of Lie type of rank $l \geq 2$

In the sequel, $G = G(q)$ denotes a group of Lie type of rank l over $GF(q)$ associated with the root system Φ. If G is of type ${}^2A_{2l}$, ${}^2A_{2l-1}$, ${}^2D_{l+1}$, ${}^2E_6$, ${}^3D_4$, then we relate to G the root system $\Phi = B_l, C_l, B_l, F_4, G_2$ respectively. Therefore the element c and the set A from Section 3 are defined naturally in these cases also. Notation for subgroups and elements of G is standard, in particular, such as in [29]. We note only that we will denote by n_w one of preimages of $w \in W$ in the monomial subgroup N of G, the coefficients of which lie in the prime subfield of $GF(q)$.

For example, if the base field is finite, Theorems 1 and 2 of [1] give

Proposition 5. *Suppose that $l \geq 3$, the type of G is different from ${}^2A_{2l}$ and ${}^2F_4$, M is a subgroup of G, and t and t^2 are generators of $GF(q)$. Suppose also that $M \cap X_r \neq 1$, $r \in \Phi$, $x_r(1) \in M$, $r \in \Pi$, and $x_s(t) \in M$ for some $s \in \Phi$. Then $M = G$.*

Considering Proposition 5 from the point of view of choosing the generators of G, we have an abundant number of generators, equal to $|\Phi|$. The author [14] proposed the following economical method for the choice of generators of G using the element $c = w_{r_1} w_{r_2} \cdots w_{r_l}$ and the subset $A \subset \Phi$ from Section 3.

Theorem 2. *Let $l \geq 2$, the type of G be different from ${}^2A_{2l}$ and ${}^2F_4$, M be a subgroup of G, a be a smallest (short) root of A and, t and t^2 be generators of $GF(q)$. Suppose that for all $r \in A$ the intersections $M \cap X_r$ are nontrivial, $n_c \in M$, $x_a(t) \in M$, and if the type of G is equal to A_l, D_{2k-1}, E_6, ${}^2A_{2l-1}$, ${}^2D_{2k-1}$, or ${}^2E_6$, we have $x_{-a}(t) \in M$. Then $M = G$.*

Proof. We give only a sketch of the proof. Using the homomorphism of $SL_2(q)$ onto $\langle X_r, X_{-r} \rangle$ and Proposition 1, we have $n_r(u_r) \in M$ for $r \in A$ and some $u_r \in GF(q)$. Therefore, by Lemma 2, $n_r(u_r) \in M$ for all $r \in \Phi$. Since W acts transitively on a set of roots of equal length, all intersections $M \cap X_r$, $r \in \Phi$, are nontrivial. Apply Proposition 5 to finish the proof. □

Remark 1. Usually, the groups ${}^2A_{2l}(q)$ and ${}^2F_4(q)$ are required to be considered exceptionally by virtue of a more complex structure of their root subgroups. So they are excluded in Proposition 5 and in Theorem 2 as well. However, using in addition analogs of the Dickson theorem for the groups $SU_3(q)$ and ${}^2B_2(q)$, we can obtain analogs of Proposition 5 also for the groups ${}^2A_{2l}(q)$ and ${}^2F_4(q)$. (For groups ${}^2A_{2l}(q)$ this statement was obtained in [2].) We can consider the following statement as an analog of Theorem 2 for these exceptional groups: the groups ${}^2A_{2l}(q)$ and ${}^2F_4(q)$ are generated by the elements $x_{r_1}(1,t)$ and n_c (for notation see [24]).

The following theorem gives pairs of generators of G, one of which is monomial and the other unipotent of order p or p^2, where $q = p^n$, p prime; the orders of these generators depend on p and l but not on n, in contrast to [7].

Theorem 3 [14]. *Let G be a group of Lie type of rank $l \geq 3$ over $GF(q)$, t and t^2 be generators of $GF(q)$, $c = w_{r_1}w_{r_2}\dots w_{r_l}$. Then G is generated by elements n_c and α, where*

$$\alpha = \begin{cases} x_{r_1}(t)x_{-r_1-r_2-\dots-r_l}(1) & \textit{if the type of } G \textit{ is } A_l, D_l, E_6, \textit{or } {}^2A_{2l-1}, {}^2D_{l+1}, {}^2E_6, \\ x_{r_1}(t)x_{-r_l}(1) & \textit{if } p=2 \textit{ and the type of } G \textit{ is } B_l, C_l, \textit{ or } F_4, \\ x_{r_1}(t) & \textit{if the type of } G \textit{ is } E_7, E_8 \textit{ or } p>2 \\ & \quad \textit{and the type of } G \textit{ is } B_l, C_l, F_4, \\ x_{r_1}(1,t) & \textit{if the type of } G \textit{ is } {}^2A_{2l}. \end{cases}$$

The proof is reduced to verification of the conditions of Theorem 2.

5. Generating triples of involutions

In this section we concentrate on Problem 5 from Section 1. A group is said to be $(2\times 2, 2)$-*generated* if Problem 5 is solved positively for it.

We begin with groups of Lie type over a field of characteristic 2. In this case, the following theorem gives an exhaustive answer.

Theorem 4 [24]. *A simple group of Lie type over a finite field of characteristic* 2 *is $(2\times 2,2)$-generated iff its type is different from A_2, ${}^2A_2$, A_3, ${}^2A_3$.*

In [24], generating triples of involutions are indicated explicitly. For example, if the type of G is E_7 or E_8, then G is generated by involutions

$$\alpha = x_{r_1}(t), \qquad \beta = n_{\Pi_1}n_{w_0}, \qquad \gamma = n_{w_0}n_{\Pi_2},$$

where t is a generator of $GF(q)$, Π_i as in Section 3, $w_0(\Pi) = -\Pi$, and, moreover $\alpha\beta = \beta\alpha$. Really, let w be as in Lemma 1. Then $\alpha^{n_w} = \alpha$, $(\beta\gamma)^{n_w} = n_c$. The only thing we should do is to apply Theorem 3. It is the easiest case, but the concept of the proof is remained the same.

Just as in the proof of Theorem 4 we can show that the groups $A_2(q)$, $A_3(q)$, ${}^2A_3(q)$ are generated by the three involutions $x_{r_1}(t)$, n_{Π_1}, n_{Π_2} and the group

${}^2A_2(q)$ by the involutions $x_r(0,1)$, $x_{-r}(0,1)$, $n_r^{x_r(1,t)}$ (none two of which commute). Thus, we have

Corollary 1. *Every simple group of Lie type over a finite field of characteristic* 2 *is generated by three involutions.*

In characteristic $p > 2$, the following incomplete result was obtained (see [4, 13]), its proof, like Theorem 4, was based on the method of choosing the generators, mentioned above in Sections 3 and 4.

Theorem 5. *Let G be a simple group of Lie type of rank l over a finite field of odd characteristic. Then*

1) *if $l \geq 4$, then G is $(2 \times 2, 2)$-generated;*

2) *if $l = 1$, then G is $(2 \times 2, 2)$-generated iff G is different from $PSL_2(7)$, $PSL_2(9)$, and $PSU_3(9)$.*

Note that three exceptional groups in (2) are isomorphic to $PSL_3(2)$, A_6, $G_2(2)'$ respectively.

From the result of M. D. E. Conder [31], for sufficiently large n the alternating groups A_n are $(2 \times 2, 2)$-generated. In general case, for the groups A_n Problem 5 was solved in [32], namely, the following theorem is valid.

Theorem 6. *The alternating group A_n is $(2 \times 2, 2)$-generated, iff $n \neq 6, 7, 8$.*

Thus, an answer to Problem 5 is unknown only for sporadic groups and groups of Lie type of ranks 2 and 3 over a finite field of odd characteristic.

6. Inverse Galois problem and the method of rigidity

The following problem is among the well-known and oldest problems.

Inverse Galois problem. Can a finite group G be realized as a Galois group over the field of rational numbers Q?

Since the positive solution of the problem for solvable groups (I. R. Shafarevich, 1954), the simple groups are of exceptional interest. In the case, the (positive) answer has been obtained only for alternating groups (D. Hilbert, 1892), sporadic groups (except for the Mathieu group M_{23}), and some Chevalley groups of low rank l (either $l \leq 2$ or the type of G is equal to E_8) over a prime field $GF(p)$ with some additional restrictions on p. For details, see review of J.-P. Serre [33] (see also [34], p. 11–13).

Last achievements in solving the inverse Galois problem are mainly connected with the method of rigidity, which has been developed by G. V. Belyi, M. Fried, B. Matzat, and J. Thompson (see [33]). This method is applied in the next two sections, too. We explain its essence shortly.

Let $C_1, \ldots, C_k$, $k \geq 3$, be conjugacy classes of a finite group G. Denote by $P = P(C_1, \ldots, C_k)$ the set of $(g_1, \ldots, g_k) \in C_1 \times \cdots \times C_k$ such that $g_1 g_2 \ldots g_k = 1$ and $G = \langle g_1, \ldots, g_k \rangle$. The family $(C_1, \ldots, C_k)$ is said to be *rigid* if G acts transitively on P by conjugation and P is not empty.

By definition, a conjugacy class in G is said to be *rational* if the following equivalent conditions hold:

1) for any (complex) character χ of G and $c \in C$ the number $\chi(c)$ is rational;

2) if $c \in C$ and $(i, |c|) = 1$, then $c^i \in C$.

Denote by $\mathrm{Gal}_T(G)$ the property of a finite group G to be realizable as the Galois group of a regular Galois extension of the field of rational functions $Q(T)$ over Q. From the property $\mathrm{Gal}_T(G)$, inverse Galois problem is solved positively for G (see [33], where the list of simple groups G with the property $\mathrm{Gal}_T(G)$ can be found).

Now, we can formulate the main result of the method of rigidity.

Proposition 6. *Let G be a finite group with trivial center. If G has a rigid family of rational conjugacy classes, then the property $\mathrm{Gal}_T(G)$ is valid.*

For $k = 3$, a more strong statement follows from the results of G. V. Belyi [35].

Proposition 7. *Suppose that for a finite group G the following two properties hold:*

1) *G has a rigid rational triplet $(C_a, C_b, C_{b^{-1}a^{-1}})$,*

2) *$Z(G)$ is a direct factor of $N_G(\langle a \rangle)$.*

Then the property $\mathrm{Gal}_T(G)$ is valid.

Here C_x is a conjugacy class with a representative x, $N_G(A)$, $C_G(A)$ are the normalizer and the centralizer of a subgroup A in G, and $Z(G)$ is the center of G.

Verification of rigidity of some conjugacy classes (even without the condition of being rational) is a rather complicated task. (The author knows examples of rigid families for $k = 3$ only.) However, G. V. Belyi [35] offered the following elegant criterion of rigidity.

Proposition 8. *Let K be a field and G a finite group. Let the embedding $G \subset GL_{l+1}(K)$ be an irreducible representation, and the following conditions hold:*

1) *G is generated by two matrices a and b;*

2) *for some $\xi \in K$ the rank of matrix $a - E_{l+1}$ is 1;*

3) $N_{GL_{l+1}}(G) = GC_{GLl+1}(G)$.

Then the family $(C_a, C_b, C_{b^{-1}a^{-1}})$ is rigid.

Note that there are finite simple groups without rigid rational triplets, for example, the alternating group A_5. Recently, H. Völklein [36] has shown that the method of rigidity (for $k = 3$) can not be applied to the groups $PSL_2(p^n)$, $n \geq 9$.

7. Inverse Galois problem for Weyl groups

Using the method of rigidity and the results of Section 3, we prove

Theorem 7. *If the type of the Weyl group W is different from D_{2k} and F_4, then the property* $\mathrm{Gal}_T(G)$ *is valid.*

Assume, as in Theorem 7, that the type of W is different from D_{2k} and F_4. The Weyl groups of the types B_l and C_l are isomorphic, therefore we suppose that the type of W is not equal to B_l, too. Let C_1, C_2, C_3 be conjugacy classes of W with representatives w_{r_1}, c, $c^{-1}w_{r_1}$, respectively (for notation, see Section 3).

It is known that every conjugacy class of the Weyl group is rational, therefore in view of Proposition 7, Theorem 7 is a consequence of the two following lemmas.

Lemma 4. *The family (C_1, C_2, C_3) is a rigid triplet of W.*

Proof. In view of Lemmas 2 and 3, $W = \langle w_{r_1}, c\rangle$. Let $a_i \in C_i$ and $W = \langle a_1, a_2, a_3\rangle$. Without loss of generality, suppose that $a_2 = c$ and $a_1 = w_r$ for some root $r \in \Phi$. It is sufficient to show that $c^i(r) = r_1$ for some $i \in Z$. □

In view of Proposition 4, we can take the following elements as representatives of the orbit of $\langle c\rangle$ in Φ:

$$r_1 + \cdots + r_l,\ r_2 + \cdots + r_l,\ \ldots\ , r_l \quad \text{for} \quad \Phi \neq D_l\ ,\ E_l;$$
$$r_1 + r_3 + \cdots + r_l,\ r_2 + r_3 + \cdots + r_l,\ r_3 + \cdots + r_l,\ \ldots\ ,\ r_l \quad \text{for} \quad \Phi = D_l;$$
$$r_1 + \cdots + r_l,\ \ldots\ ,\ r_{l-3} + \cdots + r_l,\ r_{l-2} + r_{l-1},\ r_{l-1},\ r_l \quad \text{for} \quad \Phi = E_l.$$

Hence, we can assume that r is equal to one of the roots indicated above.

Let the root r be equal to:

$$r_1 + \cdots + r_l \quad \text{or} \quad r_l \quad \text{for} \quad \Phi = A_l;$$

$$r_1 + r_3 + \cdots + r_l \quad \text{or} \quad r_2 + r_3 + \cdots + r_l \quad \text{for} \quad \Phi = D_l \ ;$$
$$r_1 + \cdots + r_l \quad \text{or} \quad r_5 \quad \text{for} \quad \Phi = E_6,$$
$$r_1 + \cdots + r_l \quad \text{in all other cases .}$$

Then $c^i(r) = r_1$, where i is $1, 2, 1, l, 1, 6, 1$, respectively.

In the other cases, either w_r is not conjugate to w_{r_1} or $w_r c$ is not conjugate to $w_{r_1} c$. These cases are impossible. Indeed, if $r = r_i + \cdots + r_l$, where $2 \le i \le l-1$ and $i \ne 2$ for $\Phi = D_l$, or $r = r_{l-2} + r_{l-1}$ for $\Phi = E_l$, then respectively, we have

$$(w_r c)^c = w_{r_1 + \cdots + r_l} c = w_{r_1} \dots w_{r_{i-1}} w_{r_{i+1}} \dots w_{r_l},$$
$$(w_r c)^c = w_{r_1 + \cdots + r_{l-2}} c = w_{r_1} \dots w_{r_{l-3}} w_{r_{l-1}} w_{r_l}.$$

Evidently, the order of the element $(w_r c)^c$ is not equal to the order of the element $w_{r_1} c = w_{r_2} \dots w_{r_l}$ and therefore, they can not be conjugate. Finally, if $r = r_l$ for $\Phi \ne A_l$ or $r = r_{l-1}$ for $\Phi = E_7, E_8$, then the element $w_r c$ is conjugate to $c w_{r_l} = w_{r_1} \dots w_{r_{l-1}}$ and $c w_{r_{l-1}} = w_{r_1} \dots w_{r_{l-2}} w_{r_l}$ respectively. For $l \ne 2$, the elements $w_r c$ and $w_{r_1} c$ have distinct orders, since they are conjugate to Coxeter element of Weyl groups of distinct types. For $l = 2$, they coincide with reflections relative to roots of distinct lengths, and in any case they can not be conjugate. The lemma is proved.

Lemma 5. *The center $Z(W)$ is a direct factor of $N_W(\langle w_{r_1} \rangle)$ or $N_W(\langle c \rangle)$.*

Proof. If $\Phi = A_l, D_l$ (l odd), or E_6, then $Z(W) = 1$, and in this case the statement of the lemma is trivial. If $\Phi = E_7, E_8$, or G_2, then $N_W(\langle c \rangle)$ is a semidirect product, whose kernel is $\langle c^{|c|/2} \rangle \times \langle c^2 \rangle$ and the equality $\langle c^{|c|/2} \rangle = Z(W)$ holds. Hence $Z(W)$ is a direct factor of $N_W(\langle c \rangle)$. If $\Phi = C_l$, then

$$N_W(\langle w_{r_1} \rangle)$$
$$= Z(W) \times (\langle w_{2r_3 + \cdots + 2r_{l-1} + r_l}, \dots, w_{2r_{l-1} + r_l}, w_{r_l} \rangle \lambda (\langle w_{r_1} \rangle \times \langle w_{r_3}, \dots, w_{r_{l-2}} \rangle)).$$

The lemma is proved. □

In view of the isomorphisms $W/Z(W) \simeq C_3(2)$ for $\Phi = E_7$, and $W'/Z(W) \simeq D_4(2)$ for $\Phi = E_8$, and the remarks on page 275 in [35] we have

Corollary 2. *If G is the Chevalley group $C_3(2)$ or $D_4(2)$, then the property* $\mathrm{Gal}_T(G)$ *is valid.*

8. Inverse Galois problem for $PSL_n(p)$

In this section we prove a theorem which was announced in [37].

Theorem 8. *Let $G = PSL_{l+1}(p)$, where $l \ge 4$, p is an odd prime number and $(l+1, p-1) = 1$. Then the property* $\mathrm{Gal}_T(G)$ *is valid.*

In view of the restriction $(l+1,p-1)=1$ in Theorem 8, $G=PSL_{l+1}(p)=SL_{l+1}(p)$. Hence, we have an irreducible representation of G over $GF(p)$; besides, the condition 3) of Proposition 8 is true. Therefore by Propositions 7 and 8, we need only find matrices a and b such that the conditions 1) and 2) of Proposition 8 are valid and the classes C_a, C_b, C_{ab} are rational. (Note that the inclusion $x^{-1}\in C_x$ follows from a class C_x being rational.)

We take

$$a=\begin{pmatrix} 1 & 0 & 0 & \dots & 0 & 0\\ 1 & -1 & 0 & \dots & 0 & 0\\ 0 & 0 & -1 & \dots & 0 & 0\\ \vdots & \vdots & \vdots & \ddots & \vdots & \vdots\\ 0 & 0 & 0 & \dots & -1 & 0\\ 1 & 0 & 0 & \dots & 0 & -1 \end{pmatrix},\quad b=\begin{pmatrix} 0 & 0 & 0 & \dots & 0 & 1\\ 1 & 0 & 0 & \dots & 0 & 0\\ 0 & 1 & 0 & \dots & 0 & 0\\ \vdots & \vdots & \vdots & \ddots & \vdots & \vdots\\ 0 & 0 & 0 & \dots & 0 & 0\\ 0 & 0 & 0 & \dots & 1 & 0 \end{pmatrix}.$$

Further, along with the matrix notation we use the terminology of Chevalley groups, considering G as a Chevalley group of the type A_l. Since a Chevalley group of the type A_l over $GF(p)$ is generated by root elements $x_{\pm r_i}(1)$, $i=1,\dots,l$, where r_i are fundamental roots, then it is sufficient to indicate matrices corresponding to these elements. Let

$$x_{r_i}(1)=E_{i+1,i}+E_{l+1},\quad x_{-r_i}(1)=E_{i,i+1}+E_{l+1},\quad i=1,\dots,l\ ,$$

where E_{l+1} is the identity matrix of degree $l+1$, and $E_{i,j}$ is the matrix whose (i,j)-entry is 1 and all other entries are zero. As usually,

$$n_r(t)=x_r(t)x_{-r}(-t^{-1})x_r(t)\ ,\quad n_r=n_r(1)\ ,\quad h_r(t)=n_r(t)n_r(-1)\ .$$

Also set

$$h_i=\operatorname{diag}[-1,\dots,-1,1,-1,\dots,-1]\ ,$$

where 1 is situated on the ith position. With the preceding notation

$$a=h_1x_{r_1}(-1)x_{r_1+\dots+r_l}(-1)\ ,\quad b=n_{r_1}\dots n_{r_l}\ .$$

The rank of the matrix $a+E_{l+1}$ is equal to 1 and therefore Theorem 8 follows from the two following lemmas.

Lemma 6. *The classes C_a, C_b, C_{ab} are rational.*

Proof. The class C_a is rational because a is an involution.

Show that C_b is rational. Let $(i,|b|)=1$. The element b^i is conjugate in G to

$$n_{r_1}(\pm1)\dots n_{r_l}(\pm1) \tag{1}$$

since $\langle n_{r_i}\mid i=1,\dots,l\rangle/H\simeq S_{l+1}$, where $H=\langle h_{r_i}(-1)\mid i=1,\dots,l\rangle$ and S_{l+1} is the symmetric group. Conjugating (1) by a suitable element from H, we have

$$n_{r_1}(\varepsilon)n_{r_2}\dots n_{r_l}=h_{r_1}(\varepsilon)n_{r_1}n_{r_2}\dots n_{r_l}\ ,\quad \varepsilon=\pm1, \tag{2}$$

If $\varepsilon = 1$, then the lemma is proved. Let $\varepsilon = -1$. Conjugating (2) by the element $h_{r_2}(-1)h_{r_4}(-1)\dots h_{r_l}(-1)$, we have

$$h_{r_1}(-1)\dots h_{r_l}(-1)h_{r_1+\dots+r_l}(-1)n_{r_1}\dots n_{r_l} = n_{r_1}\dots n_{r_l}.$$

Hence, the class C_b is rational.

Show that C_{ab} is rational. Conjugating ab by the element $x_{r_1+\dots+r_l}(-1)$, we have

$$g = x_{-r_1}(1)h_1 n_{r_2}\dots n_{r_l}.$$ □

Remark 2. The monomial factor $n = h_1 n_{r_2}..n_{r_l}$ of the element g corresponds to the (0,1)-matrix multiplied by the scalar matrix $(-E_{l+1})$. This fact facilitates further calculations, namely, we skip the constants $\eta_{r,s} = \pm 1$, which arise from the equalities

$$n_r x_s(1) n_r^{-1} = x_{w_r(s)}(\eta_{r,s}).$$

If we conjugate the elements $x_s(1)$ by the element n, then the product of the constants $\eta_{r,s}$ is equal to 1.

The following equality easily follows from Remark 2:

$$g^l = x_{-r_1}(1)x_{-r_1-r_2}(1)\dots x_{-r_1-\dots-r_l}(1).$$

Hence, $|g| = pl$. Let $(i, |g|) = 1$. The element $n_{r_2}\dots n_{r_l}$ centralizes h_1 and l is odd, therefore

$$g^i = x h_1 (n_{r_2}\dots n_{r_l})^i \tag{3}$$

for some x from the subgroup

$$X_1 = \langle x_{-r_1}(1), x_{-r_1-r_2}(1), \dots, x_{-r_1-\dots-r_l}(1)\rangle.$$

Let $N_1 = \langle n_{r_i} \mid i \geq 2\rangle$. Conjugating (3) by a suitable element from the subgroup HN_1, we have

$$y h_1 n_{r_2}\dots n_{r_l} \tag{4}$$

for some $y \in X_1$. Conjugating (4) by a suitable $z \in X_1$, we have

$$x_{-r_1}(k)h_1 n_{r_2}\dots n_{r_l}\ , \tag{5}$$

where $k \neq 0$, since otherwise the order of the element (5) is different from pl. Conjugating (5) by the element $h_{r_1}(t)h_{r_1+r_2}(t)\dots h_{r_1+\dots+r_l}(t)$, we have

$$x_{-r_1}(t^{l+1}k)h_1 n_{r_2}\dots n_{r_l}\ . \tag{6}$$

By hypothesis of the theorem $(l+1, p-1) = 1$, therefore there exists t such that $t^{l+1}k = 1$. So the element (6) is equal to g. Hence, the class C_{ab} is rational. The lemma is proved.

Lemma 7. *The equality $G = \langle a, b\rangle$ is true.*

Proof. Put

$$z = x_{r_1+..+r_l}(-1).$$

$$f = a^z = h_1 x_{r_1}(-1) x_{r_1+\dots+r_l}(1),$$

$$g = (ab)^z = h_1 x_{-r_1}(-1) n_{r_2} \dots n_{r_l},$$

where $x^y = xyx^{-1}$. Evidently, $\langle f, g\rangle = \langle a, b\rangle^z$. Therefore, it is sufficient to show that $\langle f, g\rangle = G$. Since $l \geq 4$, then z centralizes the element

$$d = a^{b^2} = h_3 x_{r_3}(-1) x_{-r_2}(-1) = \begin{pmatrix} -1 & 0 & 0 & 0 \\ 0 & -1 & 1 & 0 \\ 0 & 0 & 1 & 0 \\ 0 & 0 & 1 & -1 \end{pmatrix} \oplus E_{l-3},$$

so that $d \in \langle f, g\rangle$. □

As we have shown in the proof of Lemma 6, we have

$$g^l = x_{-r_1}(1) x_{-r_1-r_2}(1) \dots x_{-r_1-\dots-r_l}(1),$$

or, in matrix notation,

$$g^l = \begin{pmatrix} 1 & 1 & 1 & 1 \\ 0 & 1 & 0 & 0 \\ 0 & 0 & 1 & 0 \\ 0 & 0 & 0 & 1 \end{pmatrix} \oplus E_{l-3} + E_{1,5} + \dots + E_{1,l+1} \ .$$

Calculations (essentially with matrices of degree 4) give that

$$g^{-l} d b^l d = x_{-r_1-r_2}(-4).$$

Conjugating the last element by various powers of g, we have

$$x_{-r_1-\dots-r_i}(1) \in \langle f, g\rangle \ , \qquad i = 1, \dots, l.$$

Hence $n \in \langle f, g\rangle$, where n is as in Remark 2. Taking into account Remark 2, we have

$$(f^n f)^n = x_{r_1}(1) x_{r_1+r_2}(-2) x_{r_1+r_2+r_3}(1).$$

By not so complicated calculations with matrices of degree 4, we have

$$d(f^n f)^n d = x_{r_1}(3) x_{r_1+r_2}(2) x_{r_1+r_2+r_3}(3).$$

Finally,

$$((f^n f)^n)^{-3} d(f^n f)^n d = x_{r_1+r_2}(8),$$

$$n^{-1} x_{r_1+r_2}(8) n = x_{r_1}(8).$$

Thus the elements $x_{\pm r_1}(1)$ and hence $n_{r_1} \dots n_{r_l}$ lie in $\langle f, g\rangle$. By Theorem 2, they generate G. The lemma is proved.

References

[1] Levchuk, V. M., On generating sets of roots elements of Chevalley groups over a field. Algebra i Logika 22 (1983), 526–541.

[2] Nuzhin, Ja. N., On groups lying between groups of Lie type over distinct fields. Algebra i Logika 22 (1983), 526–541.

[3] Levchuk, V. M., Note for theorem of L. Dickson. Algebra i Logika 22 (1983), 504–517.

[4] Nuzhin, Ja. N., On structure of groups of Lie type of rank 1. Matem. Zametki, 36 (1984), 149–158.

[5] Dickson, L. E., Linear Groups with an Exposition of the Galois Field Theory. Teubner 1901.

[6] Miller, G. A., Possible orders of two generators of the alternating and the symmetric group. Trans. Amer. Math. Soc. 30 (1928), 24–32.

[7] Steinberg, R., Generators for simple groups. Canad. J. Math. 14 (1962), 277–283.

[8] Aschbacher, M., Guralnick, R., Some applications of the first cohomology group. J. Algebra 90 (1984), 446–460.

[9] Albert, A. A., Thompson, J. G., Two element generation of the projective unimodular group. Illinois J. Math. 3 (1959), 421–439.

[10] Room, T. G., Smith, R. J., A generation of the symplectic group. Quart. J. Math. 9 (1958), 177–182.

[11] Room, T. G., The generation by two operators of the symplectic group over $GF(2)$. J. Austral. Math. Soc. 1 (1959), 38–46.

[12] Stanek, P., Two-element generation of the symplectic group. Trans. Amer. Math. Soc. 108 (1963), 429–436.

[13] Nuzhin, Ja. N., Generating elements of simple groups. Abstracts of the I Intern. Conf. on Algebra, Novosibirsk 1989, 88.

[14] — Generating sets of elements of Chevalley groups over a finite field. Algebra i Logika 28 (1989), 670–686.

[15] Malle, G., Saxl, J., Weigel, T., Generation of classical groups. Geom. Dedicata, to appear.

[16] Wilson, J. S., Economical generating sets for finite simple groups. In: Proc. of the Conf. Groups of Lie Type and Their Geometries (Italy, June 1993), Cambridge Univ. Press., to appear.

[17] Miller, G. A., On the groups generated by two operators. Bull. Amer. Math. Soc. 7 (1901), 424–426.

[18] Dey, I. M. S., Wiegold, J., Generators for alternating and symmetric group. J. Austral. Math. Soc. 12 (1971), 63–68.

[19] Woldar, A. J., On Hurwitz generation and genus actions of sporadic groups. Illinois J. Math. 33 (1989), 416–437.

[20] Tamburini, M. C., Wilson, J. S., On the generation of finite simple groups by pairs of subgroups. J. Algebra 116 (1988), 316–333.

[21] Di Martino, L., Tamburini, M. C., 2-Generation of finite simple groups and some

related topics, in Generators and Relations in Groups and Geometries (editors A. Barlotti et al.), Kluwer, 1991.

[22] Unsolved problems in group theory. The Kourovka Notebook (12th edn). Novosibirsk 1992.

[23] Dalla Volta, F., Grouppi sporadici generati da tre in involuzioni. RILS A 119 (1985), 65–87.

[24] Nuzhin, Ja., N., Generating triples of involutions of Chevalley groups over finite fields of characteristic 2. Algebra i Logika 29 (1990), 192–206.

[25] Wagner, A., The minimal number of involutions generating some three-dimensional groups. Boll. Un. Mat. Ital. A (5) 15 (1978), 431–439.

[26] Tamburini, M. C., Wilson, J. S., Gavioli, N., On the (2,3)-generation of some classical groups. J. Algebra, to appear.

[27] Hering, C., Kantor, W. M., Seitz, G., Finite groups with a split BN-pair of rank 1. I., J. Algebra 20 (1972), 435–475.

[28] Levchuk, V. M., Nuzhin, Ja. N., On structure of Ree groups. Algebra i Logika 24 (1985), 26–41.

[29] Carter, R. W., Simple groups of Lie type. Wiley, London 1972.

[30] Bourbaki, N., Groupes et Algebres de Lie. IV-VI, Hermann, Raris 1968.

[31] Conder, M. D. E., More on generators for alternating and symmetric groups. Quart. J. Math. Soc. 32 (1981), 137–163.

[32] Nuzhin, Ja. N., Generating triples of involutions of alternating groups. Matem. Zametki 51 (1992), 91–95.

[33] Serre, J. P., Groupes de Galois sur Q. In: Sem. Bourbaki, 1987-88, 687, Asterisque 161–162 (1988), 73–85.

[34] Gorenstein, D., Classifying the simple groups. Bull. Amer. Math. Soc. 14 (1986), 1–98.

[35] Belyi, G. V., On Galois extensions of the maximal cyclotomic field. Izv. AN SSSR 43 (1979), 267–276.

[36] Völklein, H., $PSL_2(q)$ and extensions of $Q(x)$. Bull. Amer. Math. Soc. 24 (1991), 145–153.

[37] Nuzhin, Ja. N., Generating elements of simple groups and their application for some tasks. Abstracts of the IIIrd Intern. Conf. on Algebra. Krasnoyarsk 1993, 462.

On prime QF-extensions of rings

Konstantin A. Pavlov *

Abstract. Given a prime ring R and its fixed ring R^G, where $G \leq \operatorname{Aut} R$ is a reduced-finite group of automorphisms with a quasi-Frobenius algebra, the ring R is proved to preserve some conditions (as satisfiability of polynomial identity, the properties being left Artinian, left Noetherian, left Goldie) lying on R^G.

1991 Mathematics Subject Classification: 13B05, 16N60, 16L60, 16R20, 16P20, 16P40, 16P60.

One of the goals of Galois theory is investigation of relations between a ring R and the subring of invariants R^G where $G \leqslant \operatorname{Aut} R$ stands for a group of automorphisms of R. In the recent papers of V. K. Kharchenko [1] and A. N. Koryukin [2] it was shown that the most wide class of automorphism groups acting on prime rings, for which relations between R and R^G are close enough, is the class of reduced-finite groups with quasi-Frobenius algebras (QF-algebras). In this connection, there was posed a number of questions in [1]:

1. Let R^G satisfy a nontrivial polynomial identity. Is R a PI-ring?
2. Let R^G be a left Artinian ring. Is R left Artinian?
3. Let R^G be a left Noetherian ring. Is R left Noetherian?

The answers to some of these questions can be found in this paper. As was proved in [1], if R^G satisfies a polynomial identity then R is a GPI-ring, i.e. R satisfies a nontrivial generalized polynomial identity. By [3], the ring R is not necessarily a PI-ring.

Let R be a prime associative ring with identity. Denote by C its generalized centroid and by Q the symmetric Martindale quotient ring. If $G \leqslant \operatorname{Aut} R$ is an automorphism group of R then its action can be uniquely extended over Q.

Following V. K. Kharchenko [4], consider the group

$$\mathbb{A}(R) \overset{\mathrm{df}}{=} \{g \in \operatorname{Aut} Q \mid I_1 \subseteq I^g \subseteq R \text{ for suitable nonzero ideals } I_1,\, I \text{ of } R\}.$$

It contains the group $\operatorname{Aut} R$ (for arbitrary $g \in \operatorname{Aut} R$ put $I_1 = I = R$). On the other hand, it also contains all inner automorphisms of Q. Note that all the results

*This work was supported by Russian Fund for Fundamental Research (grant 93-011-1520).

presented below remain valid for an arbitrary reduced-finite group $G \leqslant \mathbb{A}(R)$ with quasi-Frobenius algebra.

The set G_{inn} of inner (for Q) automorphisms is a normal subgroup of G. This is the set of so-called X-inner (with respect to R) automorphisms. The algebra $\mathbb{B}(G)$ of a group G is a C-vector space

$$\mathbb{B}(G) \stackrel{\text{df}}{=} \sum_{G_{\text{inn}}} C \cdot \varphi_g$$

i.e. a span (over C) of the elements $\varphi_g \in Q$ corresponding to automorphisms from G_{inn}, i.e. $x^g = \varphi_g \cdot x \cdot \varphi_g^{-1}$ for any $x \in Q$.

A group G is reduced-finite if the index of G_{inn} in G is finite and the algebra $\mathbb{B}(G)$ is finite-dimensional over C. It is clear that the class of reduced-finite groups contains that of all finite groups.

Let $\mathbb{B}(G)$ be a quasi-Frobenius algebra. We have a decomposition

$$\mathbb{B}(G) = \sum_{i=1}^{n} \lambda_i = \sum_{i=1}^{n} \rho_i,$$

where $\lambda_i \lhd_l \mathbb{B}(G)$, $\rho_i \lhd_r \mathbb{B}(G)$ and $\lambda_i \cong \rho_i^* = \operatorname{Hom}(\rho_i, C)$ for any $i \in \{1, \dots, n\}$. Let $\{a_{ij}\}_{j=1,\dots,s(i)}$, and $\{b_{i,j}\}_{j=1,\dots,s(i)}$ be dual bases of λ_i and ρ_i respectively, i.e. $(a_{ki}, b_{kj}) = \delta_{ij}$ where $(\lambda_k, \rho_k) \to C$ is a nondegenerate associative bilinear form and δ_{ij} denotes Kronecker delta. We can define a trace form $\tau_i : Q \to Q^G$ by

$$\tau_i(x) = \sum_{k=1}^{m} \left(\sum_{j=1}^{s(i)} a_{ij} x b_{ij} \right)^{n_k},$$

where $h_1, \dots, h_m$ are transversals for G_{inn} in G. Since the coefficients of τ_i belong to Q, there exists a nonzero ideal $I_i \lhd R$ such that $0 \neq \tau_i(I_i) \subseteq R$. Consequently, $\tau_i(I_i) \subseteq R^G$. It is easy to see that for any onesided ideal L of R there exists an $i \in \{1, \dots, n\}$ such that $\tau_i(L) \neq 0$ (see [4, 2.2.2] for details).

We say that an ideal I of a ring R is *locally finite* over a subring T of R if, for any $0 \neq a \in I$, there exist $r_1, \dots, r_l \in R$ such that $Ra \subseteq \sum^{l} T \cdot r_i$.

We will use the following results.

Theorem A [1]. *Let R be a prime ring and G a reduced-finite group of automorphisms of R such that the algebra $\mathbb{B}(G)$ is quasi-Frobenius. If R^G satisfies a polynomial identity then the central closure RC is a primitive ring with nonzero socle whose skew field is finite-dimensional.*

Theorem B [1]. *Let the algebra $\mathbb{B}(G)$ of a reduced-finite group G be quasi-Frobenius. There exists a locally finite over R^G ideal I of a ring R.*

Here we use the following notation. The left Goldie dimension of a ring R is denoted by Gol-dim R and the left invariant Goldie dimension of R is denoted by Gol-dim$^G R$, i.e. this is the Goldie dimension of the lattice of left invariant ideals of R. We write Gol-dim $[A,B]$ for the Goldie dimension of a sublattice $[A,B]$ of some modular lattice [0,1]. For example, Gol-dim $[O,R]$ = Gol-dim R, where $[O,R]$ is the lattice of all left ideals of R.

Theorem C [5]. *Let L be a modular lattice with the largest element 1 and the smallest element 0. If $a_1 \wedge \cdots \wedge a_n = 0$ for some $a_1, \ldots, a_n \in L$ then* Gol-dim $L \leq \sum^{n}$ Gol-dim $[a_i, 1]$.

Theorem 1. *Let R be a prime ring, $G \leq \operatorname{Aut} R$ a reduced-finite group of automorphisms with quasi-Frobenius algebra $\mathbb{B}(G)$. Then the following assertions hold:*

1) *if R^G is a left Artinian ring then $R \cong \operatorname{Matr}_n(\Delta)$ where Δ is a skew field;*
2) *if R^G is a left Goldie ring (not necessarily semiprime) and R is left-nonsingular then R is a (prime) left Goldie ring;*
3) *if R^G is a left Noetherian ring and R is nonsingular then R is also left Noetherian;*
4) *if R^G is a left Noetherian PI-ring then R a left Noetherian PI-ring;*
5) *if R^G is a left Goldie ring satisfying a polynomial identity then the same is true for R.*

Proof. The argument is based on the lemma that is of independent interest.

Lemma 1. *Let R be a prime ring, G a reduced-finite group of automorphisms of R such that the algebra $\mathbb{B}(G)$ is quasi-Frobenius. If R^G has finite left Goldie dimension then R also is finite-dimensional in the sense of Goldie.*

Proof. Since $\mathbb{B}(G)$ is quasi-Frobenius, there exist finitely many trace forms $\tau_1, \ldots, \tau_s$ such that for any onesided ideal L of R at least one of the inequalities $\tau_i(L) \neq 0$, $i \in \{1, \ldots, s\}$, holds. Each form τ_i can be presented as

$$\tau_i(x) = \sum_k d_k(i) \cdot x^{f_k(i)}$$

where $d_k(i) \in Q$ and $f_k(i) \in G$ (see the proof of Lemma 6.3.10 of [4]). Let L be an arbitrary left ideal of R. Consider the intersection

$$L_1 = \bigcap_{i,k} L^{f_k(i)^{-1}}.$$

Note that $L \bigcap R^G \neq 0$ provided that $L_1 \neq 0$. Indeed, choose a nontrivial ideal I of R such that $(d_k(i)^{f_k(i)^{-1}}) \cdot I \subseteq R$ for all i, k. Then

$$\tau_i(I) \subseteq \sum_k d_k(i) I^{f_k(i)} = \sum_k \left(d_k(i)^{f_k(i)^{-1}} \cdot I\right)^{f_k(i)} \subseteq R.$$

Therefore, if $L_1 \neq 0$ then for a left ideal $I \cdot L_1$ there exists a number $q \in \{1, \ldots, s\}$ such that

$$0 \neq \tau_q(IL_1) \subseteq \tau_q(IL_1) \bigcap R^G \subseteq L \bigcap R^G.$$

Indeed,

$$\tau_q(IL_1) \subseteq \sum d_k(q) I^{f_k(q)} \cdot L_1^{f_k(q)} \subseteq \sum d_k(q) I^{f_k(q)} \cdot L \subseteq R \cdot L \subseteq L.$$

Conversely, if $L_1 = 0$ then

$$L \bigcap R^G \subseteq \bigcap_{k,i} (L \bigcap R^G)^{f_k(i)^{-1}} \subseteq \bigcap_{k,i} L^{f_k(i)^{-1}} = L_1 = 0.$$

Consider the set

$$\mathfrak{N} \stackrel{\text{df}}{=} \{L \triangleleft_l R \mid \bigcap_{i,k} L^{f_k^{-1}(i)} = 0\}.$$

We have shown that $L \in \mathfrak{N}$ if and only if $L \bigcap R^G = 0$. The set of indices $\{i, k\}$ is finite, hence $\mathfrak{N}$ is inductive. Applying Zorn lemma we can choose a maximal on inclusion element M of $\mathfrak{N}$.

By the definition of M, the intersection $\bigcap_{i,k} M^{f_k(i)^{-1}}$ is trivial. By Theorem C we have

$$\text{Gol-dim}\, R \leq \sum_{i,k} \text{Gol-dim}\, [M^{f_k(i)^{-1}}, R].$$

Since the set $\{i, k\}$ of indices is finite, to establish finiteness of Gol-dim R it suffices to prove the inequalities

$$\text{Gol-dim}\, [M, R] \leq \text{Gol-dim}^G R < \infty.$$

The inequality $\text{Gol-dim}^G R < \infty$ is clear, for if A is a left invariant ideal of R, then for a suitable trace form τ_A and a suitable nontrivial ideal $I_A \triangleleft R$, the inclusion

$$0 \neq \tau_A(I_A \cdot A) \subseteq A \bigcap R^G$$

takes place, i.e. each left invariant ideal intersects R^G nontrivially. The Goldie dimension of R^G is finite in view of the assumptions of the lemma. This implies finiteness of the invariant left Goldie dimension of R.

We prove the inequality

$$\text{Gol-dim}\, [M, R] \leq \text{Gol-dim}^G R.$$

Let $J_1, \ldots, J_p$ be independent and nontrivial in $[M, R]$, i.e., $J_k \neq M$ for enery $k \in \{1, \ldots, p\}$ and the intersection

$$J_k \bigcap \Big(\sum_{i \neq k} J_i \Big)$$

is equal to M for every $k \in \{1, \dots, p\}$. Maximality of M in $\mathfrak{N}$ gives $J_k \notin \mathfrak{N}$, i.e., $J_k \bigcap R^G \neq 0$ for every k. Choose a nontrivial element $x_k \in J_k \bigcap R^G$ and consider the left ideals $Rx_1, \dots, Rx_p$. They are invariant and $Rx_k \subseteq J_k$ for every k. It remains to show that these left invariant ideals form a direct sum in R. Let

$$P_t = Rx_t \bigcap \Big(\sum_{k \neq t} Rx_k \Big);$$

P_t is left invariant ideal and

$$P_t \subseteq J_t \bigcap \Big(\sum_{k \neq t} J_k \Big) = M,$$

i.e.

$$P_t = \bigcap_{i,k} P_t^{f_k(i)^{-1}} \subseteq \bigcap_{i,k} M^{f_k(i)^{-1}} = 0.$$

Lemma 1 is proved.

Remark. The proof of Lemma 1 is based on an idea of [5] about action of finite groups on modular lattices. But we cannot immediately apply the results of [5] because reduced-finite groups with quasi-Frobenius algebras are not necessarily finite.

Now we complete the proof of the theorem.

1. Let R^G be a left Artinian ring. By Theorem B there exists a nonzero element $a \in R$ such that $Ra \subseteq \sum_{t=1}^{m} R^G r_t$ for suitable $r_1, \dots, r_m \in R$. Therefore Ra is an Artinian R-module. This means that R contains a minimal left ideal. Applying Lemma 1 we obtain finiteness of Gol-dim R. It is well known that a primitive ring with minimal left ideal of finite Goldie dimension is a simple Artinian ring, i.e. it is isomorphic to the ring of matrices over a skew field.

2. Let R^G be a left Goldie ring. Nonsingularity of R together with Lemma 1 implies that R is a left Goldie (prime) ring.

3. Let R^G be a left Noetherian ring. Then R is a left Goldie ring. In this case every locally finite ideal I contains a regular element $b \in I$ (in view of the well-known Goldie theorems). Therefore $R \cong R \cdot b$, and $R \cdot b$ is embeddable in a module finitely generated over a Noetherian ring. Consequently, R is left-Noetherian.

4. Let R^G be a left Noetherian PI-ring. Theorem A implies complete primitiveness of the central closure RC. Since R has finite Goldie dimension, RC has finite Goldie dimension too. Hence RC is a simple Artinian ring that is isomorphic to $\mathrm{Matr}_n(\Delta)$ where Δ is finite-dimensional over the generalized centroid C (by the Martindale theorem). Therefore R is a PI-ring. But a prime ring satisfying

polynomial identities is nonsingular. By above, R is a left Noetherian ring. The theorem is proved.

The author wishes to express his thanks to Professor Kharchenko V. K. for suggesting the problem and help during the preparation of the paper.

References

[1] Kharchenko, V. K., On invariants of reduced-finite groups with quasi-Frobenius algebra. Commun. Algebra 21:4 (1993), 1221–1228.

[2] Koryukin, A. N., To the question on the double centralizer in prime rings. Sib. Mat. Zh. 32:6 (1991), 81–86. English translation: Sib. Math. J. 32:6 (1991), 965–970.

[3] L'vov, I. V., On centralizers of finite dimensional subalgebras in the algebra of linear transformations. In: Abstracts of the 3rd Int. Conf. on Algebra in Memory of M. I. Kargapolov (Krasnoyarsk, August 23–28, 1993), 213. INOPROF, Krasnoyarsk 1993.

[4] Kharchenko, V. K., Automorphisms and derivations of associative rings. Kluwer Academic Publlisher Group, Dordrecht 1991.

[5] Grzeszczuk, P., Puczylowski, E. R., Goldie dimension and chain conditions for modular lattices with finite group actions. Can. Math. Bull. 29:3 (1986), 274–280.

Automorphism groups of free pro-p-groups of finite ranks $\geqslant 2$ are infinitely generated

Vitaliĭ A. Roman'kov

Abstract. The automorphism groups of free pro-p-groups of finite ranks 2 are proved to be infinite generated as topological groups. The automorphism groups of free metabelian pro-p-groups of finite ranks $\geqslant 2$ are proved to be infinite generated as topological groups. Some related groups of automorphisms are also proved to be finite or infinite generated.

1991 Mathematics Subject Classification: 20F28, 20E18, 22A05.

1. Automorphisms of profinite groups. Let G be a profinite group. Denote by $\operatorname{Aut} G$ the group of all topological automorphisms of G with the natural congruence topology. A base for neighbourhoods of identity is given by the subgroups

$$\Gamma(U) = \{\varphi \in \operatorname{Aut} G \mid g^{-1}g^{\varphi} \in U \text{ for any } g \in G\},$$

as U runs over the open normal subgroups of G. This makes $\operatorname{Aut} G$ into a Hausdorff topological group. In general, $\operatorname{Aut} G$ is not a profinite group. However, $\operatorname{Aut} G$ is a profinite group provided that G is a finitely generated (f. g.) profinite group. Note that any abstract automorphism of a f. g. pro-p-group G is topological.

As usual, $\operatorname{I Aut} G$ denotes the subgroup of $\operatorname{Aut} G$ which consists of all automorphisms inducing the identity map on the abealization G/G'.

The main question considered in the present paper is whether $\operatorname{Aut} G$ is finite or infinite generated as a topological group.

2. Free metabelian pro-p-groups. Let us introduce

$A_n(p)$, the free abelian pro-p-group freely generated by $a_1, \dots, a_n$;

$M_n(p)$, the free metabelian pro-p-group freely generated by $x_1, \dots, x_n$;

$F_n(p)$, the free pro-p-group freely generated by $f_1, \dots, f_n$.

$\widetilde{\Lambda}_n(p) = \mathbb{Z}_p[A_n(p)]$, the group algebra of $A_n(p)$;

$b_i = a_i - 1$, elements of $\widetilde{\Lambda}_n(p)$;

$\widetilde{\Delta}_n(p) = \sum_i b_i \widetilde{\bigwedge}_n(p)$, the fundamental ideal of $\widetilde{\bigwedge}_n(p)$;

$\bigwedge_n(p) = \mathbb{Z}_p[[b_1, \dots, b_n]]$, the algebra of power series on $b_1, \dots, b_n$ (completed group algebra of $A_n(p)$);

$\Delta_n(p) = \sum_i b_i \bigwedge_n(p)$, the maximal subalgebra of $\bigwedge_n(p)$;

T_n, the free $\bigwedge_n(p)$-module freely generated by $t_1, \dots, t_n$.

The map $x_i \to \begin{pmatrix} a_i & t_i \\ 0 & 1 \end{pmatrix}$ is uniquely extended to the embedding (see [1])

$$M_n \to \begin{pmatrix} A_n(p) & T_n \\ 0 & 1 \end{pmatrix}.$$

Define topological derivations as follows: for

$$g \to \begin{pmatrix} \overline{g} & \alpha_1 t_1 + \dots + \alpha_n t_n \\ 0 & 1 \end{pmatrix}, \quad g \in M_n(p),$$

introduce

$$\partial g / \partial x_i = \alpha_i \in \bigwedge\nolimits_n(p).$$

All properties of ordinary Fox derivations including the main identity

$$\overline{g} - 1 = \sum\nolimits_i \partial g / \partial x_i b_i,$$

are valid in the topological case too.

3. The group $\mathrm{I\,Aut}\, M_n(p)$. The topological variant of the well-known Bachmuth embedding

$$\mathrm{I\,Aut}\, M_n(p) \to GL_n(\bigwedge\nolimits_n(p))$$

is represented as follows:

$$\varphi \to J_\varphi = (\partial x_i^\varphi / \partial x_j).$$

Proposition 1 [2]. $\det \mathrm{I\,Aut}\, M_n(p) = 1 + \Delta_n(p)$.

Note that $1 + \Delta_n(p)$ is the free abelian pro-p-group freely generated by $1 + b^{\overline{\alpha}} = 1 + b_1^{\alpha_1} \dots b_n^{\alpha_n}$, $\overline{\alpha} = (\alpha_1, \dots, \alpha_n) \in \mathbb{N}^n$.

Theorem 1. $\mathrm{I\,Aut}\, M_n(p)$ *is not finitely generated for* $n \geqslant 2$.

Corollary 1. $\mathrm{I\,Aut}\, F_n(p)$ *and* $\mathrm{I\,Aut}\, F_n$ *(F_n free profinite of rank n) are not finitely generated for* $n \geqslant 2$.

4. The group Aut $M_n(p)$. It is well known that Aut $A_n(p)$ is isomorphic to the analytic group $GL_n(\mathbb{Z}_p)$ (f. g. as well). Thus, the main portion of the question on f. g. of Aut $M_n(p)$ is one about f. g. of $1 + \Delta_n(p)$ as a $GL_n(\mathbb{Z}_p)$-group.

Denote by

$$\begin{aligned} i_k(l) \ : \ & x_k \to x_k^l,\ (l,p) = 1, \\ & x_j \to x_j\ (j \neq k) \end{aligned}$$

a diagonal automorphism of $M_n(p)$ (inducing the diagonal automorphism of $A_n(p)$ given by the diagonal matrix diag$(1, \dots, l, \dots, 1) \in GL_n(\mathbb{Z}_p)$); by

$$\begin{aligned} \rho_{ij} \ : \ & x_i \to x_i x_j\ (j \neq i), \\ & x_q \to x_q\ (q \neq i) \end{aligned}$$

the Nielsen automorphism of $M_n(p)$ (inducing the Nielsen automorphism of $A_n(p)$, given by the transvection $t_{ij} \in GL_n(\mathbb{Z}_p)$).

Then $GL_n(\mathbb{Z}_p)$ is generated by the induced automorphisms $i_k(l)$, t_{ij} (we may choose only finite number of $i_k(l)$ since Aut $\mathbb{Z}_p \simeq \mathbb{Z}_p^*$ is f. g. (see [3])).

Theorem 2. *The group* Aut $M_n(p)$ *is not finitely generated for every finite* $n \geqslant 2$.

Corollary 2. *For* $n \geqslant 2$ *the groups* Aut $F_n(p)$, Aut F_n *are not finitely generated.*

The corollary gives the answer to the well-known question (posed explicitly by A. Lubotskiĭ in Oberwolfach, 1990) on f. g. of Aut $F_n(p)$.

Sketch of proof. Note that $Q_n(p) = 1 + \Delta_n(p)/(1 + \Delta_n(p))^p$ can be considered as an $\mathbb{F}_p[[GL_n(\mathbb{Z}_p)]]$-module. The ring $\mathbb{F}_p[[GL_n(\mathbb{Z}_p)]]$ is topologically Noetherian [4].

Let a group G be an extension of a normal subgroup N by a finitely presented (f. p.) group H. The group G is f. g. iff N is f. g. as a normal subgroup.

Suppose that Aut $M_n(p)$ is a f. g. Then $Q_n(p)$ is f. g. $\mathbb{F}_p[[GL_n(\mathbb{Z}_p)]]$ -module. Then every submodule of $Q_n(p)$ is f. g. Consider the module $1 + p\Delta_n(p) (\text{mod } p^2)$ as a ring with zero multiplication. Since

$$(1 + \Delta_n(p))^p \cap 1 + p\Delta_n(p) = 1 + p^2 \Delta_n(p),$$

the computations in $1 + p\Delta_n(p)$ are natural.

The automorphisms $i_k(l)$, ρ_{ij} act on $1 + \Delta_n(p)$ as the following maps:

$$\begin{aligned} \sigma_k(l) : \ & b_k \to (1 + b_k)^l - 1, \\ : \ & b_j \to b_j\ (j \neq k); \\ \tau_{ij} \quad : \ & b_i \to b_i + b_j + b_i b_j\ (j \neq i), \\ : \ & b_q \to b_q\ (q \neq i). \end{aligned}$$

The module $1 + p\Delta_n(p)(\text{mod } p^2)$ is not f. g. Hence the module $1 + \Delta_n(p)$ is not f. g. too.

5. The group Aut $M_n(p,p)$. Denote by $M_n(p,p) = M_n(p)/(M_n(p)')^p$ the free metabelian pro-p-group with derived subgroup of period p.

Proposition 2 [2]. *The group* $\mathrm{I\,Aut}\, M_n(p,p)$ *is not finitely generated for every finite* $n \geqslant 2$.

Theorem 3 [2]. *The group* $\mathrm{Aut}\, M_n(p,p)$ *is finitely generated for* $n = 2$ *or finite* $n \geqslant 4$.

Sketch of proof. Let $\bigwedge_n(p,p) = \mathbb{F}_p[[b_1, \ldots, b_n]]$, and let $\Delta_n(p,p) = \sum_i b_i \bigwedge_n(p,p)$ be the maximal subalgebra of $\bigwedge_n(p,p)$.

The Bachmuth embedding $\mathrm{I\,Aut}\, M_n(p,p) \to GL_n(\bigwedge_n(p,p))$ gives a possibility to define $\det \mathrm{I\,Aut}\, M_n(p,p) = 1 + \Delta_n(p,p)$, the free abelian pro-$p$-group freely generated by

$$1 + b^{\overline{\alpha}} = 1 + b_1^{\alpha_1} \ldots b_n^{\alpha_n}, \; \bar{\alpha} = (\alpha_1, \ldots, \alpha_n) \in \mathbb{N}^n \setminus pN^n.$$

$\mathrm{Aut}\, M_n(p,p)$ is generated by the set of the Nielsen automorphisms of $M_n(p,p)$ combined with the set of Chein automorphisms of $M_n(p,p)$, jointly with some finite set of special automorphisms [2]. A Chein automorphism is defined as

$$x_i \to v_i^{\alpha} x_i$$

where $\alpha \in \bigwedge_n(p,p)$, $v_i \in M_n(p,p)'$ is independent of x_i,

$$x_j \to x_j, \quad i \neq j.$$

Thus, $\mathrm{Aut}\, M_n(p,p)$ is f. g. for $n \geqslant 4$. The case $n = 2$ is considered separately. The case $n = 3$ is still open.

6. Some applications to discrete groups. Let N_{nl} be a free nilpotent group of class l and M_{nl}, a free metabelian nilpotent group of class l.

Theorem 4. *For* $n \geqslant 2$ *the number of generating elements of* $\mathrm{Aut}\, N_{nl}$ *as well as of* $\mathrm{Aut}\, M_{nl}$ *can not be bounded by some function of* n *only.*

References

[1] Remeslennikov, V. N., The embedding theorems for profinite groups. Izv. Acad. Nauk SSSR, Ser. Mat. 43 (1979), 399–417.

[2] Roman'kov, V. A., Generating elements of automorphism groups of free metabelian pro-p-groups. Sib. Mat. Zh. 33:5 (1992), 145–158. English translation: Sib. Math. J. 33:5 (1992), 878–890.

[3] Weil, A., Basic number theory. Springer Verlag, Berlin. Heidelberg–New York 1967.

[4] Lazard, M., Groups analytiques p-adiques. Publ. Math. IHES 26 (1965), 389–603.

Elementary theories of free algebras for varieties corresponding to non-classical logics

*Vladimir V. Rybakov**

Abstract. Questions on the recursive complexity of free algebras from varieties corresponding to different non-classical logics are studied. A uniform method for proving the undecidability of elementary theories of free algebras with representable lattice structure is given. It is proved that the free algebras of a wide class, which contains in particular the free algebras for varieties of K-, K4-, T-, S4-modal algebras, GL-diagonalizable algebras, pseudo-Boolean algebras and IL-interpretability algebras, have hereditarily undecidable first-order theories.

1991 Mathematics Subject Classification: 03B25, 03B20, 03B45, 06E25.

Introduction

Properties of free algebras for varieties corresponding to different propositional non-classical logics reflect important properties of these logics. For instance, an equational theory describes the set of theorems for the corresponding logic [1]. A quasiequational theory corresponds to the set of admissible rules of the logic [2, 3]. The results of [3–5] can be interpreted as results about decidability of universal theories of free algebras. For instance, the result [3, 4, 6] on the universal theory of the Heyting free algebra being decidable yields a positive solution to H. Friedman's problem [7, Problem 40] about recognizing admissible inference rules in Heyting intuitionistic logic. This shows the importance of studies on recursive complexity of first-order theories of free algebras corresponding to non-classical logics.

In this paper, we develop a uniform method for showing undecidability of elementary theories of free algebras with representable lattices structure, and find a number of applications. It is proved that free algebras from a wide class, containing in particular free algebras for varieties corresponding to the modal logics

*Supported by the Alexander von Humboldt Foundation in 1992/1993 during a stay at the Free University of Berlin.

$S4, T, K4$, intuitionistic propositional calculus H, provability logic GL, binary interpretability logic IL and unary interpretability logic, have hereditarily undecidable first-order theories.

1. Background

We assume the reader to be familiar with the basic concepts and definitions of first-order logic and non-classical propositional logic and propositional modal logic (see, for instance [1, 8–12]).

We start by defining the notion of a CDL-class of algebraic systems and prove a lemma on first order theories of these classes to be hereditarily undecidable.

Definition. Let a class of algebras (or models) K have the predicate of partial order $\leq$ expressible by first-order formulas. Let each member of K be a lattice with respect to this predicate (thus, $\vee, \wedge$ have their usual meanings in algebras of K). We call this class a *CDL-class* if the following conditions are true:

(a) there exists a term $f(x)$ in the signature of K such that for each number n and for any $M \in K$ for each $g_i, d \in M$, on M the following holds

$$f(d) \leq \bigvee_{1 \leq i \leq n} f(g_i) \Rightarrow \exists i (f(d) \leq f(g_i)); \tag{1}$$

(b) for each $n \in N$ and each family $X_i \subseteq \{1, \ldots, n\}$, $1 \leq i \leq n$ such that $\forall i, j \in \{1, \ldots, n\}(i \in X_j \Leftrightarrow j \in X_i)$ & $\forall i(i \notin X_i)$, there exist terms $t_{X_i}(x_1, \ldots, x_n, y)$ and an algebra $M \in K$ such that there is a tuple of elements $a_i, 1 \leq i \leq n$, of this algebra and $v \in M$ for which the elements $b_i, 1 \leq i \leq n$, of M, where $b_i := f_i(t_{X_i}(a_1, \ldots, a_n, v))$, satisfy in M:

$$b_i \wedge b_j = v \Longleftrightarrow j \in X_i \ \& \ i \in X_j, \tag{2}$$

$$b_i \not\leq b_j, b_j \not\leq b_i. \tag{3}$$

For the definition of relative elementary definability of a class of first-order models in an appropriate class of models we refer to [11]. It is well known that if a class K_1 is relatively elementary definable in a class K_2 and K_1 has a hereditarily undecidable first-order theory, then $Th(K_2)$ is hereditarily undecidable as well.

Lemma 1. *The class K_0 of finite models of the symmetric irreflexive binary predicate is relatively elementary definable in each CDL-class of algebras K.*

Proof. Let $A(x, y)$ and $B(x, y, z)$ be fixed first-order formulas in the signature of K, where $A(x, y)$ expresses the property: y is a maximal element of the form $f(z)$

contained in x; and $B(x, y, z)$ states that the intersection of x and y is equal to z, and that x and y are different.

Let $\langle\{i \mid 1 \leq i \leq n\}, P(x, y)\rangle$ be a model of K_0, $X_i := \{j \mid 1 \leq j \leq n \;\&\; P(i, j)\}$. By assumption, $P(x, y)$ is a symmetric irreflexive relation and we have $\forall i, j \in \{1, \ldots, n\}(i \in X_j \Leftrightarrow j \in X_i)$. By the conditions of the Lemma and (2), there are $M \in K$ and elements $b_i, v \in M$, where $b_i := f_i(t_{X_i}(a_1, \ldots, a_n, v))$, such that $b_i \wedge b_j = v \Longleftrightarrow (j \in X_i) \;\&\; (i \in X_j)$. Then

$$P(i, j) \Longleftrightarrow (j \in X_i) \;\;\&\;\; (i \in X_j) \Longleftrightarrow b_i \wedge b_j = v. \tag{4}$$

Consider the set $C := \{y \mid F \models A(a, y)\}$, where $a := \bigvee_{1 \leq i \leq n} b_i$. By (3) the elements b_i are incomparable. Then by (1), the set C consists of all elements b_i (and only them).

Now on the set C we define a predicate $P_1(x, y)$ as follows

$$P_1(x, y) \Leftrightarrow F \models B(x, y, v).$$

From (4) we obtain

$$F \models B(b_i, b_j, v) \Leftrightarrow (b_i \wedge b_j = v) \;\&\; (i \neq j \Leftrightarrow P(i, j)).$$

Thus $P_1(b_i, b_j) \Leftrightarrow P(i, j)$ and we have proved that the models $\langle C, P_1\rangle$ and $\langle\{i \mid 1 \leq i \leq n\}, P(x, y)\rangle$ are isomorphic. This completes the proof of Lemma 1.

2. Free algebras with representable lattice structure

We can now apply Lemma 1 to various kinds of free algebras having representable lattice structure, in particular to various free modal algebras. It is clear that all the proofs are based on successfully finding appropriate terms and checking properties from the definition of CDL-class.

For any propositional logic (modal, intermediate etc.), $\lambda\,\mathrm{Var}(\lambda)$ denotes the variety of algebras on which all theorems of this logic are true, $F_\lambda(\alpha)$ denotes the free algebra of rank α from this variety.

The *smallest provability Gödel–Löb logic*, GL, is the normal modal logic with the modal axioms $\Box(p \to q) \to (\Box p \to \Box q)$, $\Box p \to \Box\Box p$, and $\Box(\Box p \to p) \to \Box p$. Recall that a *diagonalizable* (*provability modal*) *algebra* is a Boolean algebra with an additional modal unary "necessity" operation $\Box$ which satisfies the identities $\Box(x \to y) \to (\Box x \to \Box y) = T$, $\Box x \to \Box\Box x = T$, and $\Box(\Box x \to x) \to \Box x = T$, $\Box T = T$.

We call an extension λ of GL a $GL(3, \infty)$-*logic* if all theorems of this logic are true on each frame $\sum_{1 \leq i \leq n} F_i \oplus 1$ which is a disjoint union, with the added smallest irreflexive element, of frames F_i which are three-element irreflexive transitive chains.

Theorem 2. *Let K be a class of free algebras of the variety* Var(λ) *which contains among their subalgebras all free algebras of finite rank, where λ is any $GL(3,\infty)$-logic. Then the elementary first-order theory of this class is hereditarily undecidable.*

Proof. We apply Lemma 1. For this we need a choice of some terms corresponding to the definition of CDL-class. Let

$$f(x) := \Box(\Diamond x \vee \Box \bot).$$

Prove (1). Elements of the free algebra are considered as terms which are built up on free generators. The premise of (1) in our case is transformed into

$$\Box(\Diamond d \vee \Box \bot) \leq \bigvee_{1\leq i\leq n} \Box(\Diamond g_i \vee \Box \bot).$$

If (1) is not true, then by the finite model property of provability logic GL [10], for each i there is a finite transitive irreflexive model M_i on which the formula $\Box(\Diamond d \vee \Box \bot) \to \Box(\Diamond g_i \vee \Box \bot)$ is not true. It is easy to see that there is a model K_i, based on a three-element irreflexive transitive chain, on which this formula is not true. We take the model $\sum_{1\leq i\leq n} K_i \oplus 1$ which is the direct sum (disjoint union) of the given models with the added smallest irreflexive element. Directly, on this model the formula $\Box(\ \Diamond d \vee \Box \bot\ ) \to \bigvee_{1\leq i\leq n} \Box(\Diamond g_i \vee \Box \bot)$ is not true as well. As this logic is a $GL(3,\infty)$-logic, the last formula is not a theorem of λ. Contradiction to the premise of (1). Thus, (1) holds.

We pass to the proof of (2). As elements a_i, $1 \leq i \leq n$, we take free generators of the free algebra $F_\lambda(n)$, which is a subalgebra of some free algebra $F_\lambda(\beta)$ from K. Let $v := \bot$ and

$$t_{X_i}(x_1, \dots, x_n, y) := (x_i \wedge \bigwedge_{j\in X_i} \neg\Diamond x_j)).$$

Every b_i is of the form

$$\Box(\Diamond(a_i \wedge \bigwedge_{j\in X_i} \neg\Diamond a_j) \vee \Box \bot).$$

If $j \notin X_i$ then by assumption in the definition of CDL-class we have $i \notin X_j$. Consider the three-element irreflexive transitive chain M_3 in the last element of which only the variables a_i, a_j are true. Easily, the formula $b_i \wedge b_j \equiv \Box \bot$ is not true on this model and so it is not a theorem of GL. Then $b_i \wedge b_j \neq \Box \bot$ in the free algebra $F_\lambda(\beta)$.

Now, $j \in X_i, i \in X_j$. Then $\Box\Box \bot \leq b_i \wedge b_j \leq \Box\Diamond a_i \wedge \Box\Diamond\neg\Diamond a_i \leq \Box \bot$. But $\Box\Box \bot = \Box \bot$ in this case. Thus (2) is true as well.

To finish the proof, we check (3). Let S be a three-element irreflexive transitive chain on the last element of which only the variable a_i is true. Easily, the formula $b_i \to b_j$ is not true in this model, which implies incomparability of the elements b_i and b_j (the countermodel to the opposite implication can be constructed in the same way) and (3) is true. Now by Lemma 1 we have that the first-order theory of the class K is hereditarily undecidable.

Corollary 3. *The class of all free diagonalizable algebras of finite rank, as well as the class of one free algebra of countable rank, has a hereditarily undecidable elementary theory.*

Now, we extract a similar result for free algebras of varieties corresponding to interpretability logic. Recall that (binary) interpretability logic IL (see [13]) is obtained from the logic GL by adding to the language a single binary interpretability connective $\rhd$ and some additional axioms for this connective. An IL-frame is a triple $\langle W, R, S\rangle$, where the frame $\langle W, R\rangle$ has a transitive and conversely well-founded relation (frame for GL), and $S = \{S_w \mid w \in W\}$ is a collection of binary relations on W, where S_w is the reflexive and transitive relation on elements accessible from w by R such that for every two a, b accessible from w by R, $aRb \Rightarrow aS_wb$.

The truth of formulas in such models is defined like the truth of modal formulas, assuming that $u \Vdash A \rhd B \iff \forall v(uRv \wedge v \Vdash A \Rightarrow \exists w(vS_uw \;\&\; w \Vdash B)$. It is known [13] that IL is complete with respect to finite IL-models.

We call an extension λ of IL a *$IL(3,\infty)$-logic* if all theorems of this logic are true on each IL-frame $\bigotimes_{1\leq i\leq n} F_i \oplus 1$ obtained from the IL-frames F_i, each of which is a three-element irreflexive transitive chain, as the direct sum (disjoint union) of these frames with the added smallest irreflexive element a. The relation S_a is defined to be the set-theoretic union of all corresponding relations in frames F_i with the added pairs consisting of generators of F_i and all elements accessible from them.

Corollary 4. *Let K be a class of free algebras of variety* $\mathrm{Var}(\lambda)$, *where λ is any $IL(3,\infty)$-logic, which contains among its subalgebras all free algebras of finite rank. Then the elementary first-order theory of this class is hereditarily undecidable.*

Proof. We follow the proof of Theorem 2. Let

$$f(x) := \Box(\Diamond x \vee \Box\bot).$$

Then (1) can be checked similarly to Theorem 2: in the case the premise of (1) is transformed into

$$\Box(\Diamond d \vee \Box\bot) \leq \bigvee_{1\leq i\leq n} \Box(\Diamond g_i \vee \Box\bot).$$

If (1) is not true, then by the pointed completeness theorem for IL, for each i there is a finite transitive irreflexive IL-model M_i on which the formula $\Box(\Diamond d \vee \Box\bot) \to$

$\Box(\Diamond g_i \vee \Box \perp)$ is not true. Easily, there is an IL-model K_i with the properties as in the definition of $IL(3, \infty)$-logic on which this formula is not true. We take the model $\bigotimes_{1 \leq i \leq n} K_i \oplus 1$ as above given. Directly, on this model the formula $\Box(\Diamond d \vee \Box \perp) \rightarrow \bigvee_{1 \leq i \leq n} \Box(\Diamond g_i \vee \Box \perp)$ is not true as well. As this logic is a $IL(3, \infty)$-logic, the last formula is not a theorem of λ. Contradiction with the premise of (1). Thus (1) holds.

The modal logic $\lambda_\Box$ which consists of all formulas without $\rhd$ having occurrences in λ is a $GL(3, \infty)$-logic and $\alpha \equiv \beta \in \lambda_\Box$ iff $\alpha \equiv \beta \in \lambda$ and $F_{\lambda_\Box}(m)$ is a subalgebra of $F_\lambda(m)$. Therefore, (2) and (3) follow immediately from the corresponding relations from the proof of Theorem 2. Now we use Lemma 1 and obtain the claim of the theorem.

The same result for unary interpretability logic (see [14]) and its free algebras can be obtained just by the trivial rewriting of the proof of the above placed corollary for binary interpretability logic.

A modal logic λ extending the normal modal logic $S4$ is called an $S4(1, \infty)$*-logic* if for any tuple of reflexive one-element frames S_i, all theorems of this logic are true on the frame $\sum_{1 \leq i \leq n} S_i \bigodot 1$ which is obtained as a direct sum with the added smallest reflexive element.

In the all remaining theorems, we use Lemma 1. For this reason, we do not mention it and start every proof just from a choice of the needed terms and after that, check the properties from the definition of a CDL-class.

Theorem 5. *Let λ be a $S4(1, \infty)$-logic and K be a class of free algebras of variety* $\mathrm{Var}(\lambda)$ *which contains among their subalgebras all free algebras of finite rank. Then the first-order theory of K is hereditarily undecidable.*

Let $f(x) := \Box\Diamond x$. We need to show (1). Again, we can consider all elements of the free algebra as terms which are built up on free generators. In this case, the premise of (1) has the form

$$\Box(\Diamond d) \leq \bigvee_{1 \leq i \leq n} \Box(\Diamond g_i).$$

Suppose that (1) is not true. By the finite model property of the modal logic $S4$ [10], for each i there is a finite rooted transitive reflexive model M_i on which the formula $\Box\Diamond d \rightarrow \Box\Diamond g_i$ is not true. Then it is clear that this formula is not true on some one-element model K_i. We take model $\sum_{1 \leq i \leq n} K_i \bigodot 1$ which is constructed as in the definition of the $S4(1, \infty)$-logic. It can be verified directly that the formula $\Box\Diamond d \rightarrow \bigvee_{1 \leq i \leq n} \Box\Diamond g_i$ is not true on this model as well. As this logic is a $S4(1, \infty)$-logic, the last formula is not a theorem of λ. Contradiction with the premise of (1). Thus, (1) holds.

Pass to the proof of (2). As elements a_i, $1 \leq i \leq n$, we take free generators of the free algebra $F_\lambda(n)$ which is a subalgebra of some $F_\lambda(\beta)$ from K, $v := \perp$. Let

$$t_{X_i}(x_1, \dots, x_n, y) := (a_i \wedge \bigwedge_{j \in X_i} \neg\Diamond a_j)).$$

In this case, every b_i has the form

$$\Box(\Diamond(a_i \wedge \bigwedge_{j \in X_i} \neg\Diamond a_j)).$$

If $j \notin X_i$ then by assumption in the definition of a CDL-class we have $i \notin X_j$. Consider the one-element reflexive model M_1 in whose element only the variables a_i, a_j are true. Easily, the formula $b_i \wedge b_j \equiv \perp$ is not true on this model. It is not a theorem of λ because λ is an $S4(1,\infty)$-logic. Therefore, $b_i \wedge b_j \neq \perp$ in the free algebra $F_\lambda(\beta)$.

Now, assume $j \in X_i, i \in X_j$. In this case $b_i \wedge b_j \leq \Box\Diamond a_i \wedge \Box\Diamond\neg\Diamond a_i) \leq \perp$. Thus (2) is true as well. Check (3). Let S be a one-element reflexive frame in whose element only the variable a_i is true. Easily, the formula $b_i \to b_j$ is not true in this model, which implies incomparability of the elements b_i and b_j (a countermodel to the inverse implication can be constructed in the same way) and (3) holds. Now we can use Lemma 1, which proves the theorem.

Let λ be an intuitionistic intermediate logic. We call λ an $H(1,\infty)$-*logic* if all theorems of λ are true on posets of the form $\sum_{1 \leq i \leq n} D_i \odot 1$ which are obtained as direct sums (disjoint unions) of one-element posets D_i with the added smallest reflexive element. Some proof of the theorem given below can use steps from the proof of Theorem 5, but we prefer to give a bit more direct proof.

Theorem 6. *Let λ be an $H(1,\infty)$-logic and K a class of free algebras of the variety* $\mathrm{Var}(\lambda)$, *which contains among its subalgebras all free algebras of finite rank. Then $Th(K)$ is hereditarily undecidable.*

Proof. Let $f(x) := \neg\neg x$. Check (1). All elements of the free algebra can be supposed to be terms which are built up on free generators. In this case, the premise of (1) has the form

$$\neg\neg d \leq \bigvee_{1 \leq i \leq n} \neg\neg g_i.$$

Suppose that (1) is not true. By the finite model property of Heyting intuitionistic logic, for each i there is a sharp finite poset and a model M_i based on this poset such that the formula $\neg\neg d \supset \neg\neg g_i$ is not true on M_i. Then it is clear that this formula is not true on some one-element model K_i. We take the model $\sum_{1 \leq i \leq n} K_i \odot 1$ which is constructed as in the above definition of an $H(1,\infty)$-logic. Directly, on this model the formula $\neg\neg d \supset \bigvee_{1 \leq i \leq n} \neg\neg g_i$ is not true as well. The last formula is not a theorem of λ because this logic is an $H(1,\infty)$-logic. Contradiction to the premise of (1). Thus, (1) holds.

Pass to the proof of (2). As elements a_i, $1 \leq i \leq n$, we take the free generators of the free algebra $F_\lambda(n)$, which is a subalgebra of some free algebra $F_\lambda(\beta)$ from K, $v :=\perp$. Let

$$t_{X_i}(x_1, \dots, x_n, y) := (x_i \wedge \bigwedge_{j \in X_i} \neg x_j).$$

Elements b_i have in this case the form

$$\neg\neg(a_i \wedge \bigwedge_{j \in X_i} \neg a_j).$$

If $j \notin X_i$ then by the assumption in the definition of a CDL-class we have $i \notin X_j$. Consider the one-element poset model M_1 in whose element only the variables a_i, a_j are true. Easily, the formula $b_i \wedge b_j \equiv\perp$ is not true on this model. This formula is not a theorem of λ because the latter is an $H(1,\infty)$-logic. Then $b_i \wedge b_j \neq\perp$ in the free algebra $F_\lambda(\beta)$. Now, let $j \in X_i, i \in X_j$. Then $b_i \wedge b_j \leq \neg\neg a_i \wedge \neg\neg\neg a_i \leq\perp$. Thus (2) is true as well. We can verify (3) as the corresponding part of Theorem 5. Thus, K is a CDL-class and use Lemma 1 to complete the proof of the theorem.

We call an extension λ of the smallest normal modal logic K a $K(\infty)$-*logic* if all theorems of this logic are true on the reflexive one-point frame and the following is true: for any tuple F_i of finite, but non one-point, rooted irreflexive frames holding all theorems of this logic, all the theorems are true on the frame $\biguplus_{1\leq i\leq n} F_i \oplus 1$ as well. This frame is obtained from the frames F_i as follows.

Let the tuple F_i consist of a tuple D_i of frames with reflexive roots and a tuple G_i of frames with irreflexive roots. Let each W_i be obtained from G_i by removing the irreflexive root. Then the frame $\biguplus_{1\leq i\leq n} F_i \oplus 1$ is a disjoint union of all D_i and all W_i with the added root from which only such elements are accessible: all elements of each D_i which are accessible from the root of D_i and all elements of every W_i which are accessible from the root of G_i.

Theorem 7. *Let C be a class of free algebras of the variety* Var(λ), *where λ is any $K(\infty)$-logic, which contains among its subalgebras all free algebras of finite rank. Then the elementary first-order theory of C is hereditarily undecidable.*

Proof. Let $f(x) := \Box x$. We check (1). In our case the premise of (1) is transformed into $\Box d \leq \bigvee_{1\leq i\leq n} \Box g_i$. If (1) is not true, then by the finite model property of logic K [10], for each i there is a finite, but non one-point, rooted irreflexive model M_i on which the formula $\Box d \to \Box g_i$ is not true. Take the model $\biguplus_{1\leq i\leq n} M_i \oplus 1$, which is constructed as in the definition of a $K(\infty)$-logic. Directly, on this model the formula $\Box d \to \bigvee_{1\leq i\leq n} \Box g_i$ is not true as well. The last formula is not a theorem of λ because this logic is $K(\infty)$-logic, which contradicts to the premise of (1). Thus (1) holds.

Pass to the proof of (2). As elements a_i, $1 \le i \le n$, we take the free generators of the free algebra $F_\lambda(n)$ which is an subalgebra of some algebra $F_\lambda(\beta)$ from K. Let $v := \Box \perp$ and

$$t_{X_i}(x_1, \dots, x_n, y) := (x_i \wedge \bigwedge_{j \in X_i} \neg x_j)).$$

In this case elements b_i are of the form

$$\Box(a_i \wedge \bigwedge_{j \in X_i} \neg a_j).$$

If $j \notin X_i$, then by the assumption in the definition of a CDL-class we have $i \notin X_j$. Consider the one-element reflexive transitive model M_1 on whose element only the variables a_i, a_j are true. Easily, the formula $b_i \wedge b_j \equiv \Box \perp$ is not true on this model so this formula is not a theorem of λ. Then $b_i \wedge b_j \neq \Box \perp$ in the free algebra $F_\lambda(\beta)$.

Assume $j \in X_i, i \in X_j$. In this case, $\Box \perp \le b_i \wedge b_j \le \Box a_i \wedge \Box \neg a_i \le \Box \perp$. Thus (2) is true as well. We can verify (3) as in the proof of Theorem 5. By Lemma 1, the first-order theory of our class K is hereditarily undecidable. This completes the proof.

We can extract a corollary for modal free algebras of the varieties corresponding to the smallest normal modal logic K, the modal logic $K4$ and the Von Wright modal logic T.

Corollary 8. *The free modal algebras $F_K(\infty)$, $F_{K4}(\infty)$ and $F_T(\infty)$ have hereditarily undecidable first-order theories.*

As we have seen above, all the theorems can be proved by the described uniform method. The theorems concern a wide class of free algebras for which a suitable lattice structure is (first-order) representable (see Lemma 1). The general method can be applied to other, similar classes of free algebras.

References

[1] Blok, W. J., Pigozzi, D., Algebraizable logic. Mem. Am. Math. Soc. 77:396 (1989).

[2] Rybakov, V. V., Bases of admissible rules of the modal system *Grz* and of intuitionistic logic. Mat. Sb., Nov. Ser. 128:3 (1985), 321–338. English translation: Math. USSR, Sb. 56:2 (1987), 311–331.

[3] — Problems of admissibility and substitution, logical equations and restricted theories of free algebras. In: Proc. of the 8th Int. Congr. of Logic, Method., and Phil. of Science (Elsevier Sci. Publ.), 121–139. North-Holland, Amsterdam 1989.

[4] — Problems of substitution and admissibility in the modal system Grz and in intuitionistic propositional calculus. Ann. Pure Appl. Logic 50:1 (1990), 71–106.

[5] — Logical equations and admissible rules of inference with parameters in modal provability logics. Stud. Log. 49:2 (1990), 215–239.

[6] Roziere, P., Admissible rules in intuitionistic propositional calculus. Ph. D. Thesis. Paris-7 Univ., 1992.

[7] Friedman, H., One hundred and two problems in mathematical logic. J. Symb. Log. 40:3 (1975), 113–130.

[8] Goldblatt, R. I., Metamathematics of modal logics. Rep. Math. Logic 6 (1976), 41–78 (Pt 1); 7 (1976), 21–52 (Pt 2).

[9] Rautenberg, W., Klassische und nichtklassische Aussagenlogik. Friedr. Vieweg&Sohn, Braunschweig–Wiesbaden 1979.

[10] Segerberg, K., An Essay in Classical Modal Logic 1–3 (Filosofiska Studier 13). Uppsala University, Uppsala 1971.

[11] Handbook on Mathematical Logic. North-Holland, Amsterdam 1986.

[12] Shavrukov, V. Yu., A note on the diagonalizable algebras of PA and ZF (ITLI Prepublication Ser. ML–88–03). University of Amsterdam, ML–91–09, Amsterdam 1991.

[13] De Jongh, D., Veltman, F., Provability logics for relative interpretability (ITLI Prepublication Series ML-88-03). University of Amsterdam, Enclosed in: Proc. of the 1988 Heyting Colloquium. Plenum Press, Boston 1990.

[14] De Rijke, M., Unary interpretability logic (ITLI Prepublication Series ML-90-04). University of Amsterdam, Amsterdam 1990.

The finite generation of projective modules

Iskhak I. Sakhaev

Abstract. Associative rings with unit and unitary modules are studied with emphasis on projective and flat modules. Questions on rings and modules to be finitely generated and questions on existence of special modules are discussed.

1991 Mathematics Subject Classification: 16S15, 16D40, 16L30, 16D90.

We study associative rings with unit and unitary modules. D. Lazard [1, p. 440] posed the following hypothesis.

Hypothesis A. *If the quotient module $P/PJ(R)$ of a projective right R-module P modulo its radical is finitely generated then the module P is finitely generated.*

D. Lazard confirmed Hypothesis A for the case of commutative ring. The authors of [2] stated the following hypothesis.

Hypothesis B. *If R is semilocal ring, Π is a finitely generated flat right R-module, and the quotient module $\Pi/\Pi J(R)$ is $R/J(R)$-projective then the module Π is projective.*

S. Jondrup [3] established that Hypothesis B is true if Hypothesis A is true. J. Valette [4] proved that if the quotient ring R/ρ is a Goldie ring for every prime ideal ρ then Hypothesis A is satisfied for the ring R.

The author [5] proved that Hypotheses A and B are equivalent. A number of necessary and sufficient left-right symmetric conditions on the ring under which Hypothesis A holds can be found in [6].

As was shown by V. N. Gerasimov and the author [7] there exists a semilocal ring for which Hypothesis A fails. Note that Cohn's theory of localization of associative rings [8] was further developed by V. N. Gerasimov [9]. He indicated a new interpretation of elements of the universal Σ-inverting ring $R\Sigma^{-1}$. It was very useful for the proof of finite generation of the projective module in [7].

In the sequel we denote by P^0 ($P^0(n)$) a projective right R-module P if the quotient module $P/PJ(R)$ is finitely generated (n-generated by $M = \langle u_2, u_2, \dots, u_n \rangle$) n-generated right R-module.

Let $F = \langle x_1, x_2, \dots, x_n \rangle$ be a free right R-module with basis $x_1, x_2, \dots, x_n$, and A_m ($m = 1, 2, \dots$) elements of the $n \times n$-matrix ring R_n over R such that

$$A_m = A_{m+1} A_m \qquad (m = 1, 2, \dots).$$

Denote by $P(A_m = A_{m+1} A_m)$ the module

$$\bigcup_{m=1}^{\infty} P(A_m)$$

where

$$P(A) = \langle u_1, u_2, \dots, u_{nm} \rangle$$

and the elements are defined as follows:

$$(u_{1m}, u_{2m}, \dots, u_{nm}) = (x_1, x_2, \dots, x_n) A_m \quad (m = 1, 2, \dots).$$

Lemma 1 [5]. *A projective right R-module P of type P is countable generated.*

Proof follows from [10] and [11].

Theorem 2 [12]. *If Π is a finitely generated flat right R-module, P a projective right R-module, and $\Pi/\Pi J(R) \cong P/PJ(R)$ then $\Pi \cong P$.*

Proof follows from the theorem of [13].

Theorem 3 [5]. *Let P be a projective right R-module of type $P^0(n)$. There exists a right R-module $P^* = P(A_m = A_{m+1} A_m)$, $A_m \in R_n$ ($m = 1, 2, \dots$) such that $P/PJR \cong P^*/P^*J(R)$. If the first is finitely generated then the second, too. Furthermore, P is finitely generated iff P^* is.*

Denote by $P^0R(n)$, $PR(n)$, and $\Pi^0 R(n)$ respectively the categories of all projective right R-modules P of type $P^0(n)$, n-generated projective right R-modules, and n-generated flat right R-modules Π for which the quotient modules $\Pi/\Pi J(R)$ are $R/J(R)$-projective modules.

Theorem 4 [6]. *Let n be an integer and R a ring. Then the following conditions are equivalent:*

1) *the categories $P^0R(n)$ and $PR(n)$ coincide,*

2) *$A = A \cdot E^{-1} \cdot A$ whenever two elements A and E of the ring R_n satisfy the conditions $1 - E \in J(R_n)$ and $A^2 = E \cdot A$,*

2′) $A_* = A_* E_*^{-1} \cdot A_*$ *whenever two elements* A_*, *and* E_* *of the ring* R_n *satisfy the conditions* $1 - E_* \in J(R_n)$ *and* $A_*^2 = A_* E_*$,

3) *the categories* $\Pi^0 R(n)$ *and* $PR(n)$ *coincide.*

The following theorem generalizes the result of [7].

Theorem 5 [13]. *For any* $n \geqslant 2$, *there exists a semilocal* $(n-1)$-FI-*algebra* S *such that* $P^0 S(n-1) = PS(n-1)$, $P^0 S(n) \neq PS(n)$.

Theorem 6. *Let* $\mathfrak{A}$ *be an ideal of a ring* R, $\mathfrak{A} \leqslant J(R)$, *and the idempotents modulo* $\mathfrak{A}_n$ *be liftable. If the quotient module* $P/P \cdot \mathfrak{A}$ *of a projective right* R-*module* P *is* n-*generated then it is true for the* R-*module* P.

Proof follows from Theorem 4 and 5.

Theorem 7. *Let* n *be an integer,* R^0 *a subring,* $\mathfrak{A}$ *an ideal of a ring* R, *and* $R = R^0 \oplus \mathfrak{A}$, $\bigcap_{p=1}^{\infty} \mathfrak{A}^p = 0$. *If every right* R^0-*module of the type* $P(A^0 = A^0_{m+1} \cdot A^0_m)$ *is finitely generated then every right* R-*module of the type* $P(A_m = A_{m+1} \cdot A_m)$ *is, too; here* $A^0_m \in R^0_n$ *and* $A_m \in R_n$ $(m = 1, 2, \dots)$.

Proof. Assume that

$$P = P(A_m = A_{m+1} \cdot A_n), \quad A_m \in R_n, \quad A_m = A^0_m + A^*_m,$$

$$A^0_m \in R_n, \quad A^*_m \in \mathfrak{A}_n \qquad (m = 1, 2, \dots).$$

Since $A^0_m = A^0_{m+1} \cdot A^0_m$ by the assumption, the R^0-module $P(A^0_m = A^0_{m+1} \cdot A^0_m)$ is finitely generated. Hence $A^0_m = A^0_{m'} \cdot A^0_m$ for $m, m' \geqslant q$ with a fixed number q. Further,

$$P = P(A_q) + P\mathfrak{A} = P(A_q) + P\mathfrak{A}^l,$$

$$A_m - A_{m'} \cdot A_m \equiv 0 \qquad (\operatorname{mod} \mathfrak{A}^l_m) \qquad (l = 1, 2, \dots).$$

By the condition we have

$$\bigcap_{l=1}^{\infty} \mathfrak{A}^l = 0.$$

This implies $A_m = A_{m'} \cdot A_m$.

Theorem 8. *Let* n *be an integer,* $\hat{R} = R\langle Y \rangle$ *a polynomial ring in* y_α $(\alpha \in I)$, $Y = \{y_\alpha, \alpha \in I\}$. *If every right* R-*module of type* $P(A_m = A_{m+1} \cdot A_m)$ *is finitely generated then every right* $\hat{R}$-*module* $P(\hat{A}_m = \hat{A}_{m+1} \cdot \hat{A}_m)$ *is, too; here* $A_m \in R_n$, $\hat{A}_m \in \hat{R}_n$ $(m = 1, 2, \dots)$.

Proof follows from Theorem 7.

Theorem 9. *Let $Y = \{y_\alpha\}$ $(\alpha \in I)$ be an alphabet. If $R = R^0\langle Y \rangle$ is a polynomial ring over a ring R^0 in y_α $(\alpha \in I)$ then every projective right R-module of type $P^0(n)$ is finitely generated.*

Proof. Denote by $\mathfrak{B}$ the ideal of the ring R generated by commutators $[y_\alpha, y_\beta]$ $(\alpha, \beta \in I)$. By Theorem 1 of [14] $J(\overline{R}) = \mathfrak{N}^0[\overline{y}]$ where $\overline{R} = R/\mathfrak{B}$ is the quotient ring, $\mathfrak{N}^0$ is the absolute nil ideal of the ring R^0. Assume that $P = P(A_m = A_{m+1} \cdot A_m)$, $A_m = A_m^0 + A^*$, $A_m^0 \in R_n^0$, $A_m^* \in \mathfrak{A}_n$. Denote by $\mathfrak{A}$ the ideal generated by y_α $(\alpha \in I)$. Then the right R^0-module $P^* = P(A_m^0 = A_{m+1}^0 \cdot A_m^0)$ is a finitely generated module $\mathfrak{A}_n^0$ and P^* is finitely generated in view of Theorem 6. Therefore every right R-module of type $P^0(n)$ is finitely generated.

Theorem 10. *If R is an algebraic algebra over a field then every projective right R-module of type $P^0(n)$ is finitely generated.*

Proof. By Theorem 4, it suffices to show that every projective right R-module of a type $P(A_m = A_{m+1} \cdot A_m)$ is finitely generated; here $A_m = E^{-m-1} \cdot A \cdot E^{-n}$, $A \in R_n$, $1 - E \in J(R_n)$, $A^2 = E \cdot A$. We now proceed by induction on n and l where l designates the number of non-nilpotent elements on the diagonal of the matrix A_1.

1. Let the element a_{11} of A_1 be invertible. Since $m > 1$, there exist the invertible matrices U, V such that

$$A_1^* = U \cdot A_1 \cdot V = \begin{pmatrix} 1 & | & 0 \\ - & - & - \\ 0 & | & A_{11} \end{pmatrix}, \quad A_{11} \in R_{n-1},$$

$$A_m^* = U A_m U^{-1} = \begin{pmatrix} 1 & | & * \\ - & - & - \\ 0 & | & A_{11}^m \end{pmatrix}, \quad A_{11}^m \in R_{n-1}.$$

Hence we can construct a right R-module $P^* = P(A_{11}^{(m)} = A_{11}^{(m+1)} \cdot A_{11}^{(m)})$ that is finitely generated by the induction.

2. We now proceed by induction on l. In view of item 1 we may suppose that the element $a = a_{11}$ of the matrix A_1 is non-nilpotent. Hence, there exists an $e = a \cdot d$ such that $e = e^2$ and $a \cdot e = e \cdot a$. Therefore the element

$$E = A_1 \cdot d \cdot e_{11} = \begin{pmatrix} 1 & | & 0 \\ - & - & - \\ 0 & | & A_{11} \end{pmatrix}$$

is idempotent, where e_{11} is the matrix unit. We set $\widetilde{A}_m = (1-E)A_m$. Then the number of non-nilpotent elements on the diagonal of the matrix $\widetilde{A}_{11}$ is less than l, and by induction the right R-module $\widetilde{P} = P(\widetilde{A}_{m+1} + \widetilde{A}_m)$ is finitely generated. Consequently, P is also finitely generated.

We now consider the question on projective modules of the type $P(A_m = A_{m+1} + A_m)$ over PI-ring to be finitely generated.

Theorem 11 [15]. *Let R be a PI-ring. If elements a_m $(m = 1, 2, \dots)$ of R satisfy the condition $a_m = a_{m+1} \cdot a_m$ then there exist elements d_m $(m = 1, 2, \dots)$ of R such that $d_m = d_{m+1} \cdot d = d_{m+2} \cdot d_m$ and $P(a_m = a_{m+1} \cdot a_m) = P(d_m = d_{m+1} \cdot d_m)$.*

Theorem 12 [15]. *Let n be an integer, R a PI-ring. Then the following conditions are equivalent:*

1) *every right R-module $P(A_m = A_{m+1} \cdot A_m)$ is finitely generated; here $A_m \in R_n$ $(m = 1, 2, \dots)$;*
2) *every left R-module $P(A_m^* = A_m^* \cdot A_{m+1}^*)$ is finitely generated; here $A_m^* \in R_n$ $(m = 1, 2, \dots)$.*

Theorem 13 [15]. *Let $\mathfrak{A}$ be an ideal of a ring R, $\mathfrak{A}$ be a PI-ring, the quotient module $P/P\mathfrak{A}$ be finitely generated. If the right R-module P is of type $P^0(n)$ then P is a finitely generated R-module.*

We say that a right R-module M satisfies ACCS$_n$ (DCCS$_n$) if there no infinite properly ascending (descending) chain

$$M_1 \leqslant M_2 \leqslant \cdots \leqslant M_m \dots$$

$$(M_1 \geqslant M_2 \geqslant \cdots \geqslant M_m \dots)$$

of n-generated submodules $M_m = \langle U_{1m}, U_{2m}, \dots, U_{nm}\rangle$ $(n = 1, 2, \dots)$ of the module M, where

$$(U_{1m}, U_{2m}, \dots, U_{nm}) = (U_{1m+1}, U_{2m+1}, \dots, U_{nm+1})$$

$$((U_{1m+1}, U_{2m+1}, \dots, U_{nm+1}) = (U_{1m}, U_{2m}, \dots, U_{nm}) \cdot B_{m+1})$$

and $A_m = A_{m+1} \cdot A_m$, $B_{m+1} = B_m \cdot B_{m+1}$, $A_m, B_m \in R_n$ $(m = 1, 2, \dots)$.

Theorem 14. *Let n be an integer, R a PI-ring, $R \leqslant \Lambda_q$, $\Lambda = \prod_{\alpha \in I} \mathbf{H}^\alpha$ the direct product of algebraically closed fields $\mathbf{H}^\alpha (\alpha \in I)$. Then the following conditions are equivalent:*

(a) *the right R-module Λ_q satisfies* ACCS$_n$,

(b) *the right R-module Λ_q satisfies* DCCS$_n$,

(c) *every right R-module of a type* $P(A_m = A_{m+1} \cdot A_m)$ *is finitely generated where* $A_m \in R_n$ $(m = 1, 2, \dots)$.

Proof. (a)$\Rightarrow$(c) Let P be a right R-module of type $P(A_m = A_{m+1} \cdot A_m)$ where $A_m \in R_n$ $(m = 1, 2, \dots)$. Since the field $\mathbf{H}^\alpha$ is algebraically closed, there exists an invertible matrix $U = (U^\alpha)$ $\alpha \in I$, $U^\alpha \in \mathbf{H}^\alpha_{nq}$, $U \in \Lambda_{qn}$, such that

$$\widetilde{A}^\alpha_m = U^\alpha A^\alpha_m U^{\alpha-1} = \begin{pmatrix} J^\alpha_m & | & 0 \\ - & - & - \\ 0 & | & 0 \end{pmatrix},$$

where $A_m = (A^\alpha_m)$, $\alpha \in I$, $A^\alpha_m \in \mathbf{H}^\alpha_{nq}$, and J^α_m is a Jordan box. We can assume that J^α_m is determined by the Jordan boxes corresponding to the elements $h^\alpha \in \mathbf{H}^\alpha$, $h^\alpha \neq 0$. We denote

$$\widetilde{R}_n = U R_n U^{-1}, \quad \widetilde{A}_m = U A_m U^{-1} = (\widetilde{a}_{i\,jm}),$$

$$(\widetilde{a}_{i\,1m}, \widetilde{a}_{i\,2m}, \dots, \widetilde{a}_{i\,nm}) = (0, \dots, \varepsilon_{im}, \delta_{im}, 0, \dots, 0).$$

The elements are determined by the Jordan matrix $\widetilde{A}^\alpha_m$ and belong to $\mathbf{H}^\alpha_q$ $(i = 1, 2, \dots, n;\ m = 1, 2, \dots)$. Therefore, we have

$$\langle 0, \dots 0, \varepsilon^\alpha_{iq}, \delta^\alpha_q, 0, \dots, 0\rangle = \langle 0, \dots 0, \varepsilon^\alpha_{im}, \delta^\alpha_{im}, 0, \dots, 0\rangle \tag{1}$$

$$r_{A^\alpha} = r_{A^\alpha_m} \qquad (m = q+1, \dots, \alpha \in I).$$

Moreover, $A^\alpha_m = A^\alpha_{m+1}$ $(\alpha \in I)$ and $\widetilde{A}_m = \widetilde{A}_{m+1}$, $A_m = A_{m+1}$ $(m = q, q+1, \dots)$. Consequently, the R-module P is finitely generated. By Theorem 12 we can assume that $B_{m+1} = B_m \cdot B_{m+2} = B_{m+2} \cdot B_m$, hence the equivalence (b)$\Leftrightarrow$(a) follows from (1) and the equality $A_m = A_{m+1} \cdot A_m$ where $A_m = 1 - B_m$.

(c)$\Rightarrow$(a) Since the R-module $P(A_m = A_{m+1} \cdot A_m)$ is finitely generated then $A_m = A_q \cdot A_m = A_{m'} \cdot A_m$ for $m, m' \geqslant q$. Therefore the condition is satisfied.

Theorem 15. *There exists a PI-algebra R such that every right R-module* $P(a_m = a_{m+1} \cdot a_m)$, $a_m \in R$ $(m = 1, 2, \dots)$ *is finitely generated, and there exists a right R-module* $P(A_m = A_{m+1} \cdot A_m)$, $A_m \in R_n$, $n \geqslant 2$, *that is not finitely generated.*

Proof. Let K be a field $\operatorname{Char} K \neq 2$, and $R = K\langle a, b, c \mid A = A^2, bc = cb\rangle$ be a K-algebra where

$$A = \begin{pmatrix} a & b \\ c & a \end{pmatrix}.$$

We have $a^2 + bc = a$, $b = ab + ba$, $c = ac + ca$. It is easy to show that $(x_1 \cdot x_2 - x_2 \cdot x_2)^2 \cdot x_3 - x_3 \cdot (x_1 \cdot x_2 - x_2 \cdot x_2)^2 = 0$ is an identity on R. Hence R is a PI-algebra.

We write

$$E_m = \begin{pmatrix} a^{(m)} & b^{(m)} \\ c^{(m)} & a^{(m)} \end{pmatrix}$$

and denote by

$$\begin{aligned} S = K\langle a^{(m)}, b^{(m)}, c^{(m)} \ (m = 1, 2, \dots) \mid E_m = E_m^2,\ b^{(m)}c^{(m)} = c^{(m)}b^{(m)}; \\ x^{(m)}x^{(m')} = x^{(m')}x^{(m)} = 0;\ x^{(m)} = a^{(m)},\ b^{(m)},\ c^{(m)}; \\ x^{(m')} = a^{(m')},\ b^{(m')},\ c^{(m')};\ (m \neq m'\ m, m' = 1, 2, \dots)\rangle \end{aligned}$$

a K-algebra. It is obvious that S is a PI-algebra. We can prove that the algebra

$$S^{(m)} = K\langle a^{(m)},\ b^{(m)},\ c^{(m)} \mid E_m = E_m^2, b^{(m)}c^{(m)} = c^{(m)}b^{(m)}\rangle$$

has no zero divisors. If $P = P(d_t = d_{t+2}d_t)\ d_t \in S\ (t = 1, 2, \dots)$ then

$$d_t = d_{t+1}d_t, \quad d_t = \sum_m d_t^{(m)} + a_t^0,\ a_t^0 \in K, \quad d_t^{(m)} = f_t(a^{(m)}, b^{(m)}, c^{(m)}).$$

The last expression implies either $d_t = 1$ or $d^t = 0$ $(t = 2, 3, \dots)$ Therefore, the S-module P is finitely generated. But the right S-module $P^* = P(A_m = A_{m+1}A_m)$ is not finitely generated; here $A_m = \sum_{i=2}^{m} \cdot E_i$.

References

[1] Lazard, D., Liberté des gros modules projectifs. J. Algebra 31 (1974), 437–451.

[2] Sakhaev, I. I., Čirkov, G. V., Projectivity of finitely generated plane modules over a similocal ring with a single primitive ideal. Izv. Vyssh Uchebn. Zaved., Mat. 97:6 (1970), 106.

[3] Jondrup, S., Projective modules. Proc. Am. Math. Soc. 59:2 (1976), 217–221.

[4] Valette, J., Sur les modules projectifs. C. R. Acad. Sci., Paris, Sér. A–B, 282 (1976), 821–823.

[5] Sakhaev, I. I., Finite generation of projective modules. Izv. Vyssh. Uchebn. Zaved., Mat. 9 (1977), 69–79. English translation: Sov. Math. (Iz. VUZ) 21:9 (1977), 55–62.

[6] — Lifting the finite generation of a projective module modulo its radical. Mat. Zametki 49:3 (1991), 97–108. English translation: Math. Notes 49:3 (1991), 291–301.

[7] Gerasimov, V. N., Sakhaev, I. I., A counterexample to two hypotheses on projective and flat modules. Sib. Mat. Zh. 24:6 (1984), 31–33. English translation: Sib. Math. J. 24:6 (1984), 855–859.

[8] Cohn, P. M., Free Rings and Their Relations. Mir, Moscow 1975. Original: Lond. Math. Soc. Monogr. 2. Academic Press, London–New York 1971.

[9] Gerasimov, V. N., Localizations in associative rings. Sib. Mat. Zh. 23:6 (1982), 36–54. English translation: Sib. Math. J. 23:6 (1982), 788–804.

[10] Kaplansky, I., Projective modules. Ann. Math., II. Ser. 68:2 (1958), 372–377.

[11] Bass, H., Finitistic dimension and a homological generalization of semi-primary rings. Trans. Am. Math. Soc. 95:3 (1960), 466–488.

[12] Sakhaev, I. I., Projectivity of finitely generated flat modules over semilocal rings. Mat. Zametki 37:2 (1985), 152–161. English translation: Math. Notes 37:2 (1985), 85–90.

[13] Sakhaev, I. I., On BI-rings and the Lazard conjecture. Math. Nachr. 144 (1989), 87–98.

[14] Amitsur, S. A., Radicals of polynomial rings. Can. J. Math. 8:3 (1956), 355–361.

[15] Sakhaev, I. I., On projective modules over rings with polynomial identities to be finitely generated. Izv. Vyssh. Uchebn. Zaved., Mat. 8 (1993), 65–75.

Determination of abelian groups by their endomorphism rings

Anatoliĭ M. Sebel'din

Abstract. The well-known result of Baer–Kaplansky on two periodic abelian groups with isomorphic rings of endomorphisms to be isomorphic has given rise to numerous studies. The purpose is to give a survey of the obtained results.

1991 Mathematics Subject Classification: 20K30, 20K15.

The well-known result of Baer–Kaplansky [1, Section 108] on two periodic abelian groups with isomorphic rings of endomorphisms to be isomorphic has given rise to numerous studies. The purpose of this paper is to give a survey of the obtained results.

Denote by $\mathfrak{X}$ a class of abelian groups and by $\mathfrak{U}$ the class of all abelian groups.

We say that a group G from a class $\mathfrak{X}$ is *determined by its endomorphism ring* $E(G)$ *in* $\mathfrak{X}$ if there is no group H in $\mathfrak{X}$ such that $H \ncong G$ and $E(H) \cong E(G)$. Denote by $\mathfrak{X}(E)$ the subclass of $\mathfrak{X}$ which is composed of groups determined by their endomorphism rings in $\mathfrak{X}$. Problem 43 of [2] can be reformulated as follows: "In which cases $E(G) \cong E(H)$ implies $G \cong H$?"

The above-mentioned problem can be viewed as determination of an abelian group by its endomorphism ring in the class $\mathfrak{U}$, i.e., we have the description of $\mathfrak{U}(E)$.

A class $\mathfrak{X}$ is called an *E-class* if any group from $\mathfrak{X}$ is determined by its endomorphism ring in $\mathfrak{X}$, i.e., $\mathfrak{X}(E) = \mathfrak{X}$, a *NE-class* if there is no group in $\mathfrak{X}$ which is determined by their endomorphism rings in $\mathfrak{X}$, i.e. $\mathfrak{X}(E) = \emptyset$.

Denote by $\mathfrak{X}(EI)$ the subclass of $\mathfrak{X}(E)$ which consists of groups G such that any ring isomorphism $\alpha : E(G) \to E(H)$ is induced by a group isomorphism $\beta : G \to H$ $(\alpha : \gamma \to \beta\gamma\beta^{-1})$.

A class $\mathfrak{X}$ is called an *EI-class* if $\mathfrak{X} = \mathfrak{X}(EI)$. In particular, the class $\mathfrak{T}$ of periodic abelian groups is an EI-class [1, Section 108].

Similar definitions can be introduced for the multiplicative semigroup $E^{\cdot}(G)$, the additive group $E^{+}(G)$, and the group of invertible elements $\mathrm{Aut}\,(G)$ of the endomorphism ring of an abelian group G.

By [3], a periodic abelian group $G = \bigoplus_{p \in \operatorname{supp} G} G_p$ is determined by its endomorphism ring in the class of all abelian groups $\mathfrak{U}$ iff either G is bounded or $\operatorname{supp} R(G) \cap \operatorname{supp} D(G) \neq \emptyset$ where $R(G)$ stands for the reduced part of G and $D(G)$ denotes a divisible part of G. Thus, the construction of groups from $\mathfrak{F} \cap \mathfrak{U}(E)$ is known.

It is clear that the intersection $\mathfrak{F} \cap \mathfrak{U}(E)$ is not empty. Here and in the sequel we denote by $\mathfrak{F}$ the class of torsion-free abelian groups. Indeed, every divisible torsion-free abelian group belongs to $\mathfrak{U}(E)$. The class $\mathfrak{U}(E)$ includes reduced torsion-free abelian groups [4, 5]. A nonreduced completely decomposable torsion-free abelian group of finite rank belongs to $\mathfrak{U}(E)$ iff it is almost divisible, i.e. the set $P(G) = \{p : p \text{ is prime } pG \neq G\}$ is finite [6].

To construct mixed abelian groups from $\mathfrak{U}(E)$ we shall make the following remark. If $G \in \mathfrak{U}$, $G = F(G) \oplus T(G)$, $0 \neq F(G) \in \mathfrak{F}(E)$, $0 \neq T(G) \in \mathfrak{U}(E)$, where $F(G)$ is the torsion-free part of a group G and $T(G)$ a periodic one, then $G \notin \mathfrak{U}(E)$ iff $F(G)$ is fully invariant reduced almost divisible algebraically compact and $T(G)$ is a bounded subgroup of G.

Suppose that $G \notin \mathfrak{U}(E)$. There is an abelian group H such that $H \not\cong G$ and $E(H) \cong E(G)$. In this case we have $H = H_1 \oplus H_2$ where $E(F(G)) \cong E(H_1)$ and $E(T(G)) \cong E(H_2)$. Then $T(G) \cong H_2$. It is clear that $T(H_1) \neq 0$ (otherwise $G \cong H$). Since $\operatorname{Hom}(H_2, H_1) = 0$, we have $\operatorname{supp} T(H_1) \neq \operatorname{supp} H_2$. Consequently, $T(G) \not\cong T(H)$. Taking into account [3] and [4] we obtain the required assertion.

On the contrary, if $F(G)$ is a fully invariant almost divisible subgroup then $P(F(G))$ is a finite set and $P(F(G)) \cap \operatorname{supp} T(G) = \emptyset$. Therefore, there is a periodic abelian group H such that $E(G) \cong E(H)$ [3, 4].

Let $\mathfrak{U}_A$ be a class of abelian groups $G = R(G) \oplus D(G)$ such that $R(G)$ is included in $D(G)$ and $D(G)$ has $\mathbb{Z}(p^\infty)$ for any prime p. As was shown in [7], $\mathfrak{U}(EI) \supset \mathfrak{U}_A$.

A torsion-free abelian group belongs to $\mathfrak{U}(E)$ iff it belongs to $\mathfrak{F}(E)$ and is not reduced algebraically compact [8]. Thus, determination of torsion-free abelian groups by their endomorphism rings in the class of abelian groups can be reduced to the same problem in the class of torsion-free abelian groups. If $G \in \mathfrak{F}$ and $R(G) \in \mathfrak{F}(E)$ then $G \in \mathfrak{F}(E)$. The converse assertion is wrong (for example see [6]). Torsion-free reduced abelian groups rarely belong to $\mathfrak{F}(E)$. Due to the well-known result of Corner (see [1, Section 110]), the class of countable reduced torsion-free abelian groups is a NE-class. Summing up this result, Dugas and Göbel have shown the class of all reduced torsion-free abelian groups with no subgroups $\mathbb{J}_p$ for all prime p to be a NE-class [9]. On the other hand, the class $\mathfrak{F}_{AC}$ of algebraically compact torsion-free abelian groups is an E-class and a group from $\mathfrak{F}_{AC}$ belongs to $\mathfrak{F}(E)$ iff it is almost divisible [4].

The following condition necessary for $G \in \mathfrak{F}(E)$ was found in [8]:

(α) every pure subgroup of rank 1 is almost divisible.

A class $\mathfrak{X}$ is called an *E_γ-class* if a condition (γ) for $G \in \mathfrak{X}$ is necessary and sufficient for $G \in \mathfrak{X}(E)$.

List some examples of E_α-classes:

the class $\mathfrak{F}_{CD}$ of completely decomposable torsion-free abelian groups [10, 11],

the class $\mathfrak{F}_{HS}$ of homogeneous separable torsion-free abelian groups [12],

the class $\mathfrak{F}_{DS}$ of homogeneously decomposable separable torsion-free abelian groups [8],

the class $\mathfrak{F}_S$ of separable torsion-free abelian groups [13].

Note that the condition (α) for the groups of the above-mentioned classes can be written as follows:

(β) every direct summand of rank 1 is almost divisible.

The class $\mathfrak{F}_{RV}$ of reduced vector groups, whose cardinality does not exceed the first cardinal of non-zero measure, is a E_β-class [14].

The author [15] pointed out EI- and E_α-classes. The class of subgroups of a direct power of the group $\mathbb{Z}$ is an EI-class [16]. W. May [17] announced that the class of mixed torsion-free abelian groups of finite rank with totally projected periodic part and which are $\mathbb{Q}_P^*$-modules is an EI-class. He also established the existence of an example of a NE-class consisting of mixed abelian groups of torsion-free rank 1 with fixed separable p-primary periodic parts and which are $\mathbb{Q}_P^*$-modules. There is a construction of a NE-class of mixed abelian groups with fixed p-primary periodic part such that the factor groups of groups of the class by pointed parts are divisible groups of finite rank more than 1 (see [18]). The negative solution of Problem 87 of book [1], examples of E-classes of mixed abelian groups of torsion-free rank 1 were given in [19, 20].

Summarizing the results of I. Kaplansky [21] and K. Wolfson [22] we can assert that if R is a complete discrete valuation ring and both R-modules A and B are either periodic or torsion-free, then each isomorphism $E_R(A) \cong E_R(B)$ is induced by an isomorphism of R-modules A and B.

For mixed R-modules, W. May [23] proved that the isomorphisms are induced if B is an arbitrary R-module, a R-module A is mixed with additional conditions. An example of an EI-class [24] is pointed in the class of endocyclic torsion-free abelian groups.

Under the condition (ZFC+GCH), the class $\mathfrak{T}(E^+)$ is described in [25] and the problem on a reduced primary group to be in the class $\mathfrak{U}(E^+)$ is solved there, too. Under the same condition, S. Ya. Grinshpon and the author proved that if $G = R(G) \oplus D(G)$ is a nonreduced periodic abelian group from the class $\mathfrak{T}(E^+)$ then $E^+(G) \cong E^+(H)$, where $H = R(G) \oplus H'$ and H' is a torsion-free abelian group. Thus, the class $\mathfrak{T} \cap \mathfrak{U}(E^+)$ is described in the axiomatic system (ZFC+GCH).

Clearly, torsion-free divisible groups of finite rank, mixed groups with finite periodic part and with torsion-free part being divisible of finite rank belong to $\mathfrak{U}(E^+)$. The class $\mathfrak{U}(E^+)$ has also nondivisible torsion free groups [5]. The classes $\mathfrak{F}_{RV}(E^+)$ and $\mathfrak{F}_{CD}(E^+)$ were described in [14, 26], and the description of the last class depends on the choice of axiomatic system. For instance, it formally coincides with the description of the class $\mathfrak{F}_{RV}(E^+)$ in the axiomatic system (ZFC+GCH) because the classes satisfy the condition (β) and the following condition

(β') if the set of direct summands of rank 1 of the same type of fixed decomposition is finite then the set of direct summands of rank 1 of a smaller type is also finite.

Both the conditions are necessary and sufficient. In axiomatic system (ZFC+GCH) the class $\mathfrak{F}_D$ of divisible torsion-free abelian groups is a subclass of $\mathfrak{U}(E^+)$ and without GCH the inclusion $\mathfrak{F}_D \subset \mathfrak{U}(E^+)$ is equivalent to the following proposition: *for any cardinals* $\mathfrak{M}$ *and* $\mathfrak{N}$, *the equality* $2^{\mathfrak{M}} = 2^{\mathfrak{N}}$ *implies* $\mathfrak{M} = \mathfrak{N}$.

A reduced torsion-free abelian group belongs to $\mathfrak{U}(E^+)$ iff it belongs to $\mathfrak{F}(E^+)$ and is not algebraically compact [5]. Under the condition (ZFC+GCH), a nonreduced torsion-free abelian group belongs to $\mathfrak{U}(E^+)$ iff it belongs to $\mathfrak{F}(E^+)$ and its reduced part is not a direct sum of its two fully invariant subgroups, one of which is non-zero algebraically compact [6]. Under the same condition, the class of reduced algebraically compact torsion-free abelian groups is an E^+-class and a nonreduced algebraically compact group belongs to the class $\mathfrak{F}_{AC}(E^+)$ iff the rank of its reduced part is strictly less than that of its divisible part [5]. Thus, the class $\mathfrak{F}_{AC}(E^+)$ is described in the axiomatic system (ZFC+GCH).

The class of periodic abelian groups is an $E^{\cdot}$-class [27]. Generalizing May's result ($\mathfrak{U}_A \subset \mathfrak{U}(E)$) P. Puusemp [28] proved that $\mathfrak{U}_A \subset \mathfrak{U}(E^{\cdot})$. The class $\mathfrak{F}_{CD}(E^{\cdot})$ is described in [29]. A nonreduced torsion-free abelian group G belongs to $\mathfrak{X}(E^{\cdot})$ if its reduced part $R(G)$ belongs to $\mathfrak{X}(E^{\cdot})$ [29]. Let G be a nonreduced torsion-free abelian group. Suppose that the rank of the reduced part of G does not exceed that of the divisible part. Then G belongs to $\mathfrak{X}(E^{\cdot})$ iff $G \in \mathfrak{X}(E)$ [29]. In particular, divisible torsion-free abelian groups belong to $\mathfrak{U}(E^{\cdot})$. Using Lemma 6.1 of [29] it can be proven that $\mathfrak{F}_{AC}$ is an $E^{\cdot}$-class and a group from $\mathfrak{F}_{AC}$ belongs to $\mathfrak{F}(E^{\cdot})$ iff it is almost divisible. In [30], the author announced that a separable torsion-free abelian group G of rank more than 1 and having the smallest type in the set of all different types of direct summands of rank 1 belongs to $\mathfrak{F}_S(E^{\cdot})$ iff G satisfies the condition (α). Interesting $E^{\cdot}$-classes of abelian groups can be obtained from module results of [31] and [32]. Let $A = B \oplus C$ be an abelian group such that $E(B)$ is a ring of unique addition. Then $E(A)$ is a ring of unique addition if one of the following conditions is valid: 1) $E(C)$ is a ring of unique addition; 2) C is isomorphic to a subgroup of B; 3) C is an epimorphic image of B.

Denote by $\mathfrak{F}_{AC}^R$ the class of reduced groups from $\mathfrak{F}_{AC}$. The following equalities hold: $\mathfrak{F}_{AC}(E^{\cdot}) = \mathfrak{F}_{AC}$, $\mathfrak{F} \cap \mathfrak{U}(E^{\cdot}) = \mathfrak{F}(E^{\cdot}) \setminus \mathfrak{F}_{AC}$, $\mathfrak{F}_{AC} \cap \mathfrak{F}(E^{\cdot}) = \mathfrak{F}_{AC} \cap \mathfrak{F}(E)$, and $\mathfrak{T} \cap \mathfrak{U}(E^{\cdot}) = \mathfrak{T} \cap \mathfrak{U}(E)$.

In conclusion we give some results on automorphism groups. H. Leptin [33] for $p \geqslant 5$ and W. Liebert [34] for $p \geqslant 3$ have shown that the class of primary abelian groups is an Aut-classe. On the other hand, V. A. Nikiforov [35] proved that the class of torsion-free abelian groups with finite automorphism group is a NAut-class. A NAut-class of countable torsion-free abelian groups with countable divisible automorphism group is pointed out in [36]. It is proved in [37] that if $A \in \mathfrak{X}(E)$ ($A \in \mathfrak{X}(\mathrm{Aut})$) then $A \in \mathfrak{X}(EI)$ ($A \in \mathfrak{X}(\mathrm{Aut}\, I)$) iff every automorphism of the ring $E(A)$ (of the group $\mathrm{Aut}\,(A)$) is inner. The class of p-adic ($p \neq 2$) algebraically compact torsion-free abelian groups is an Aut-class [38].

References

[1] Fuchs, L., Infinite abelian Groups 1,2 (Pure Appl. Math. 36, 36–II). Academic Press, New York–London 1970, 1973.

[2] — Abelian Groups. Hungarian Academy of Sciences, Budapest 1958.

[3] May, W., Toubassi, E., Endomorphisms of abelian groups and the theorem of Baer and Kaplansky. J. Algebra 43:1 (1976), 1–13.

[4] Krylov, P. A., Sebel'din, A. M., On determination of a torsion-free abelian group by its endomorphism ring. In: Proc. Ist Conf. of Young Scientists, 30–33. Tomsk 1975.

[5] Sebel'din, A. M., On determination of a torsion-free abelian group by its endomorphism ring or its endomorphism group. In: Groups and Modules, 78–85. Tomsk. Gos. Univ., Tomsk 1976.

[6] — Definability of a torsion-free unreduced abelian group by its group of endomorphisms. In: Abelian Groups and Modules, 102–108. Tomsk. Gos. Univ., Tomsk 1980.

[7] May, W., Endomorphism rings of abelian groups with ample divisible subgroups. Bull. Lond. Math. Soc. 10:3 (1978), 270–272.

[8] Sebel'din, A. M., Torsion-free abelian groups with isomorphic endomorphism rings. In: Abelian Groups and Modules, 165–170. Tomsk. Gos. Univ., Tomsk 1979.

[9] Dugas, M., Göbel, R., Every cotorsion-free ring is an endomorphism ring. Proc. Lond. Math. Soc., III. Ser. 45:2 (1982), 319–336.

[10] Sebel'din, A. M., Condition for the isomorphism of completely decomposable torsion-free abelian groups with isomorphic rings of endomorphisms. Mat. Zametki 11:4 (1972), 403–408. English translation: Math. Notes 11:4 (1972), 248–250.

[11] — Complete direct sums of torsion-free abelian groups of rank 1 with isomorphic groups or rings of endomorphisms. 1. Mat. Zametki 14:6 (1973), 867–878. English translation: Math. Notes 14:6 (1973), 1057–1063.

[12] Hauptfleisch, G. J., Torsion-free abelian groups with isomorphic endomorphism rings. Arch. Math. 24:3 (1973), 269–273.

[13] Bazzoni, S., Metelli, C., On abelian torsion-free separable groups and their endomorphism rings. In: Gruppi abeliani e Loro Relazioni con la Teoria dei Moduli (Dicembre 12–16, 1977) (Symp. Math. 23), 259–285. Academic Press, London–New York 1979.

[14] Sebel'din, A. M., Complete direct sums of torsion-free abelian groups of rank 1 with isomorphic groups or rings of endomorphisms. II. In: Abelian Groups and Modules, 159–164. Tomsk. Gos. Univ., Tomsk 1979.

[15] — Definability of abelian groups by their endomorphism rings. In: Abelian Groups and Modules 8, 113–123. Tomsk. Gos. Univ., Tomsk 1989.

[16] Mikhalev, A. V., Isomorphisms of endomorphism rings of modules that are close to free. Vestn. Mosk. Univ. Ser. I 2 (1989), 20–27. English translation: Mosc. Univ. Math. Bull. 44:2 (1989), 28–38.

[17] May, W., Endomorphism rings of mixed abelian groups. Contemp. Math. 87 (1989), 61–74.

[18] Göbel, R., May, W., The construction of mixed modules from torsion-free modules. Arch. Math. 48:6 (1987), 476–490.

[19] May, W., Toubassi, E., Isomorphisms of endomorphism rings of rank one mixed groups. J. Algebra 71:2 (1981), 508–514.

[20] — A result on problem 87 of L. Fuchs. In: Lect. Notes Math. 616, 354–367. Springer-Verlag, Berlin etc. 1977.

[21] Kaplansky, I., Infinite Abelian Groups. Michigan University Press, Ann Arbor 1954.

[22] Wolfson, K., Isomorphisms of the endomorphism rings of torsion-free modules. Proc. Am. Math. Soc. 13:5 (1962), 712–714.

[23] May, W., Isomorphism of endomorphism algebras over complete discrete valuation rings. Math. Z. 204:4 (1990), 485–499.

[24] Krylov, P. A., Irreducible torsion-free abelian groups and their endomorphism rings. In: Abelian Groups and Modules 4, 73–100. Tomsk. Gos. Univ., Tomsk 1986.

[25] Grinshpon, S. Ya., Primary abelian groups with isomorphic endomorphism groups. Mat. Zametki 14:5 (1973), 733-740. English translation: Math. Notes 14:5 (1973), 979–982.

[26] Sebel'din, A. M., Completely decomposable torsion-free abelian groups with isomorphic endomorphism rings. In: Post-Graduate Student's Works. 28–34. Tomsk. Gos. Univ., Tomsk 1974.

[27] Puusemp, P., On the determination of a torsion abelian group by its semigroup of endomorphisms in the class of all periodic abelian groups. Eesti NSV Tead. Akad. Toimetised Füüs.-Mat. 29:3 (1980), 246–253.

[28] — On a theorem of May. Eesti NSV Tead. Akad. Toimetised Füüs.-Mat. 38:2 (1989), 139–145.

[29] Sebel'din, A. M., On the determination of abelian groups by their semigroups of endomorphisms. In: Abelian Groups and Modules, 125–133. Tomsk. Gos. Univ., Tomsk 1991.

[30] — On determination of separable abelian groups by their semigroups of endomorphisms. In: VI Symp. on Ring Theory. Abstracts. 111. L'vov 1990.

[31] Mikhalev, A. V., Isomorphisms of semigroups of endomorphisms of modules. Algebra Logika 5:5 (1966), 59–67.

[32] Mikhalev, A. V., Isomorphism of semigroups of endomorphisms of modules. II. Algebra Logika 6:2 (1967), 35–47.

[33] Leptin, H., Abelsche p-Gruppen und ihre Automorphism-engruppen. Math. Z. 73 (1960), 235–253.

[34] Liebert, W., Isomorphic automorphism groups of primary abelian groups. In: Abelian Group Theory, Proc. 3rd Conf. (Oberwolfach, August 11–17, 1985), 9–31. Cordon and Breach, New York 1987.

[35] Nikiforov, V. A., Torsion-free abelian groups with finite groups of automorphisms. Mat. Zametki 39:5 (1986), 641–646. English translation: Math. Notes 39:5 (1986),

350–353.

[36] Dixon, M. R., Evans, M. J., Divisible automorphism groups. Q. J. Math., Oxf. II. Ser. 41:162 (1990), 179–188.

[37] Grinshpon, S. Ya., In: XIX All-Union Algebra Conf. (L'vov, September 9–11, 1987). Abstracts. Pt 1, 74. Institut Prykladnykh Problem Mekhaniki i Matematiki AN SSSR, L'vov 1987.

[38] Solonina, A. G., On determination of p-adic modules by automorphism groups. Deposited in VINITI, No 2660–75.

Fréchet algebras of global dimension zero

Yuriĭ V. Selivanov *

Abstract. The structure of a Fréchet algebra, i.e. a Fréchet space turned into an algebra by means of jointly continuous multiplication, of global dimension zero is investigated here.

1991 Mathematics Subject Classification: 46H20, 46H15.

Introduction

Let A be a complex Fréchet algebra, that is, a Fréchet space turned into an algebra by means of jointly continuous multiplication (everywhere in this work, A is assumed to be unital). Let A-mod be the category of (unital left) Fréchet A-modules, and let $\mathrm{dh}_A X$ ($X \in A$-mod) be the *homological dimension* of a Fréchet A-module X (for the definition of this dimension see [1]). The *global dimension* of A is, by definition, the number $\mathrm{dg}A = \sup\{\mathrm{dh}_A X : X \in A\text{-mod}\}$.

The present paper is devoted to the study of the structure of Fréchet algebras of global dimension zero. The analogous question posed for Banach algebras was considered earlier by the author in [2]. It transpired that a Banach algebra A with $\mathrm{dg}A = 0$ is isomorphic to the direct product of a finite number of full matrix algebras (and, as a consequence, it is finite-dimensional) provided that some spaces related to A have the (Grothendieck) approximation property. However outside the framework of Banach structures it is possible to find interesting examples of infinite-dimensional algebras with $\mathrm{dg}A = 0$. These examples (not necessarily with a metrizable topology) include:

(i) the algebra of all complex-valued functions on an arbitrary set (with the topology of pointwise convergence) [1, Theorem IV.5.27];

(ii) the topological product of an arbitrary family of full matrix algebras;

(iii) the algebra of distributions on a compact Lie group [3, Section 7].

We recall [1, Chapter III, Section 5] that the (Fréchet) A-modules X with $\mathrm{dh}_A X = 0$ are precisely the *projectives* ones; as regards the latter, they can be described in the following way.

* Supported by the Russian Fund for Fundamental Research (grant 93-011-156).

The *canonical morphism* for an A-module X means the morphism $\pi : A\widehat{\otimes}X \to X$ defined by the formula $\pi(a \otimes x) = a \cdot x$ $(a \in A,\ x \in X)$, where $\widehat{\otimes}$ denotes the projective tensor product of locally convex spaces [4], and the module operation $A\widehat{\otimes}X \in A$-mod is defined by $a \cdot (b \otimes x) = ab \otimes x$ $(a, b \in A, x \in X)$. We recall [1, Chapter IV,Section 1] that an A-module X is projective if and only if the canonical morphism π has a right inverse morphism ρ; in this case ρ is said to be a *coretraction of the projective A-module X*.

If S is an arbitrary subset of A, then $\operatorname{lan}(S)$ and $\operatorname{ran}(S)$ denote the *left* and *right annihilators* of S respectively [5, Section 30]. We recall that an algebra A is said to be *semiprime* if it contains no bi-ideals, $I \neq 0$, with $I^2 = 0$, and *right annihilator* if $\operatorname{ran}(J) \neq 0$ for each closed left ideal J, $J \neq A$.

The *topological radical* of a Fréchet algebra A is the intersection of all its closed maximal left ideals, or A itself if there are no such ideals. It is easy to see that the topological radical is also the intersection of the annihilators of all irreducible Fréchet A-modules and is therefore a closed bi-ideal. An algebra A is *topologically semisimple* if its topological radical is 0. It is clear that topologically semisimple algebras are semisimple (and, in particular, semiprime). The converse is false, since the closed maximal left ideals form only a part of the totality of maximal left ideals [6, Example 12.1].

We recall that the *left socle*, $\operatorname{soc}(A)$, of an algebra A is the sum of all its minimal left ideals. The left socle $\operatorname{soc}(A)$ is well known to be a bi-ideal of A. Finally, we recall the following definition: a Fréchet space E has the *approximation property* (AP for short) if for each compact set $K \subset E$ and any neighborhood U of zero in E there exists a continuous finite-dimensional operator $\varphi : E \to E$ such that $x - \varphi(x) \in U$ for all $x \in K$.

1. Homologically trivial Fréchet algebras

We shall prove the following theorem.

Theorem 1. *Let A be a semiprime Fréchet algebra with the approximation property and suppose that $\operatorname{dg}A = 0$. Let R be the topological radical of this algebra, and $\{I_\nu : \nu \in \Lambda\}$ the collection of finite-dimensional minimal bi-ideals of A such that $I_\nu \cap R = 0$; we put $P_\nu = \operatorname{lan}(I_\nu)$ $(\nu \in \Lambda)$. Then:*

(i) *A is the closure of the direct sum of R and all the I_ν, and moreover, for each ν, $A = I_\nu \oplus P_\nu$;*

(ii) *for each $\nu \in \Lambda$ the algebra I_ν is isomorphic to some full matrix algebra;*

(iii) *any closed bi-ideal of A containing R is the intersection of the bi-ideals P_ν those contain it.*

We preface to the proof of Theorem 1 a number of preparatory results. The following assertion is well known for the case of Banach spaces [4, Proposition 36].

Lemma 1. *Let E and F be Fréchet spaces, at least one of which has AP. Then for any $u \in E\widehat{\otimes}F$, $u \neq 0$, there exist continuous linear functionals $f \in E^*$ and $g \in F^*$ such that $(f\widehat{\otimes}g)(u) \neq 0$.*

Proof. Suppose that F has AP. Let $u \in E\widehat{\otimes}F$, $u \neq 0$. By [7, Lemma 1], for any neighborhood U of zero in $E\widehat{\otimes}F$ there exist a (continuous) finite-dimensional operator $\varphi : F \to F$ such that $u - (1_E\widehat{\otimes}\varphi)(u) \in U$. It follows that, for some finite-dimensional operator φ_0 on F, $v - (1_E\widehat{\otimes}\varphi_0)(u) \neq 0$ and v belongs to the algebraic tensor product $E \otimes F \subset E\widehat{\otimes}F$. Using [1, Proposition II.1.5] and the Hahn–Banach theorem, it is easy to find $f \in E^*$ and $h \in F^*$ such that $(f\widehat{\otimes}h)(v) \neq 0$. Putting $g = h\varphi_0$, we get$(f\widehat{\otimes}g)(u) \neq 0$. □

Lemma 2. *Let A be a Fréchet algebra, X a projective Fréchet A-module, and at least one of them have* AP. *Then for each $x \in X$, $x \neq 0$, there exists a (continuous) A-module morphism $\psi : X \to A$ such that $\chi(x) \neq 0$.*

Proof. Let $\rho : X \to A\widehat{\otimes}X$ be a coretraction of the module X. We consider $u = \rho(x) \in A\widehat{\otimes}X$. Since $\pi(u) = x \neq 0$, then $u \neq 0$. By Lemma 1, the approximation property for A or X implies the existence of two functionals $f \in A^*$ and $g \in X^*$ such that $(f\widehat{\otimes}g)(u) \neq 0$. The latter quantity is equal to $f(a)$, where $a = (1_A\widehat{\otimes}g)\rho(x) \in A\widehat{\otimes}\mathbf{C} = A$. Since $f(a) \neq 0$, then $a \neq 0$. Let $\chi = (1_A\widehat{\otimes}g)\rho$. Then $\chi : X \to A$ is an A-module morphism and $\chi(x) = a \neq 0$. □

Lemma 3. *Let A be a Fréchet algebra, I a closed left ideal in A, and at least one of the spaces A and A/I have* AP. *If the A-module A/I is projective, then* $\mathrm{lan}(\mathrm{ran}(I)) = I$.

Proof. It is clear that $I \subset \mathrm{lan}(\mathrm{ran}(I))$. Suppose that there exists $a \in \mathrm{lan}(\mathrm{ran}(I))$ such that $a \notin I$. We consider $x = a + I \in A/I$. By Lemma 2, $\chi(x) \neq 0$ for some A-module morphism $\chi : A/I \to A$. Let $b = \chi(1 + I)$. For any $c \in I$, we have $cb = \chi(c + I) = 0$. Hence $b \in \mathrm{ran}(I)$. Since $a \in \mathrm{lan}(\mathrm{ran}(I))$, then $ab = 0$, that is, $\chi(x) = \chi(a \cdot (1 + I)) = ab = 0$. We obtain a contradiction. □

Corollary 1. *Let A be a Fréchet algebra having AP and* $\mathrm{dg}A = 0$. *Then A is right annihilator.*

Proposition 1. *Let A be a semiprime Fréchet algebra, I a closed maximal left ideal in A, and at least one of the spaces A and A/I have* AP. *Then the A-module A/I is projective if and only if there exists a closed left ideal J in A such that $A = I \oplus J$.*

Proof. Sufficiency is obvious; we turn to necessity. By T denote the linear space, generated by the images of all possible A-module morphism from A/I to A. It is easy to see that T is a bi-ideal in A. Consequently, if $T \subset I$ then $a \cdot x = 0$ for all $a \in T$ and $x \in A/I$, and hence $T^2 = 0$; the latter implies that $T = 0$, since A is semiprime. According to Lemma 2, $T \neq 0$. Therefore there exists a morphism $\chi : A/I \to A$ whose image is not entirely contained in I. We denote this image by J. Since I is maximal, $I + J = A$. Obviously, the A-module A/I is irreducible; therefore J is a minimal left ideal in A. Hence $I \cap J = 0$. So we have shown that $A = I \oplus J$. Since J is of the form Ae, where e is an idempotent in A [5, Lemma 30.2], it follows that J is closed. □

Corollary 2. *Let A be a semiprime Fréchet algebra having AP, $\mathrm{dg}A = 0$, R its topological radical. Then:*

(i) *every irreducible Fréchet A-module is topologically isomorphic to a minimal left ideal (of the form Ae, where $e^2 = e \in A$);*

(ii) $\mathrm{lan}(\mathrm{soc}(A)) = R$.

Lemma 4. *Let A be a right annihilator, semiprime Fréchet algebra, and I a bi-ideal in A. Then* $\mathrm{lan}(I) = \mathrm{ran}(I)$, $I \cap \mathrm{ran}(I) = 0$, *and* $\overline{I \oplus \mathrm{ran}(I)} = A$.

The proof of this lemma is analogous to the proof of [5, Lemma 32.4].

The following corollary follows from Lemmas 3, 4 and Corollaries 1, 2.

Corollary 3. *Let A be a semiprime Fréchet algebra having* AP, $\mathrm{dg}A = 0$, *R its topological radical. Then $A = \overline{I \oplus R}$, where I is a closed bi-ideal in A such that $I = \overline{\mathrm{soc}(A)}$ and $I = \mathrm{lan}(R)$.*

Lemma 5. *Let A be a semiprime Fréchet algebra and J a minimal left ideal in A. Then for each idempotent e in A with $Ae = J$ there exists $f \in A^*$ such that $eae = f(a)e$ $(a \in A)$.*

Proof. Suppose that e is an idempotent in A with $Ae = J$. We consider the set $eAe = \{a \in A : eae = a\}$. Obviously, eAe is a closed subalgebra of A and therefore a Fréchet algebra with unit element e. According to [5, Lemma 30.2], eAe is a division algebra. Hence $eAe = \mathbf{C}e$ [6, Theorem 9.4]. Thus there exists a linear functional f on A with $eae = f(a)e$ $(a \in A)$. It is clear that $f \in A^*$. □

We recall that a *minimal closed bi-ideal* (of A) is a closed bi-ideal $I \neq 0$ that contains no closed bi-ideals other than 0 and I.

Lemma 6. *Let A be a semiprime Fréchet algebra, $I = \overline{\mathrm{soc}(A)}$, $\{I_\nu : \nu \in \Lambda\}$ the family of minimal closed bi-ideals of A such that $I_\nu \subset I$. Then I is the closure* *of the direct sum of all the* I_ν *and moreover, for each* ν, *the closed bi-ideal* I_ν *is generated by an idempotent* $e_\nu \in A$ *such that for some* $f_\nu \in A^*$, $e_\nu a e_\nu = f_\nu(a)e_\nu$ $(a \in A)$.

The proof of this lemma, which uses Lemma 5 and arguments borrowed from the theory of annihilator Banach algebras [8, Section 25], is analogous to the proof of [9, Lemma 1.10].

For a Fréchet space E, we shall consider the space E^*, dual to E, as a locally convex space with the strong topology (that is, the topology of uniform convergence on bounded sets).

Lemma 7 [5, Lemmas 32.8 and 32.19]. *Let* A *be a right annihilator, semiprime Fréchet algebra,* I *a minimal closed bi-ideal in* A, *which is generated by an idempotent* $e \in A$ *such that for some* $f \in A^*$, $eae = f(a)e$ $(a \in A)$. *Then:*

(i) $\operatorname{ran}(eA) \cap I = 0$;

(ii) $\operatorname{ran}(J) \cap I \neq 0$ *for each closed left ideal* J *in* A *such that* $J \subset I$ *and* $J \neq I$;

(iii) *the operator* $\Psi : Ae \to (eA)^*$, *given by* $x \longmapsto \psi_x$ *where* $\psi_x(y) = f(yx)$ $(y \in eA)$, *is a continuous linear bijection of* Ae *onto* (eA^*).

Proof. (i) Since $I = \overline{AeA}$, we have $\operatorname{ran}(eA) = \operatorname{ran}(I)$, and so (i) follows from Lemma 4.

(ii) Let J be a closed left ideal such that $J \subset I$ and $\operatorname{ran}(J) \cap I = 0$. Hence $\operatorname{ran}(J)I \subset \operatorname{ran}(J) \cap I = 0$ and therefore $\operatorname{ran}(J) \subset K$, where $K = \operatorname{lan}(I)$. We put $S = \overline{J + K}$. By Lemma 4,

$$\operatorname{ran}(S) = \operatorname{ran}(J) \cap \operatorname{ran}(K) \subset K \cap \operatorname{ran}(K) = 0.$$

Since S is a closed left ideal and A is right annihilator, then $S = A$. By Lemma 4, $K = \operatorname{ran}(I)$, hence $IK = 0$ and so $I = IA = IS \subset J$. Therefore $J = I$.

(iii) It is easy to see that Ψ is a continuous linear operator from Ae to $(eA)^*$, and $\operatorname{Ker}\Psi \subset Ae \cap \operatorname{ran}(eA) = 0$, by (i). We shall prove that $\operatorname{Im}\Psi = (eA)^*$.

Indeed, let $\varphi \in (eA)^*$, $\varphi \neq 0$. Set $M = \operatorname{Ker}\varphi$ and set $J = \{a \in I : \varphi(eAa) = 0\}$. It is easy to show that $M \subset J$, and J is a closed left ideal in A such that $J \subset I$ and $J \neq I$. By (ii), $\operatorname{ran}(J) \cap I \neq 0$, that is we may choose $b \in I$, $b \neq 0$, such that $Jb = 0$ and therefore $Mb = 0$. Choose $u \in eA$ with $\varphi(u) = 1$. Let $y \in eA$. Since $y - \varphi(y) \in M$, then $(y - \varphi(y)u)b = 0$, and so $yb = \varphi(y)ub$. By (i), we have $y_0 b \neq 0$ for some $y_0 \in eA$. Hence $ub \neq 0$. Since eA is a minimal right ideal in A [5, Proposition 30.6], we have $ubA = eA$. Choose $c \in A$ with $ubc = e$. Then $ybce = \varphi(y)ubce = \varphi(y)e$ $(y \in eA)$, and so $\varphi = \Psi(x)$, where $x = bce \in Ae$. □

Proposition 2. *Let A be a right annihilator, semiprime Fréchet algebra, I a minimal closed bi-ideal in A, which is generated by an idempotent $e \in A$ such that, for some $f \in A^*$, $eae = f(a)e$ $(a \in A)$. Then I is finite-dimensional and is isomorphic, as an algebra, to a full matrix algebra.*

Proof. By associating to each $a \in A$ the operator $m_a : eA \longrightarrow eA$ given by $y \longmapsto ya$, we obviously get a continuous linear operator from A into $\mathcal{B}(eA)$ (the space of all continuous linear operators on eA with the topology of uniform convergence on bounded sets). It is clear that m_a is a finite-dimensional operator if $a \in AeA$, and that $m_a = 0$ if $a \in \operatorname{ran}(I)$. Since $I = \overline{AeA}$ and, by Lemma 4, $A = \overline{I \oplus \operatorname{ran}(I)}$, then for each $a \in A$ the operator m_a is approximated in the topology of the space $\mathcal{B}(eA)$ by finite-dimensional operators. In particular, the identity operator on eA (the image of the unit element of A) is the limit of some net of finite-dimensional operators on eA. Thus for any bounded set $S \subset eA$ and neighborhood U of zero in eA, there exists a finite-dimensional operator $\varphi : eA \longrightarrow eA$ such that $x - \varphi(x) \in U$ for all $x \in S$; since the set $\varphi(S)$ is precompact, the set $S \subset \varphi(S) + U$ is precompact, too. Consequently, the Fréchet space eA is a Montel space. The latter implies [10, Chapter IV] that the space $(eA)^*$ is a Montel space, too and, in particular, barreled.

Let Ψ be as in Lemma 7. We have proved in Lemma 7 that Ψ is a continuous linear bijection of (the Fréchet space) Ae onto $(eA)^*$. Since the space $(eA)^*$ is barreled, the open mapping theorem [10, Chapter VI, Theorem 7 and Corollary 2] implies that Ψ is a topological isomorphism. It follows that the (strong) dual space $(eA)^*$ of the Fréchet space eA is a Fréchet space, too. By [10, Chapter VI, Proposition 15], eA is normable. Since eA is a Montel space, it is finite-dimensional. It follows that the space Ae is finite-dimensional. It remains to take advantage of the regular representation of the algebra I on Ae. □

From Lemma 6 and Proposition 2 we obtain the following corollary.

Corollary 4. *Let A be a right annihilator, semiprime Fréchet algebra, $I = \overline{\operatorname{soc}(A)}$, $\{I_\nu : \nu \in \Lambda\}$ the collection of finite-dimensional minimal bi-ideals of A such that $I_\nu \subset I$. Then I is the closure of the direct sum of all the I_ν, and moreover, for each ν, the bi-ideal I_ν is isomorphic, as an algebra, to some full matrix algebra.*

Proof of Theorem 1. We set $I = \overline{\operatorname{soc}(A)}$. By Corollary 3, $A = \overline{I \oplus R}$ and $I = \operatorname{lan}(R)$. Let J be a finite-dimensional minimal bi-ideal in A. If $J \subset I$ then clearly $J \cap R = 0$. On the other hand, if $J \cap R = 0$ then $JR \subset J \cap R = 0$, and therefore $J \subset \operatorname{lan}(R) = I$. Using Corollary 4, we obtain that (i) and (ii) hold.

For (iii), suppose that K a closed bi-ideal of A containing R. Choose $a \in \cap\{P_\nu : P_\nu \subset K\}$. Since A is unital, to show $a \in K$ it suffices to prove $Aa \subset K$.

Let $\nu \in \Lambda$. Then either $I_\nu \subset K$ or $I_\nu \cap K = 0$. In the first case, $I_\nu a \subset I_\nu \subset K$. In the second, $KI_\nu \subset K \cap I_\nu = 0$ and consequently, $K \subset \operatorname{lan}(I_\nu) = P_\nu$. Hence $a \in P_\nu$ and $I_\nu a \subset I_\nu P_\nu = 0 \subset K$. Thus $I_\nu a \subset K$ for any $\nu \in \Lambda$. Furthermore,

$Ra \subset R \subset K$. By (i), the sum of R and the bi-ideals I_ν is dense in A. Thus $Aa \subset K$ and consequently, (iii) holds. □

The following is a direct consequence of Theorem 1.

Theorem 2. *Let A be a Fréchet algebra with* AP. *Suppose also that there exists a faithful irreducible Fréchet A-module. Then* $\mathrm{dg}A = 0$ *iff A is isomorphic to a full matrix algebra $M_n(\mathbf{C})$ for some integer n.*

Theorem 3. *Let A be a topologically semisimple Fréchet algebra with* AP, $\mathrm{dg}A = 0$. *Then A is the closure of the algebraic direct sum of a family of full matrix algebras. Furthermore, any closed bi-ideal of A is the closure of the direct sum of all finite-dimensional minimal bi-ideals of A contained in it.*

Proof. The first assertion of the theorem is a consequence of Theorem 1. For the second assertion, suppose that K is a closed bi-ideal of A and $\{I_\nu : \nu \in \Lambda\}$ the collection of finite-dimensional minimal bi-ideals of A. We set

$$\Lambda_1 = \{\nu \in \Lambda : I_\nu \subset K\},$$

$$K_1 = \overline{\bigoplus_{\nu \in \Lambda_1} I_\nu}.$$

As we noted in the proof of Theorem 1(iii), $K \subset P_\nu$ for each $\nu \in \Lambda \setminus \Lambda_1$, and hence $K_1 \subset P_\nu$. For any $\nu \in \Lambda_1$ we have $K_1 \not\subset P_\nu$, since otherwise $I_\nu \subset P_\nu$ and $I_\nu = I_\nu \cap P_\nu = 0$. By Theorem 1(iii),

$$K_1 = \bigcap_{\nu \in \Lambda \setminus \Lambda_1} P_\nu.$$

Hence $K \subset K_1$, and so $K_1 = K$. □

2. The Arens–Michael hull

We recall that for an arbitrary polynormed algebra one can define, in a reasonable way, the Arens–Michael algebra (another terminology is the complete locally multiplicatively-convex algebra) that is "closest to it". This is the so-called *Arens–Michael hull* of the given algebra [11, V.2.21]. We shall prove the following theorem.

Theorem 4. *Let A be a semiprime Fréchet algebra with* AP, $\mathrm{dg}A = 0$. *Let R be the topological radical of this algebra, and $\{I_\nu : \nu \in \Lambda\}$ the collection of finite-dimensional minimal bi-ideals of A such that $I_\nu \cap R = 0$. Then the Arens–Michael* *hull $\widetilde{A}$ of the algebra A is topologically isomorphic to the Cartesian product of all the I_ν, and hence $\widetilde{A}$ is the Cartesian product of full matrix algebras.*

To prove Theorem 4, we need a number of lemmas. Suppose first that A and B are two Fréchet algebras connected by a (unital continuous) homomorphism $\alpha : A \to B$, and that X is a Fréchet B-module. Then obviously, the action of B on X induces a continuous action of A on X defined by $ax = \alpha(a) \cdot x$ ($a \in A$, $x \in X$). In particular, B is a Fréchet A-module with the action $a \cdot b = \alpha(a)b$ ($a \in A, b \in B$); in this case, the operator $\alpha : A \to B$ is an A-module morphism.

Lemma 8. *Let $\alpha : A \to B$ be a Fréchet algebra homomorphism with a dense image and X a Fréchet B-module which is projective as an A-module. If A is semiprime and has* AP *then for any $x \in X$, $x \neq 0$, there exists a Fréchet B-module morphism $\chi : X \to B$ such that $\chi(x) \neq 0$ and $(\operatorname{Im}\chi)^2 \neq 0$.*

Proof. Let $x \in X$, $x \neq 0$, and A be semiprime with AP. By Lemma 2, there exists a (Fréchet) A-module morphism $\varphi : X \to A$ such that $\varphi(x) \neq 0$. Let $\chi = \alpha\varphi$. Then $\chi : X \to B$ is an A-module morphism. Since $\alpha : A \to B$ has dense image, an A-module morphism between B-modules is also a B-module morphism. It follows that $\chi : X \to B$ is a B-module morphism. We set $I = \operatorname{Ker}\alpha$. Then $I \cdot X$, and hence $\varphi(x) \in \operatorname{ran}(I)$. If $\chi(x) = 0$ then $\varphi(x) \in I \cap \operatorname{ran}(I) = 0$ (since A is semiprime). Consequently, $\chi(x) \neq 0$. Let $J = \operatorname{Im}\varphi$ and $K = \operatorname{Im}\chi$. Then J is a left ideal of A, and K is a left ideal of B; in particular, J is a subalgebra of A, and K is a subalgebra of B. If $\alpha_0 : J \to K$ is the restriction of α to J, then

$$\operatorname{Ker}\alpha_0 \subset I \cap \operatorname{Im}\varphi \subset I \cap \operatorname{ran}(I) = 0,$$

$$\operatorname{Im}\alpha_0 = K.$$

Thus α_0 is an (algebraic) isomorphism of the algebras J and K. Since A is semiprime, we have $J^2 \neq 0$, and so $K^2 \neq 0$. □

Lemma 9. *Let A be a semiprime Fréchet algebra having* AP, $\mathrm{dg}A = 0$, *and B an arbitrary Fréchet algebra. If there exists a Fréchet algebra homomorphism from A to B with a dense image, then:*

(i) $\operatorname{lan}(\operatorname{ran}(I)) = I$ *for each closed left ideal I of B (and, as a consequence, the algebra B is right annihilator)*;

(ii) for any closed maximal left ideal J of B there exists a closed left ideal K of B such that $B = J \oplus K$.

The proof of this lemma, which uses Lemma 8, repeats in the main the proofs of Lemma 3 and Proposition 1.

Lemma 10. *Let A be a semiprime Fréchet algebra having* AP, $\mathrm{dg}A = 0$, *and R its topological radical. Then the image of a (continuous) homomorphism of A into a Banach algebra is isomorphic to the Cartesian product of a finite number of full matrix algebras (and, is in particular, finite-dimensional), and the kernel contains R.*

Proof. Let α be a homomorphism from A to a Banach algebra B. We set $B_0 = \overline{\operatorname{Im}\alpha}$ and $I = \operatorname{soc}(B_0)$. Suppose that $I \neq B_0$. Then $I \subset J$ for some maximal left ideal J of B_0. By Lemma 9, there exists a minimal left ideal K of B_0 such that $B_0 = J \oplus K$. We have $K \subset I \subset J$, and so $B_0 = J$. This contradiction shows that $\operatorname{soc}(B_0) = B_0$ and consequently, the Banach algebra B_0 is *classically semisimple.* Using the Wedderburn–Artin theorem [12, Section 3.4, Proposition 6] and the Gel'fand–Masur theorem, we obtain that B_0 is isomorphic to the Cartesian product of a number of full matrix algebra, with complex coefficients; in particular, B_0 is finite-dimensional and coincides with $\operatorname{Im}\alpha$. It is clear that any minimal left ideal of the algebra $B_0 = \operatorname{Im}\alpha$ is an irreducible Fréchet A-module. Since $\operatorname{soc}(B_0) = B_0$, we have $R \cdot B_0 = 0$, and hence $R \subset \operatorname{Ker}\alpha$. □

Proof of Theorem 4. For each $\nu \in \Lambda$ we put $P_\nu = \operatorname{lan}(I_\nu)$. We consider the quotient algebra $A_\nu = A/P_\nu$ and the natural epimorphism $\tau_\nu : A \to A_\nu$. By Theorem 1, $A = I_\nu \oplus P_\nu$, and the algebra I_ν is isomorphic to a full matrix algebra. The restriction of τ_ν to I_ν is an isomorphism of algebras. This enables us to identify I_ν with A_ν.

Let now $\|\cdot\|$ be some continuous *multiplicative* [11] seminorm on A. We put

$$I = \{a \in A : \|a\| = 0\};$$

obviously this is a closed bi-ideal in A. Let B be the completion of the quotient algebra A/I in the norm $\|a + I\|^\wedge = \|a\|$, so B is a Banach algebra. Consider the homomorphism $\alpha : A \to B$ given by $a \longmapsto a + I$; it is continuous. By Lemma 10, $R \subset \operatorname{Ker}\alpha = I$, and $\operatorname{Im}\alpha$ is finite-dimensional. It follows that the space A/I is also finite-dimensional. We put

$$\Lambda_1 = \{\nu \in \Lambda : I \subset P_\nu\}$$

$$J = \overline{\oplus_{\nu\in\Lambda_1}} I_\nu.$$

By Theorem 1,

$$I = \bigcap_{\nu\in\Lambda_1} P_\nu = \operatorname{lan}(J),$$

and so $A = \overline{I \oplus J}$, in view of Corollary 1 and Lemma 4. Since A/I is finite-dimensional, we obtain that J is finite-dimensional, and $A = I \oplus J$. Moreover, the set Λ_1 is finite, and

$$J = \bigoplus_{\nu\in\Lambda_1} I_\nu.$$

Consequently, up to isomorphism, $A/I = \underset{\nu\in\Lambda_1}{\times} A_\nu$. Obviously this isomorphism is topological. By A_0 denote the multinormed algebra with the same underlying pure algebra as A, but with the system of seminorms consisting of all continuous multiplicative seminorms on A [11, Chapter V, Section 2]. Further, we take the Cartesian product, $\underset{\nu\in\Lambda}{\times} A_\nu$, of all the A_ν, and we introduce the map $\rho : A_0 \to \underset{\nu\in\Lambda}{\times} A_\nu$ taking $a \in A_0$ into "row" $\{\tau_\nu(a) \in A_\nu; \nu \in \Lambda\}$. It is clear that ρ is a continuous algebra homomorphism; its kernel is R and coincides with the closed bi-ideal in A_0 consisting of all the points "tied to zero" (the intersection of all the neighborhoods of zero in A_0). We denote the (Hausdorff) quotient algebra A_0/R by A_1; according to [11, V.2.24], the completion, $\widetilde{A}$, of A_1 is the Arens–Michael hull of the algebra A. Let $\rho_1 : A_1 \to \underset{\nu\in\Lambda}{\times} A_\nu$ be the operator generated by ρ; it is a continuous embedding. It follows from Theorem 1 that the image of the operator ρ_1 is dense in the space $\underset{\nu\in\Lambda}{\times} A_\nu$. It follows from the previous paragraph that the systems of seminorms of the spaces A_1 and $\operatorname{Im}\rho_1$ are equivalent. Consequently, ρ_1 realizes a topological isomorphism between dense subspaces in $\widetilde{A}$ and $\underset{\nu\in\Lambda}{\times} A_\nu$. □

We recall [1, Chapter II, Section 5.2] that every Fréchet A-bimodule may be considered as an object in A^{env} – mod, where $A^{env} = A\widehat{\oplus}A^{op}$ is the *enveloping algebra* of the algebra A, and that the definition of a projective module is automatically extended to it. We recall also that a (unital) Fréchet algebra A is said to be *contractible* if the A-bimodule A is projective. It is known [11, VII.1.72] that an algebra A is contractible iff it has a *diagonal*, that is an element $d \in A\widehat{\oplus}A$ such that $a \cdot d = d \cdot a$ for all $a \in A$ and $\pi(d)$ is the identity element in A. The simplest example of a contractible algebra is the algebra $M_n(\mathbf{C})$ of all $n \times n$-matrices (the full matrix algebra). J. L. Taylor [13] is the first to note that the direct topological product of an arbitrary family of full matrix algebras is a contractible algebra. The fact can be obtained from the following lemma, too.

Lemma 11. *The Cartesian product of a family of contractible* (*polynormed*) *algebras is a contractible algebra.*

Proof. Let $A = \underset{\nu\in\Lambda}{\times} A_\nu$ and for each ν, d_ν be a diagonal for A_ν. By [1, Theorem II.5.19], the space $A\widehat{\oplus}A$ can be identified with $\underset{\nu,\mu\in\Lambda}{\times} (A_\nu\widehat{\otimes}A_\mu)$. It is easy to see that the "row" $d = \{x_{\nu,\mu} \in A_\nu\widehat{\otimes}A_\mu; \nu,\mu \in \Lambda\}$, where $x_{\nu,\mu} = 0$ for $\nu \neq \mu$ and $x_{\nu,\mu} = d_\nu$ for $\nu = \mu$, is a diagonal for A. □

Theorem 5. *Let A be a metrizable Arens–Michael algebra* (= *a multinormed Fréchet algebra*), *A semiprime with* AP. *Following conditions are equivalent:*

(i) *A is contractible;*

(ii) $\mathrm{dg}A = 0$;

(iii) *A is topologically isomorphic to the Cartesian product of at most countable family of full matrix algebras.*

Proof. Suppose that (i) holds. Let X be an arbitrary Fréchet A-module. Since A is a projective A-bimodule, the A-module $A\widehat{\otimes}_A X$ is projective. But this module is isomorphic to X. Consequently, X is projective, and so (ii) holds.

The implication (ii) $\Rightarrow$ (iii) follows from Theorem 4, [14, 4.2.4] and the fact that for an Arens–Michael algebra A, its Arens–Michael hull $\widetilde{A}$ is isomorphic to A itself.

The implication (iii) $\Rightarrow$ (i) follows from [11, Proposition VII.1.73] and Lemma 11. □

We note that, in the special case where A is commutative, Theorem 5 follows from [1, Theorem IV.5.27].

It remains to remark that Theorem 5 implies the following results which is related to [2, Theorem 1].

Theorem 6. *Let A be a semiprime Banach algebra with* AP. *Then the following conditions are equivalent:*

(i) *A is contractible*;

(ii) $\mathrm{dg} A = 0$;

(iii) *A is isomorphic to the Cartesian product of a finite number of full matrix algebras.*

References

[1] Helemskiĭ , A. Ya., The Homology of Banach and Topological Algebras. Kluwer Academic Publishers, Dordrecht 1989.

[2] Selivanov, Yu. V., Banach algebras of small global dimension zero. Usp. Mat. Nauk 31:2 (1976), 227–228.

[3] Taylor, J. L., A general framework for a multi-operator functional calculus. Adv. Math. 9 (1972), 183–252.

[4] Grothendieck, A., Produits tensories topologiques et espaces nucleaires. Mem. Am. Math. Soc. 16 (1955).

[5] Bonsall, F. F., Duncan, J., Complete Normed Algebras. Springer-Verlag, Berlin 1973.

[6] Zelazko, W., Metric generalizations of Banach algebras. Rozprawy Mat. 47 (1965).

[7] Selivanov, Yu. V., Projective Fréchet modules with the approximation property. Usp. Mat. Nauk 49 (1994) (to appear).

[8] Naimark, M. A., Normed Rings. Nauka, Moscow 1968.

[9] Selivanov, Yu. V., Biprojective Banach algebras. Izv. Akad. Nauk SSSR, Ser. Mat. 43 (1979), 1159–1174. English translation: Math. USSR-Izv. 15 (1980), 387–399.

[10] Robertson, A. P., Robertson, W., Topological Vector Spaces. Cambridge University Press, Cambridge 1964.

[11] Helemskiĭ, A. Yu., Banach and Polynormed Algebras: General Theory, Representations, Homology. Oxford University Press, Oxford 1993.

[12] Lambek, J., Lectures on Rings and Modules. Blaisdell, Waltham, Mass. 1966.

[13] Taylor, J. L., Homology and cohomology for topological algebras. Adv. Math. 9 (1972), 137–182.

[14] Engelking, R., General Topology. PWN, Warsaw 197?.

Development of the theory of layer-finite groups

Vladimir I. Senashov *

Abstract. In the survey, results on the so-called layer-finite groups, forming a subclass of the class of locally finite groups, are considered.

1991 Mathematics Subject Classification: 20F50, 20E15, 20E22.

Layer-finite groups appeared for the first time in S. N. Chernikov's paper [1] in 1945. This class of groups was mentioned there but without any name. S. N. Chernikov used the name "layer-finite" in some of his later works.

A group is called *layer-finite* if the set of its elements of any given order is finite.

Investigations of properties of layer-finite groups were carried out by S. N. Chernikov, R. Baer, Kh. Kh. Mukhamedzhan in 1945–1960. The basic properties were described in different journals and remained in this form until 1980, when S. N. Chernikov's book [2] was published. In this book, S. N. Chernikov collected his results in one section. The author's book [3] includes almost all of the properties of layer-finite groups.

When a great number of the properties of layer-finite groups were described, the question about the place of this class among other groups arose. The first such characterization of layer-finite groups was the establishing of interconnection with the nearest class of groups, locally normal groups. The idea of this connection appeared in the articles of S. N. Chernikov, R. Baer, Kh. Kh. Mukhamedzhan [1, 4–7]. The strongest result in the field is that of Chernikov: the class of layer-finite groups coincides with the class of locally normal groups whose Sylow subgroups satisfy the minimal condition.

A group G is called a *Chernikov group* if it is a finite extension of a direct product of a finite number of groups of type C_{p^∞}. In 1960 Ya. D. Polovitskiĭ [8] introduced the class of the so-called *layer-Chernikov groups.* It consists of the groups in which every set of elements of any given order generates a Chernikov subgroup. These groups are very close to layer-finite groups in their properties.

Between 1960 and 1975 there was a period when scientists forgot this class of group. But now, interest to layer-finite groups grows up simultaneously with

* Supported by the Krasnoyarsk Region Fund (grant 1F0140).

expansion of research on groups with the min-p condition. Interest in layer-finite groups ebbs after the completion of their description in the fifties. Recently, different methods of studying groups with different finiteness conditions have appeared. About 1975, layer-finite groups were used in the descriptions of groups with various finiteness conditions. To the author's knowledge, the first result in this direction may be found in V. P. Shunkov's article [9]. It contains methods of studying of such groups and groups with min-p condition. For more details about Shunkov's methods, see [10, 11].

It is very natural to use the theory of layer-finite groups in investigations on these groups, because groups with the conditions of a periodic type and almost local solvability are extensions of layer-finite groups by layer-finite groups. Particularly, using these methods, a characterization of layer-finite groups in the class of periodic groups with min-p condition was obtained. As we have remarked above, the first characterization of layer-finite groups was made by S. N. Chernikov:

Theorem (Chernikov [2]). *The class of layer-finite groups coincides with the class of locally normal groups, whose Sylow subgroups satisfy the minimal condition.*

After 1980, further characterizations in other classes of groups, e.g. periodic groups and binary soluble groups, were obtained in Shunkov's scientific school. These investigations represent a major advance in the research of properties of periodic groups using properties of layer-finite groups.

Consider the latest results on layer-finite groups.

Definition. A group G satisfies the *layer minimal condition* if in every descending chain

$$G_1 > G_2 > \cdots > G_i > G_{i+1} > \cdots \tag{1}$$

of subgroups for every natural n and for every prime p only finite number of differences

$$G_i \backslash G_{i+1} \tag{2}$$

contain elements of order p^n.

Theorem (Polovitskiĭ [12]). *A locally finite group satisfies the layer minimal condition if and only if it is a group with Chernikov layers.*

Corollary. *In the class of locally finite groups the conditions of layer minimality and primary minimality are equivalent.*

Theorem (Polovitskiĭ [12]). *A periodic group satisfies the layer-minimal condition if and only if for every natural n in every descending chain* (1) *of subgroups only finite number of differences* (2) *contain elements of order n.*

Definition. By a *locally graded group* we mean a group in which every non-trivial finitely generated subgroup contains a proper subgroup of finite non-unite index.

Theorem (Chernikov [13]). *Let G be a locally graded group factorizable by a pair of locally finite subgroups A and B such that A and B satisfy the minimal condition on primary subgroups and the quotient-groups $A/J(A)$ and $B/J(B)$ are almost locally normal. Then the group G is locally normal satisfies the minimal condition for primary subgroups, and $J(G) = J(A)J(B)$.*

Corollary. *Let G be a locally graded group factorizable by a pair of layer-finite subgroups A and B. Then the group G is locally finite, satisfies the minimal condition on primary subgroups and $J(G) = J(A)J(B)$.*

Theorem (Chernikov [13]). *Let G be a locally soluble-by-finite group factorizable by a pair of almost locally normal subgroups A, B and π an arbitrary non-empty set of prime numbers. Then the group G is locally finite and furthermore, each π-subgroup of G is either*

1) *trivial, or*

2) *finite, or*

3) *satisfies the minimality condition*

if and only if the same holds for all π-subgroups of A and B.

Corollary. *Let G be a locally almost soluble group factorizable by a pair of layer-finite subgroups A and B. Then the group G is locally finite and all Sylow subgroups of G are Chernikov.*

Definition. The group G is called *F^*-group* if for every subgroup H of G, every finite subgroup K from H, and any elements a, b of prime order from $T = N_H(K)/K$ there exists an element c of T such that $\mathrm{gr}\,(a, c^{-1}bc)$ is finite.

Theorem (Sedova [14]). *A periodic residually finite F^*-group, in which every locally finite subgroup is layer-finite, is a layer-finite group.*

Theorem (Sedova [14]). *A periodic group is a locally soluble layer-finite group if and only if it is binary soluble and every its locally soluble subgroup is layer-finite.*

The next theorem reinforces the previous theorem in characterization of layer-finite groups in the class of periodic binary soluble groups: here the condition of layer-finiteness is satisfied only for subgroups containing some specific element a.

Theorem (Gomer [15]). *Let G be a periodic binary soluble group and a its element of prime order p such that:*

1) *in* $C_G(a)$ *every locally finite subgroup is layer-finite and has finite Sylow* q*-subgroups for all primes* q;

2) *every locally finite subgroup containing the element* a *is layer-finite.*

Then G *is locally soluble (locally finite) layer-finite group.*

We can characterize layer-finite groups in the class of locally finite groups in the next form:

Theorem (Senashov [16]). *A locally finite group is layer-finite if and only if every its locally soluble subgroup is layer-finite.*

Theorem (Ivko, Senashov [17]). *A periodic locally soluble group* G *is layer-finite if and only if for some finite subgroup* B *of* G *the normalizer of each non-trivial* B*-invariant finite subgroup of* G *is layer-finite.*

Definition. A group G is called *conjugately biprimitively finite* if for any finite subgroup H of G the subgroup generated by two arbitrary conjugate elements of prime order of the quotient-group $H_G(H)/H$ is finite.

This class of groups was introduced by V. P. Shunkov. These conditions turned out to be very useful in the proofs of many theorems of more than twenty years old.

Theorem (Senashov [16]). *A periodic group whose finite subgroups are soluble is layer-finite if and only if it is conjugately biprimitively finite and all its locally finite subgroups are layer-finite.*

Theorem (Senashov [16]). *A periodic group without involutions is layer-finite if and only if it is conjugately biprimitively finite, and each of its locally finite subgroup is layer-finite.*

The requirement of conjugately biprimitively finiteness can never be dropped in the above two theorems, as Novikov–Adjan and Ol'shanskiĭ [18, 19] showed in their examples. Moreover the next example shows that this requirement cannot even be replaced by the condition of weak conjugately biprimitively finiteness implying that any two conjugate elements of prime order of a group generate a finite subgroup.

Let $A = \mathrm{gr}(b, c)$, where $b^p = c^p = d$, for some prime number p, A is a torsion-free group, and $A/(d)$ is a Novikov–Adjan group of period p. Then the group $B = A/(h)$, where $h = d^p$, is a group we are interesting for.

The next theorem gives a complete characterization of layer-finite groups in the class of periodic groups:

Theorem (Senashov [20]). *A periodic group is layer-finite if and only if it is conjugately biprimitively finite and each of its locally finite subgroups is layer-finite.*

The condition of conjugately biprimitive finiteness in the theorem is necessary because of the examples of Novikov–Adjan group and Ol'shanskiĭ group, in which all conditions of the theorem are valid except for the condition of finiteness, but they are not layer-finite.

This year new results for layer-finite groups were published:

Definition. Let G be a group and i an involution, satisfying the following conditions:

1) all subgroups of the form $\mathrm{gr}(i, i^g)$, $g \in G$, are finite,

2) $C_G(i)$ is infinite and has layer-finite periodic part,

3) $C_G(i) \neq G$ and $C_G(i)$ is not contained in another subgroup of G with a periodic part,

4) if K is a finite subgroup of G, not contained in $C_G(i)$, and $V = K \cap C_G(i) \neq 1$, then K is a Frobenius group with complement V.

The group G, satisfying the conditions 1)–4) is called an *F-group.*

This class of groups have been introduced by V. P. Shunkov.

Theorem (Senashov [17]). *Let G be a group and a an involution from G satisfying the next conditions:*

1) *all subgroups of the form $gr(a, a^g)$, $g \in G$, are finite,*

2) *the normalizer of any non-trivial (a)-invariant finite subgroup has a layer-finite periodic part.*

Then either all elements of finite order form a layer-finite subgroup of G or G is an F-group.

References

[1] Chernikov, S. N., On the theory of infinite p-groups. Dokl. Akad. Nauk SSSR 50 (1945), 71–74.

[2] — Groups with Given Properties of a System of Subgroups. Nauka, Moscow 1980.

[3] Senashov, V. I., Layer-finite Groups. Nauka, Novosibirsk 1993.

[4] Mukhamedzhan, Kh. Kh., On groups with an ascending central series. Mat. Sb., Nov. Ser. 28:1 (1951), 185–196.

[5] — On groups posessing increasing invariant sequence. Mat. Sb., Nov. Ser. 39:2 (1956), 201–218.

[6] Chernikov, S. N., On the centralizer of a complete Abelian normal divisor in an infinite periodic group. Dokl. Akad. Nauk SSSR 72:2 (1950), 243–246.

[7] Baer, R., Finiteness properties of groups. Duke Math. J. 15:4 (1948), 1021–1032.

[8] Polovitskiĭ, Ya. D., Layerwise groups. Dokl. Akad. Nauk SSSR 134:3 (1960), 533–535. English translation: Sov. Math., Dokl. 1 (1960), 1112–1113.

[9] Shunkov, V. P., Sufficient conditions for the existence in a group of infinite locally finite subgroups. Algebra Logika 15:6 (1976), 716–737. English translation: Algebra Logic 15:6 (1976), 445–462.

[10] — M_p-groups. Nauka, Moscow 1990.

[11] — On Embedding of Prime Orders Elements in a Group. Nauka, Novosibirsk 1992.

[12] Polovitskiĭ, Ya. D., Groups with layer-minimal condition. Ukr. Mat. Zh. 44:6 (1992), 847–851. English translation: Ukr. Math. J. 44:6 (1992).

[13] Chernikov, N. S., Groups that decompose into a product of permutable subgroups. Naukova Dumka, Kiev 1987.

[14] Sedova, E. I., Periodic F^*-groups with additional finiteness conditions. Ph. D. Thesis. Krasnoyarsk 1985.

[15] Gomer, V. O., On groups with elements of finite ranks. Ph. D. Thesis. Novosibirsk 1992.

[16] Senashov, V. I., Characterization of layer-finite groups in the class of periodic groups. Algebra Logika 24:5 (1985), 608–617. English translation: Algebra Logic 24:5 (1985), 402–408.

[17] Ivko, M. N., Senashov, V. I., On a new class of infinite groups. Preprint Computing Center of Siberian Division of Russian Science Academy, No 1, 30–44. Krasnoyarsk 1993.

[18] Adjan, S. I., The Burnside Problem and Identities in Groups. Nauka, Moscow 1975.

[19] Ol'shanskiĭ, A. Yu., Geometry of Defining Relations in Groups. Nauka, Moscow 1989.

[20] Senashov, V. I., Characterization of layer-finite groups. Algebra Logika 28:6 (1989), 687–704. English translation: Algebra Logic 28:6 (1989).

Lie groups and mechanics of deformable solid

Sergeĭ I. Senashov *

Abstract. Algebraic methods are considered for investigation and solution of differential equations of mechanics (mechanics of deformable solid). Algebraic problems, whose solution would, from the author's point of view, contribute to the progress in investigation of differential equations of mechanics, are formulated.

1991 Mathematics Subject Classification: 73A05, 35Q72, 35G25, 35G30, 16W10, 22E05.

In this paper several aspects of using algebraic methods for investigation and solution of differential equations of mechanics (mechanics of deformable solid) are considered. Some algebraic problems, the solution of which, from the authors point of view, will allow to make progress in investigation of differential equations of mechanics, are formulated.

Mechanics of deformable solid investigates tension and deformation conditions of solids, which are in elasticity, plasticity and other conditions. For determination of main characteristics of this processes it is necessary to solve nonlinear differential equations with initial boundary conditions. Nowadays the greater part of these problems cannot be solved by means of direct analytic and numeric methods. That is why inverse methods play the main role in research related to nonlinear equations of mechanics, the most effective of which have close connection to the notion of symmetry.

Let us mention some definitions which are necessary for understanding the present paper. Let $x = (x_1, \dots, x_n)$ be independent and $u = (u^1, \dots, u^m)$ dependent variables. Consider the system of differential equations

$$F = 0 \tag{1}$$

where

$$F = (F_1, \dots, F_r), \quad F_i = F_i(x, u, \dots, u_{(s)})$$

and $u_{(l)}$ is totality of all partial derivatives of order l of functions u^i in variables x_j.

*Supported by the Russian Fund for Fundamental Research (grant 93–013–1657).

Consider the infinite-dimensional space J^∞ with coordinates x, u, p^i_σ, where the symbol p^i_σ corresponds to the derivative

$$\frac{\partial^{|\sigma|}u^i}{\partial x_1^{i_1}\dots\partial x_n^{i_n}}, \quad |\sigma| = i_1+\dots+i_n, \quad \sigma=(i_1,\dots,i_n)$$

By a *differentiable function* on J^∞ we mean a differentiable function in finite number of coordinates on J^∞. The algebra of all such functions will be denoted by $\mathcal{F}$.

The *total derivative operator* with respect to x_j is

$$D_j = \frac{\partial}{\partial x_j} + \sum_{\sigma,i} p^i_{\sigma j}\frac{\partial}{\partial p^i_\sigma},$$

where $\sigma j = (i_1,\dots,i_j+1,\dots,i_n)$ if $\sigma=(i_1,\dots,i_n)$.

Let $D_\sigma = D_1^{i_1}\circ\dots\circ D_n^{i_n}$ if σ is as above. The universal linearization operator for the system (1) is defined as follows:

$$l_F = \begin{pmatrix} \sum\limits_\sigma \frac{\partial F_1}{\partial p^1_\sigma}D_\sigma & \dots & \sum\limits_\sigma \frac{\partial F_1}{\partial p^m_\sigma}D_\sigma \\ \vdots & & \vdots \\ \sum\limits_\sigma \frac{\partial F_r}{\partial p^1_\sigma}D_\sigma & \dots & \sum\limits_\sigma \frac{\partial F_r}{\partial p^m_\sigma}D_\sigma \end{pmatrix}$$

This is a matrix differential operator. The system (1) defines the submanifold A_∞ in J^∞ which is given by means of the following infinite system of equations:

$$D_\sigma(F_i) = 0 \quad \forall\sigma, i. \tag{2}$$

Via the system (2) a part of the coordinate functions P^i_σ on A_∞ can be expressed through the others which will be supposed functionally independent. The latter part of coordinates is called *internal* (on A_∞) while the first is called *external.* Of course, decompositions of variables into internal ones may be made in many different ways.

A function g on J^∞ may be restricted to A_∞. To perform this, external variables are to be replaced by suitable expressions in internal variables. The restriction of g to A_∞ is denoted by $\widetilde{g}$. Similarly the restriction $\overline{D_j}$ of operator D_j is defined by the formula

$$\overline{D_j} = \frac{\partial}{\partial x_j} + \widetilde{\sum} \bar{p}^i_{\sigma j}\frac{\partial}{\partial p^i_\sigma},$$

where $\widetilde{\ }$ means that summing is only over the internal variables p^i_σ. Let also

$$\widetilde{D_\sigma} = \widetilde{D}_1^{i_1}\circ\dots\circ\widetilde{D}_n^{i_n}, \quad \sigma=(i_1,\dots,i_n),$$

$$\bar{l}_F = \begin{pmatrix} \widetilde{\sum} \frac{\partial F_1}{\partial p^1_\sigma} \overline{D}_\sigma & \dots & \widetilde{\sum} \frac{\partial F_1}{\partial p^m_\sigma} \overline{D}_\sigma \\ \vdots & & \vdots \\ \widetilde{\sum} \frac{\partial F_r}{\partial p^1_\sigma} \overline{D}_\sigma & \dots & \widetilde{\sum} \frac{\partial F_r}{\partial p^m_\sigma} \overline{D}_\sigma \end{pmatrix}$$

Let φ be a symmetry of system (1) if

$$\bar{l}_F(\overline{\varphi}) = 0. \tag{3}$$

Symmetries of system (1) generate the Lie algebra $\operatorname{sym} F$ with respect to the brace

$$\{f, g\} = l_g f - l_f g.$$

Let

$$A_i = \sum a^i_\sigma D_\sigma$$

be some differential operator on J^∞. Then

$$A^*_i = \sum_\sigma (-1)^{|\sigma|} D_\sigma \circ a^i_\sigma$$

is formally conjugate to A_i. The function φ generates a conservation law if

$$\bar{l}^*_F(\overline{\varphi}) = 0, \tag{4}$$

where l^*_F is formally conjugate to l_F.

Note that not every solution φ of system (4) is a generating function of a conservation law. In order for it to be the case it is necessary and sufficient that the following representation holds:

$$\bar{l}_\varphi + \overline{A}^* = \overline{B} \circ \bar{l}_F, \tag{5}$$

where the operator A satisfies the equality $A(F) = l^*_F(\varphi)$, and $B = B^*$.

We refer the reader to [1, 2] for technical and conceptual details of the theory.

1. Let $\varphi_1, \dots, \varphi_k \in \operatorname{sym} F$, i.e. they are symmetries of the system (1). In this case a solution of the system

$$F = 0,$$

$$\varphi_1 = 0, \dots, \varphi_k = 0$$

(if it exists) is called an *invariant solution of the system* (1).

Knowing the algebra $\operatorname{sym} A_\infty$, it is possible to make the catalogue of all invariant solutions of the system (1). This catalogue will be useful for all studying the system (1).

An algebraic problem arises from the above consideration.

Problem 1. How to enumerate, up to conjugation, the subalgebras of a given Lie algebra $\operatorname{sym} F$ over the field of real numbers R?

Remark 1. Effective ways of solving the problem have already been formed and they are being realized for Lie symmetries [3]. The author does not know anything about similar solutions of problems for "higher" symmetries.

Remark 2. The solution of the problem is complicated by the fact that Lie algebras $\operatorname{sym} A_\infty$, as a rule, are infinite-dimensional.

2. The second problem is in solution of linear system of equations (3). This system is linear, but nevertheless it is difficult (and often impossible) to write out its general solution.

That is why it is important to make some simplifications. It turns out that sometimes it is possible to construct a linear matrix-operator T such that $T' l_F T = l_D$, where T' is the transposition of T and l_D is a diagonal matrix operator.

Problem 2. For a linear matrix differential operator l_F, choose a linear matrix operator T so that the operator $T' l_F T$ would have a diagonal form. As far as the choice of operator T is not ambiguous, then, in order to choose T, it is necessary to proceed from a maximal restoration of l_F by $\ker T' l_F T$.

An example of the above-mentioned construction is given in [4].

3. Symmetries are useful for constructing exact solutions of differential equations. However they are local by their nature. In mechanics, solutions of boundary-value problems are of particular interest because it is easy to solve these problems by means of conservation laws.

A conservation law is determined by its generating function φ, which is a solution of the equation (4). Unfortunately, the totality of generating functions does not possess the structure of a Lie algebra. This fact makes it difficult to describe.

Problem 3. Elucidate the algebraic structure of $\ker \bar{l}_F^*$.

4. Consider nonlinear elastic solid in which the law of Hooke has the form

$$\varepsilon_{ij} = f(\varepsilon_{ij}^0, \sigma_{ij}) \quad i, j = 1, 2, 3, \tag{6}$$

where ε_{ij} and σ_{ij} are the components of deformation and stress tensors, ε_{ij}^0 is the initial deformation. Transformations of type (6) under some natural restrictions on f, form a local Lie group [3]. Moreover, the components of the tensor σ_{ij} are considered as parameters of this group. Consequently, we obtain a 6-parametric Lie group with a function F from the space $R^6(\varepsilon_{ij})$. Classical connections between Lie groups and Lie algebras permit us to classify (up to isomorphism) the form of functions (6) and, consequently, all possible elastic materials with conservation law of Hooke of the form (6).

Unfortunately, the above-mentioned method and the plasticity method are not used for plastic materials where stresses and deformations are linked by (6). Consequently, in this case transformations of (6) form only a semigroup with identity. But it is not known to the author by what method it is possible to classify these transformations. For practical goals it is sufficient to consider only dimensions 1–3, because nowadays, pure technical, experiments with large number of parameters are not performed.

Problem 4. Classify one, two, and three parametric semigroups of transformations with identity of the form (6).

References

[1] Krasil'shic, I. S., Lychagin, V. V., Vinogradov, A. M., Geometry of Jet Spaces and Nonlinear Partial Differential Equations. Gordon Breach, New York 1986

[2] Senashov, S. I., Vinogradov, A. M., Symmetries and conservation laws of 2-dimensional ideal plasticity. Proc. Edinburgh Math. Soc., 31 (1988), 415–439.

[3] Ovsjannikov, L. V., Group Analysis of Differential Equations. Nauka, Moscow 1982.

[4] Ostrosablin, N. I., Senashov, S. I., General solutions and symmetries of linear elasticity equations. Dokl. Akad. Nauk SSSR 322:3 (1992), 513–515.

Recent investigations in group theory at the Krasnoyarsk State University

Vladimir P. Shunkov

Abstract. The paper describes scientific advances in group theory obtained during the last five years at the chair of algebra and logic at the Krasnoyarsk State University and, also, the history of foundation and development of the chair.

1991 Mathematics Subject Classification: 20E34, 20F50.

A negative solution of a great problem in mathematics always initiates a dramatic situation of choosing the further way of development. This happened also in the case of the well-known Burnside problem on groups of finite period. The negative solution to this problem obtained by Moscow mathematicians P. S. Novikov and S. I. Adyan allowed us to look into the unbounded cosmos of infinite groups. In this cosmos, with the help of the theorem discussed below, we can win an important part called "T_0-groups" for further detailed research.

This theorem is our main achievement for the last five years. It will become a base of a book that is now in preparation. An important event in the scientific life of our chair of algebra and mathematical logic was a successful defence of a doctoral dissertation on Chevalley groups by V. M. Levchuk. Moreover, five members of the chair (A. V. Timopheenko, V. O. Gomer, M. N. Ivko, G. A. Troyakova, and A. A. Cherep) got their candidate degrees. Three books on group theory were published [1–3].

The international conference on algebra dedicated to the memory of M. I. Kargapolov approved high standards of scientific achievements of Krasnoyarsk algebraists, in particular, in the field of group theory. So it is a suitable place to mention the years of establishing and reinforcement of the chair of algebra and mathematical logic at the Krasnoyarsk State University. The chair was organized in 1964 on the initiative of academician A. I. Mal'cev and correspondent member of the Academy of Sciences M. I. Kargapolov within the Krasnoyarsk subsidiary of the Novosibirsk State University.

A scientific support given by these mathematicians to the members of the chair was inspiring and effective. As early as the beginning of seventies, lecturers of the chair Yu. M. Gorchakov and V. P. Shunkov defended their doctoral dissertations.

After that, there was a constant increase in the number of research publications by the members of the chair (mainly, in the journal "Algebra and Logic" founded by A. I. Mal'cev), and in the number of researchers trained at the chair. In seventies N. M. Busarkin was the supervisor of 6 postgraduates that successfully defended candidate dissertations; one of them became a doctor of sciences later, and Yu. M. Gorchakov was that of 5 and one respectively. During the period 1978–1994 V. P. Shunkov was the supervisor of 17 candidate dissertations, and V. M. Levchuk was the supervisor of one.

Of course, these scientific achievements gained much from the creative atmosphere that was formed at the chair in these years (1970–1980) of its growth. At present, the chair of algebra and mathematical logic is headed by doctor of sciences V. V. Rybakov, who is an established specialist in the area of the theory of algebraic systems and constructive logic.

To summarize, the author would like to notice that scientific and pedagogical activities at the chair has always been characterized by some rush forward, and he thinks that it will retain this quality in the future.

Now we consider the above-mentioned theorem. First we define the class of T_0-groups and expose some properties of T-groups.

Let G be a group with involutions and i its involution such that

1) all subgroups of the form $\operatorname{gr}(i, i^g)$, $g \in G$, are finite,

2) Sylow 2-subgroups of G are either cyclic or generalized quaternion groups,

3) $C_G(i)$ is infinite and possesses a finite periodic part,

4) the normalizator of any nontrivial (i)-invariant finite subgroup of G is contained in $C_G(i)$ or possesses the periodic part which is a Frobenius group with Abelian kernel and finite noninvariant complement of even order,

5) $C_G(i) \neq G$ and, for every strictly real relative to i element $c \in G \backslash C_G(i)$, in $C_G(i)$ there exists an element s such that the subgroup $\operatorname{gr}(c, c^s)$ is infinite.

A group G with involution i is called a *T_0-group* if it satisfies the conditions 1–5.

We give an example of a T_0-group. Let $A = \operatorname{gr}(b, c)$, where $b^n = c^n = d$ is a torsion-free group and $A/(d)$ is a free Burnside group of period $n \geq 665$ [4]. Consider the group $B = A \wr (x) = (A \times A)\lambda(x)$,where x denotes an involution. Take an element $v = (d, d^{-1})$ of $A \times A$. It is evident that $v \in Z(A \times A)$ and $v^x = v^{-1}$. Using the abstract properties of the group A, it easy to prove that the group $G = B/(v)$ and its involution $i = x(v)$ satisfy the conditions 1–5 of the definition of a T_0-group.

Consequently, $G = B/(v)$ is a T_0-group. Moreover, in the group $G = B/(v)$, any maximal periodic subgroup containing the involution $i = x(v)$ is the dihedral group of order $2n$.

Let V be a T_0-group and i its involution. By the definition of T_0-groups the following assertions hold.

1. *A subgroup of V containing involutions either possesses almost Abelian periodic part or is a T_0-group.*
2. *If i and j are involutions of V then* $\mathrm{gr}\,(i, j, j^s)$ *is a T_0-group for some element s of $C_V(i)$.*
3. $V = C_V(i)\mathfrak{M}$, *where*

$$\mathfrak{M} = \{x \in V \backslash C_V(i) \mid c \text{ is an element of finite order and } c^i = c^{-1}\}.$$

4. *If* $V = \mathrm{gr}\,(i^g \mid g \in V)$ *and $L \lhd V$, where L is a torsion-free group, then $L \leq Z(V)$.*

The main result of the article is presented by the following theorem.

Theorem. *Let G be a group and a its element of prime order p such that*

1) *the subgroups of the form* $\mathrm{gr}\,(a, a^g)$ *are finite and almost all are solvable for $p \neq 2$;*
2) *in the centralizer $C_G(a)$, the set of all elements of finite order is finite;*
3) *in G, the normalizer of any nontrivial (a)-invariant finite subgroup has periodic part;*
4) *for $p \neq 2$, any (a)-invariant elementary Abelian subgroup of G is finite.*

Then either the set composed by all elements of G of finite order generates a periodic nilpotent-by-finite subgroup or G is a T_0-group and $p = 2$.

The proof is not trivial. Its full exposition will be presented in the author's book "T_0-groups." The most important part of the proof was already published [2, Chapter 4; 5–7].

The partial cases of the theorem are presented (explicitly or not) in many articles by the author and his pupils (M. N. Ivko, A. N. Izmaĭlov, A. N. Ostylovskiĭ, I. I. Pavljuk, A. M. Popov, E. I. Sedova (Chubarova), V. I. Senashov, A. I. Sozutov, N. G. Suchkova, A. A. Shafiro, A. K. Shlepkin). Groups satisfying the conditions similar to those of the theorem were also studied (mainly in the classes of locally finite groups, binary finite groups, and locally solvable groups) by A. O. Asar, V. V. Belyaev, D. I. Zaĭtsev, O. H. Kegel, S. P. Strunkov, B. Hartley, E. I. Khukhro, N. S. Chernikov, P. V. Shumyatskiĭ, and others.

Certainly, in proving the theorem we may not use the well-known Novikov–Adyan theorem on a free Burnside group of odd period [4]. But in this case we would obtain a completely different result. In other words, to finish the proof of the theorem we have to justify the existence of T_0-group. But, as was noted above, such justification is based on the properties of a free Burnside group of odd period ≥ 665 [4].

Here we discuss only two corollaries of the theorem. But they have the principal character.

What is the place of finite groups in the class of periodic groups? In other words, under what finiteness conditions a group becomes finite? The question was always topical because of the well-known problems of Burnside [8]. At present the question becomes even more topical. Indeed, after solving the above-mentioned problems, the question arises what role does a finite group play in the class of periodic groups in comparison with its role in the class of locally finite groups. Perhaps, the following assertion can be a consolations in this indeterminate situation.

Corollary 1. *A nontrivial periodic finitely generated group is finite if and only if there exists an element of prime order p such that assumptions 1, 2, and 4 of the theorem are satisfied.*

It is easy to prove that the following assertion is equivalent to the theorem and gives an abstract characterization of a T_0-group in the class of all groups.

Corollary 2. *Let G be a group and a its element of prime order p. Then G is a T_0-group and $p = 2$ if and only if the pair (G, a) satisfies the assumptions 1–4 of the theorem and the subgroup $\operatorname{gr}(a^g \mid g \in G)$ is not nilpotent-by-finite.*

Earlier, introducing a class of groups, the author was guided by utilitar argument. For example, it was the case, when the author was proving the well-known Blackburn Theorem in the class of biprimitive finite p-groups. This class includes, in particular, arbitrary 2-groups and is larger than the class of locally finite p-groups. The author knew that Blackburn Theorem is not valid in general (Novikov–Adyan Theorem [4]). The same can be said about the positive solution of the well-known Chernikov problem on groups with the minimal condition for (abelian) subgroups in the class of biprimitive finite groups (V. P. Shunkov, A. N. Ostylovskiĭ, N. G. Suchkova, A. K. Shlepkin). Note that this problem in general was solved in the negative (P. S. Novikov, S. I. Adyan, A. Yu. Ol'shanskiĭ [4, 9]).

As for T_0-groups, their appearance is unexpected but, simultaneously, natural for the author since this notion was born in proving the theorem. A similar situation is observed in research in finite groups when the generalized quartenions group suddenly arises and leads to essential correctives in previous plans of a researcher.

By virtue of the theorem and Corollary 2, we see that the key to answers to some general questions about the placement and inclusion of elements of prime orders in a group lies in the depth of the structure of a T_0-group. But yet we have a vague knowledge about the structure of these groups except for a few elementary properties of T_0-groups listed above. Hence it is a topical problem to elaborate approaches to the study of T_0-groups.

For example, we pose two questions.

1. Let G be a T_0-group, i an involution in G, and $G = \operatorname{gr}(i^g \mid g \in G)$. What can be said on the structure of the centralizer $C_G(i)$?

2. Does a simple T_0-group exist?

References

[1] Shunkov, V. P., M_p-groups. Nauka, Moscow 1990.

[2] Shunkov, V. P., On embedding primary elements into a group. Nauka, Novosibirsk 1992.

[3] Senashov, V. I., Layer-finite groups. Nauka, Novosibirsk 1993.

[4] Adyan, V. P., The Burnside problem and identities in groups. Nauka, Moscow 1975.

[5] Shunkov, V. P., On disposition of involutions in a group, Sib. Mat. Zh. 34:2 (1993), 208–219. English translation: Sib. Math. J. 34:2 (1993), 385–393.

[6] Shunkov, V. P., T_0-groups and elements of prime order in a group. Computer Center of Siberian Branch of Russian Academy of Sciences, Krasnoyarsk, 1993 [Preprint].

[7] Shunkov, V. P., On the proof of the theorem on T_0-group. Computer Center of Siberian Branch of Russian Academy of Sciences, Krasnoyarsk, 1994 [Preprint].

[8] Kargapolov, M. I., Merzlyakov, Yu. I., Fundamentals of group theory. Nauka, Moscow 1982.

[9] Ol'shanskiĭ, A. Yu., Geometry of defining relations in groups. Nauka, Moscow 1982.

Universal derivations and subalgebras of free algebras

Ualbaĭ U. Umirbaev

Abstract. The paper is a survey of results, mainly due to the author, on the following topics: (1) the structure of subalgebras of a free algebra; (2) the occurence problem for subalgebras of a free algebra; (3) characterization of automorphisms of free algebras of finite rank.

1991 Mathematics Subject Classification: 16S10, 16W10, 16W25, 17B35, 03D40.

Algebraists are working on the problem of description of subalgebras of free algebras for a long time. A classical example is Schreier's theorem [1] stating that subgroups of free groups are free. Now varieties of algebras with this property are called *Schreier*. The variety of Lie algebras was proved to be Schreier by A. I. Schirschov [2] and E. Witt [3]. In the case of associative algebras, the analog of Schreier's theorem is false. Some necessary and sufficient conditions on subalgebras of free associative algebras to be free are given in P. Cohn's book [4]. A real analog of Schreier's theorem is Cohn's theorem [5] stating that a submodule of a free module over a free associative algebra is free.

Nielsen transfomations [6, chapter 3], which were used to prove Schreier's theorem, give two important theorems of group theory: (a) the occurence problem for finitely generated subgroups of free groups is decidable; (b) automorphisms of free groups of finite rank are tame. A close connection of these problems to linear algebras is demonstrated by Levin's theorem [7] on equivalence of Nielsen and Schreier properties for uniform varieties of algebras.

In 1958 K. A. Mikhaĭlova [8] proved undecidability of the occurence problem for direct products of free algebras. This result is easily adapted for any variety of algebras with an undecidable word problem. Thus, it is necessary to study the occurence problem for free algebras. In this formulation, it is concerned with the general problem of description of subalgebras of free algebras. An appropriate solution of the occurence problem gives important information on automorphisms of free algebras.

In 1953 R. X. Fox [9] began to study differential calculus in group rings of free groups, which are widely applied in combinatorial group theory. In particular, the new proof of Schreier's theorem [10] clarified the connections between subgroups and submodules of the fundamental ideal. The exact formulation of the connection was given in [11]. J. S. Birman [12] obtained a characterization of automorphisms of free groups through Fox derivations.

Corresponding derivations of free associative algebras are called universal derivations [13]. Unfortunately, universal derivations of free algebras are not applicable to the investigation of arbitrary subalgebra, as was the case for groups. It turned out that the structure of subalgebras of a free algebra depends on the structure of the universal multiplicative enveloping algebra.

The present survey is an expanded version of the author talk on the IIIrd International Conference on Algebra dedicated to the memory of M. I. Kargapolov (Krasnoyarsk, 1993). The results given below, mainly due to the author, are concerned with the following topics:

(1) the structure of subalgebras of a free algebra;

(2) the occurence problem for subalgebras of a free algebra;

(3) characterization of automorphisms of free algebras of finite rank.

In the first section, basic definitions are given and methods of investigations are described. In the second section, these methods are applied to prove some classical theorems of the theory of Lie algebras and some new results on Schreier varieties of algebras. In the third section, results on the occurence problem for Lie algebras and groups are given. The occurence problem and algebraic dependence of elements for free associative algebras are considered in the fourth section. The methods for proving undecidability are described in more detail here. In the last section, theorems concerned with characterization of automorphisms and primitive elements of free algebras of finite rank via Fox derivations are formulated.

1. Universal derivations

Fix a field F and a uniform variety M of algebras over the field F. Let $A = F_M\langle X\rangle$ be the free algebra of the variety M with the base $X = \{x_1, x_2, \dots, x_n, \dots\}$. The universal multiplicative enveloping algebra of the algebra A in the variety M is denoted by $U(A) = U_M(A)$ (cf. [14]).

Give a definition of the algebra $U(A)$, convenient for our purposes. Let B be the free algebra of the variety M with the base $\{y\} \cup X$. The multiplication algebra of the algebra B is denoted by $M(B)$ (cf. [15]). Recall that $M(B)$ is generated by operators of left and right multiplication $L_b, R_b, b \in B$, acting on B. Then the algebra $U(A)$ is a subalgebra (with the identity 1) of the algebra $M(B)$ generated by the operators $L_a, R_a, a \in A$.

Let

$$I_A = y_1U(A) \oplus y_2U(A) \oplus \cdots \oplus y_nU(A) \oplus \cdots$$

be the free right $U(A)$-module with the base $y_1, y_2, \dots, y_n, \dots$.

Then the derivation

$$D = D_A\colon A \to I_A, \quad D(x_i) = y_i, i \geq 1$$

$$D(ab) = D(a)R_b + D(b)L_a, a, b \in A$$

is called a *universal M-derivation of the algebra A.*

For a subalgebra H of the algebra A, J_H denotes the $U(A)$-submodule of the module I_A which is generated by the elements $D(h), h \in H$. The variety M has the *property of differential separation of algebras* if for any subalgebra H of A and any $a \in A$, $D(a) \in J_H$ iff $a \in H$.

Lemma 1. *The following varieties have the property of differential separation of subalgebras:*

(1) *the variety of all algebras;*

(2) *the variety of commutative algebras* (char $F \neq 2$);

(3) *the variety of anticommutative algebras* ($x^2 = 0$);

(4) *the variety of Lie algebras;*

(5) *the variety of algebras with nonzero multiplication.*

The author was the first who noticed the property of differential separation of subalgebras for Lie algebras [16,17]. Later, independent proofs appeared in [18,19]. For Lie superalgebras the analogous statement was proved in [20]. Other statements of Lemma 1 were proved in [21].

Thus, Lemma 1 reduces the investigation of a subalgebra H of A to that of the submodule J_H of I_H. But the structure of J_H depends essentially on the structure of the algebra $U(A)$. Note that for varieties of algebras considered in Lemma 1, the algebra $U(A)$ is a free associative algebra. In the sequel we will need the following results from the theory of free associative algebras [5].

Cohn's Theorem. *A free associative algebra has a weak algorithm and is an FI-ring, i.e. submodules of a free module over the algebra are free and have the same rank.*

Levin's Theorem (an analog of Schreier's formula). *Let A be the free associative algebra of rank n, I be a right ideal of the algebra A such that* $\dim_F(A/I) = k < \infty$. *Then I is a free right A-module of rank $r = k(n-1)+1$.*

2. Schreier varieties of algebras

In this section for any Lie algebra L, $U(L)$ denotes the universal enveloping algebra of L. The algebra L is identified with the corresponding subalgebra of the algebra $U(L)^{(-)}$. For $M \subseteq L$, $\langle M \rangle$ denotes the subalgebra of L generated by the set M. For a subalgebra H of L, denote by J_H the right ideal of the algebra $U(L)$ generated by the set H. For Lie algebras, the property of differential separation of subalgebras stated in Lemma 1 is formulated more precisely in the following form:

$$H = L \cap J_H. \tag{2.1}$$

Consider the system of elements

$$h_1, h_2, \dots, h_k, \dots \tag{2.2}$$

of L. Denote $H = \langle h_1, h_2, \dots, h_k, \dots \rangle$.

Lemma 2. *The subalgebra H is the free Lie algebra with the base (2.2) iff J_H is the free right $U(L)$-module with the base (2.2).*

Immediate. □

In the sequel, $L = \mathrm{Lie}\langle X \rangle$ denotes the free Lie algebra with the base X. The system (2.2) of elements of L is said to be irreducible if there is no k such that $\overline{h}_k \in \langle \overline{h}_i \mid i \neq k \rangle$, where $\overline{h}$ denotes the leading part of the element h with respect to X-degree. For the free Lie algebra with the base (2.2), the system (2.2) is said to be algebraically independent.

Lemma 3. *Every irreducible system of elements of a free Lie algebra is algebraically independent.*

Algebraic dependence of the system (2.2) written in the additive form gives the equality

$$\sum_{i=1}^{k} h_i a_i = 0, a \in U(L).$$

Considering only the leading parts of additives and omitting the details, we may suppose that

$$\sum_{i=1}^{k} \overline{h}_i \overline{a}_i = 0.$$

Since $U(L)$ is a free associative algebra, the weak Cohn algorithm gives the equality

$$\overline{h}_i = \sum_{j \neq i} \overline{h}_j \overline{b}_j.$$

By (2.1), we have $\overline{h}_i \in \langle \overline{h}_j \mid j \neq i \rangle$, contrary to irreducibility of the system (2.2). □

Recall that a variety of algebras, in which the statement of Lemma 3 holds, is called Nielsen. The next theorem is an immediate consequence of the lemma.

Theorem (Shirshov–Witt [2, 3]). *A subalgebra of a free Lie algebra is free.*

Lemma 4. *Let* (2.2) *be an irreducible system of elements and* $h \in H$. *Then* $\overline{h} \in \langle \overline{h}_i \mid i \geq 1 \rangle$.

Lemma 4 easily implies the decidability of the occurence problem for free Lie algebras and the next theorem.

Theorem (Cohn [4]). *Automorphisms of free Lie algebras are tame.*

Every statement in this section becomes true for Lie p-algebras if we consider $U(L)$ as the designation for the bounded universal enveloping algebra of L.

Theorem (Kukin [22]). *Let* L *be the free Lie* p*-algebra of rank* n *and let* H *be its* p*-subalgebra of codimension* $k < \infty$. *Then* H *is the free Lie* p*-algebra of rank* $r = p^k(n-1)+1$.

Here, the module J_H has codimension p^k in $U(L)$. By Levin theorem, J_H is of rank $r = p^k(n-1)+1$. By lemma 2, r is equal to the rank of H. □

There are five well known Schreier varieties of (usual) linear algebras: namely, that of all algebras (A. G. Kurosh [23]), of commutative and of anticommutative algebras (A. I. Shirshov [24]), of Lie algebras (A. I. Shirshov [2], E. Witt [3]), and of algebras with zero multiplication. The varieties of Lie p-algebras [3], of Lie superalgebras [25,26], and of Lie p-superalgebras [27] are Schreier too. The problem of description of Schreier varieties of algebras (and of existence of Schreier varieties different from the five pointed above) is raised in [23, problem 1.179]. We mention some theorems that belong to the subject.

Theorem 1 [21]. *For a uniform Schreier variety* M, *the following conditions hold:*

(1) *for a free algebra* A *from* M, $U_M(A)$ *is a free associative algebra;*

(2) *for any uniform subalgebra* H *of a free algebra* A *from* M, $U_M(A)$ *is a free right* $U_M(H)$*-module.*

A variety for which the conditions 1,2 are satisfied is Schreier.

Corollary (Bakhturin [29]). *A uniform Lie variety is Schreier iff it coincides with either the variety of all Lie algebras or the variety of all algebras with zero multiplication.*

A method to test that a variety is Schreier is given by the next theorem.

Theorem 2 [21]. *Let M be a variety such that the following conditions are satisfied:*

(1) *for a free algebra A from M, $U_M(A)$ is a free associative algebra;*

(2) *the variety M has the property of differential separation of subalgebras.*

Then M is Schreier.

By Theorem 2, the varieties considered in lemma 1 are Schreier. Theorem 2 implies also the next

Theorem 3 [21]. *The variety of algebras defined by the identity*

$$xx^2 = 0$$

is Schreier.

We formulate some open questions related to last results.

Question 1. Is M a Schreier variety provided that $U_M(A)$ is a free associative algebra for every free algebra A of M?

Question 2. Is a variety with the property of differential separation of subalgebras Schreier?

Question 3. Let M be a Schreier variety and assume that universal derivations of free algebras of M are one-to-one. Does M have the property of differential separation of subalgebras?

3. The occurence problem for Lie algebras and groups

By (2.1) the decidability of the occurence problem for a finitely generated subalgebra H of a Lie algebra L is reduced to decidability of the occurence problem for the finitely generated right ideal J_H of $U(A)$. For free Lie algebras the latter is solved easily. Thus the next theorem is valid.

Theorem 4 [17]. *A free Lie algebra is residually finite with respect to occurence in finitely generated subalgebras.*

Theorem 4 answers G. P. Kukin's question [28, Question 2.61]. An analog of the theorem for Lie superalgebras was proved by A. A. Mikhalev [20].

Since there exist right ideals of $U(L)$ different from J_H, the solvability of the occurence problem for L does not imply the solvability of the problem for right ideals of $U(L)$. The difference between these algorithmic problems is shown by the next theorem.

Theorem 5 [30]. *Let L be a free metabelian Lie algebra of sufficiently large rank. Then in the algebra $U(L)$, the occurence problem for finitely generated right ideals is undecidable.*

This means that solvability of linear equations of the type

$$a_1x_1 + a_2x_2 + \cdots + a_nx_n = b, \quad a_i, b \in U(L),$$

is algorithmically undecidable. On the contrary, for Lie algebras of finite rank the problem is decidable. The solvability of a system of linear equations is reduced to the solvability of a single equation by enlargement of the rank of L.

Free metabelian Lie algebras are residually finite with respect to occurence in finitely generated subalgebras (for example, [31]). S. A. Agalakov [32] proved that free solvable Lie algebras (groups) of solvability length ≥ 3 aren't residually finite with respect to occurence in finitely generated subalgebras (subgroups). The question of decidability of the occurence problem was open in this case. It turns out that the next theorem holds.

Theorem 6 [30]. *The occurence problem for free solvable Lie algebras of solvability length ≥ 3 is undecidable.*

There is a well-known theorem of K. A. Mikhaĭlova [33] in group theory, which states decidability of the occurence problem for direct product $G_1 * G_2$ if the problem is decidable for groups G_1, G_2. For Lie algebras, the next theorem holds.

Theorem 7 [30]. *If L is a free metabelian Lie algebra of sufficiently large rank and L_1 is a one-dimensional Lie algebra, then the occurence problem in the algebra $L * L_1$ is undecidable.*

Theorem 7 answers G. P. Kukin's question [28, Question 1.86]. Another (more simple) example answering the same question was given independently by S. A. Agalakov [34]. The Agalakov theorem and Theorem 5 are proved by interpretation of two-tape Minsky machines [35]. In the author paper [30], Theorems 6, 7 are shown to be consequences of Theorem 5.

Lemma 5. *Let L_1 be a one-dimensional Lie algebra and L any Lie algebra. If the occurence problem is decidable in the algebra $L * L_1$, then the occurence problem for right ideals is decidable in $U(L)$, too.*

Theorem 7 is an immediate consequence of Theorem 5 and Lemma 5.

The next queston is still open.

Question 4 (Kukin [28]). Is the occurence problem in a Lie algebra with one identity decidable?

By Lemma 5, in any case we need to solve the occurence problem for right ideals of $U(L)$.

Question 5. Is the occurence problem for differentiation Lie algebras $W_n, n = 2, 3$, of a polynomial ring of char $= 0$ decidable?

For sufficiently large n, undecidability of the occurence problem in algebra W_n follows from Theorem 6.

For groups, the analogs of Theorem 5, 6 hold.

Theorem 8. *The occurence problem for right ideals of the integer group ring ZG of a free metabelian group G (of sufficiently large rank) is undecidable.*

Theorem 9. *The occurence problem for free solvable groups of solvability length ≥ 3 is undecidable.*

This answers Kargapolov's question [36, Question 21].

Undecidability of the word problem in the variety of solvable Lie algebras (groups) of solvability length ≥ 3 is well known [37-40]. By the proofs of Theorems 6, 9, there are finitely presented Lie algebras (groups) with undecidable word problem and with identities from the last commutant in the varieties. For a Lie algebra (group) with a single identity of this kind, the word problem is decidable [41,42].

4. Algebraic dependence of elements of a free associative algebra

Let $A = \text{Ass}\,\langle X \rangle$ be the free associative algebra with a base X. For the algebra A' antiisomorphic to A ($L_a = a' \oplus 1, R_a = 1 \oplus a, \quad a \in A$), $U(A) = A' \oplus_F A$. Let I_A be the kernel of the mapping $\lambda \colon U(A) \to A$ defined by the rule $\lambda(\sum a_i' \oplus b_i) = \sum a_i b_i$. The module I_A is a free right $U(A)$-module with the base $x_i' \oplus 1 - 1 \oplus x_i, x_i \in X$.

The universal derivation

$$D\colon A \to I_A$$

is specified by

$$D(A) = a' \oplus 1 - 1 \oplus a, a \in A.$$

The variety of associative algebras does not have the property of differential separation of subalgebras. Nevertheless, many subalgebras of the algebra A are differentially separated. For these subalgebras the occurence problem is reduced to the occurence problem for submodules of I_A.

To prove undecidability, the majority of authors on algorithmic themes use four elements a, b, c, d satisfying the identities

$$ac = ca, ad = da, bc = cb, bd = db. \tag{4.1}$$

These are the first identities for the semigroup of G. S. Tseitin [43]. In $U(A)$, the elements $L_{x_1}, L_{x_2}, R_{x1}, R_{x_2}$ satisfy them, i.e. the identities (4.1) can be used without additional suppositions on A. Thus, it is natural to expect undecidability of any algorithmic question concerning $U(A)$.

Let M be a two-tape Minsky machine (for definitions see [35]) with internal states $q_0, q_1, \dots, q_n$ (q_0 is final) and instructions of the form

$$q_i\varepsilon\delta \to q_j T_\alpha T_\beta, \tag{4.2}$$

$1 \leq i \leq n$, $0 \leq j \leq n$, ε, $\delta = 0, 1$, α, $\beta = -1, 0, 1$, $\alpha \geq 0$ for $\varepsilon = 1$ and $\beta \geq 0$ for $\delta = 1$.

If the machine M scanes the cell with a number t_k on a tape k ($k = 1, 2$) in the internal state q_i, then we say that M is in the configuration $[i, t_1, t_2]$; the notation $[i, t_1, t_2] \to [j, s_1, s_2]$ means that M moves from first to second configuration in one step.

Let P be a semigroup with generators

$$X = \{q_0, q_1, \dots, q_n, x, y, z, t\}$$

and with identities

$$x^\varepsilon y^{2-\varepsilon} q_i z^{2-\delta} t^\delta = x^\varepsilon y^{2-\varepsilon+\alpha} q_j z^{2-\delta+\beta} t^\delta, \tag{4.3}$$

which correspond to commands (4.2) of M.

Associate with every word of the form

$$u = xy_{s+1} q_i z^{r+1} t \tag{4.4}$$

the cofiguration $[u] = [i, s, r]$.

Say that v is obtained from u by right elementary transformation (cf. [35]) if u has a subword identical to the left hand part of an identity of type (4.3) and v is the result of the substitution of this subword in u by the right hand part of the identity.

Lemma 6. *Let u, v be words of type* (4.4). *The word v is obtained from u by elementary transformation iff* $[u] \to [v]$.

Thus (with an appropriate choice of the machine M), the semigroup P has an undecidable word problem. Unlike other semigroups with the property, P does not admit redundant elementary transformations which does not correspond to instructions of M. A semigroup P with additional identities occured in [38].

If we add the identities

$$ab = 0, \quad a, b \in X,$$

where $ab \neq xy, y^2, yq_i, z, z_2, zt$, to the identities (4.3), then the identity

$$x[y,z][t,k]l = 0 \tag{4.5}$$

holds on the semigroup algebra.

Thus the statement of Lemma 6 is true.

Theorem 10 [44]. *Let M be a variety of associative algebras defined by* (4.5). *Then the occurence problem for free algebras of M is undecidable.*

The analogous statement holds for the variety considered in [39].

Corollary [45, 46]. *The occurence problem for free associative algebras is undecidable.*

Earlier, S. A. Agalakov [47] proved finite inseparability of finitely generated subalgebras of free associative algebras.

Theorem 11 [46]. *The occurence problem for free Jordan algebras is undecidable.*

In connection with Cohn's theorem [4], the problem of description of free subalgebras of free associative algebras is raised. In this direction, the next theorem seems to be interesting.

Theorem 12 [44]. *Algebraic dependence of a finite system of elements of a free associative algebra is algorithmically undecidable.*

The theorem answers the question of A. T. Kolotov [28, Question 2.51]. The author does not know whether there exists another variety of algebras for which an analog of Theorem 12 holds.

Question 6. Is algebraic dependence of a a finite system of elements of free special Jordan algebra algorithmically undecidable?

Now we show how to use a Minsky machine to prove Theorem 12.

Denote by $g(i,\varepsilon,\delta)$ the element of $A = \text{Ass}\,\langle X\rangle$ which is the difference of the left and right sides of the identity (4.3). We say that the machine M has a cycle if there exists a configuration $[i, t_1, t_2]$ such that M, starting in it, returns to it in a finite non-zero number of steps. In [44] the algorithmic undecidability of the property of a machine to have a cycle is proved.

Proposition. *The system of elements*

$$x, y, z, t, g(i,\varepsilon,\delta),\ 1 \neq i \neq n,\ \varepsilon,\delta = 0,1,$$

of A is algebraically independent iff M has a cycle.

Methods of the proof of Theorem 12 give an opportunity to prove the next theorem.

Theorem 13 [44]. *There is a finitely presented associative algebra with the undecidable word problem whose module of identities is free.*

Question 7. Is there a finitely presented Lie algebra with the undecidable word problem whose module of identities is free?

The notion of a module of identities is investigated intensively in group theory [48].

Note that the occurence problem for the polynomial ring $F(x_1, x_2, \dots, x_n)$ is decidable.

Question 8. Is the occurence problem for the algebra $M_2(F[x_1, x_2, \dots, x_n])$ decidable?

5. A characterization of automorphisms and primitive elements of free algebras

We use the notation of section 1. The partial derivatives $\frac{\partial f}{\partial z_i}$ (Fox derivatives) of an element f of a free algebra A of a variety M are uniquely defined by the the formula

$$D(f) = \sum_i y_i \frac{\partial f}{\partial x_i}, \quad \frac{\partial f}{\partial x_i} \in U(A).$$

Let $\mid X \mid = n$. Denote by $\partial(f)$ the column-vector $(\frac{\partial f}{\partial x_1}, \frac{\partial f}{\partial x_2}, \dots, \frac{\partial f}{\partial x_n})^T$, where the symbol T denotes a transposition. The Jacobian matrix of an endomorphism $\varphi\colon A \to A,\quad \varphi(x_i) = f_i, 1 \le i \le n$, is the matrix

$$J(\varphi) = (\partial(f_1), \partial(f_2), \dots, \partial(f_n)).$$

The Jacobian hypothesis for the variety M states that the matrix $J(\varphi)$ is invertible over $U(A)$ iff φ is an automorphism. In the case of a variety of associative commutative algebras over a field of zero characteristic, the latter coincides with the classical Jacobian conjecture (Kähler's problem [28]). For free groups it was proved by J. S. Birman [12]. By Lemma 1, the Jacobian hypothesis is valid for varieties of all algebras, of ε-algebras [51], and of Lie algebras [52,18,19]. The same is true for Lie superalgebras [20] and associative algebras [53,54]. Characterizations of generators of Lie algebras of type L/R^2 are given in [19,55]. The last one is an analog of A. F. Krasnikov's [56] result.

Let G be the free group with a base $x_1, x_2, \dots, x_n$. Recall that a system of elements of a group G is primitive if it can be extended to a base of G.

Theorem 14 [57]. *A system of elements $g_1, g_2, \dots, g_k$ from G is primitive iff the matrix $(\partial(g_1), \partial(g_2), \dots, \partial(g_k))$ is left invertible over ZG.*

Theorem 14 is an easy consequence of Theorem D from [11], which is proved applying the developed theory of ends of groups. But the statement of the theorem has a huge methodological value. By the primitive element hypothesis for a variety M we mean the statement of Theorem 14 for free algebras of M.

The truth of the primitive element hypothesis for free metabelian groups was proved by V. A. Roman'kov [58] and E. I. Timoshenko [59]. An analogous statement is valid for metabelian Lie algebras [55]. Give an exact formulation of the result.

In the sequel, L is the free Lie algebra with a base $x_1, x_2, \dots, x_n$. Let $R = [L, L]$. Then $\tilde{M} = L/R$ is the abelian Lie algebra with the base $\tilde{x}_1, \tilde{x}_2, \dots, \tilde{x}_n$ and $\bar{L} = L/[R, R]$ is the free metabelian Lie algebra with the base $\bar{x}_1, \bar{x}_2, \dots, \bar{x}_n$. The kernel of the map

$$\varrho \circ \partial \colon L[U(L)^n]^T[U(\tilde{L})^n]^T$$

coincides with $[R, R]$ [60], which defines the map

$$\bar{\partial} \colon \bar{L} \to [U(\tilde{L})^n]^T.$$

The algebra $U(\tilde{L})$ is a polynomial ring.

Theorem 15 [55]. *Let $g_1, g_2, \dots, g_k$ be a system of elements of $\bar{L}$. Then the following conditions are equivalent:*

(1) *the system $g_1, g_2, \dots, g_k$ is primitive;*

(2) *the matrix $(\bar{\partial}(\bar{g}_1), \bar{\partial}(\bar{g}_2), \dots, \bar{\partial}(\bar{g}_k))$ is left invertible over $U(\tilde{L})$;*

(3) *k-minors of $(\bar{\partial}(\bar{g}_1), \bar{\partial}(\bar{g}_2), \dots, \bar{\partial}(\bar{g}_k))$ generate the trivial ideal of $U(\tilde{L})$.*

Corollary. *Primitiveness of a finite system of elements of a free metabelian Lie algebra over a recursive field is algorithmically decidable.*

A. A. Mikhalev, A. A. Zolotykh [61] have proved a very strong result on the rank of an element of a free Lie algebra over a field of characteristic zero (or Lie p-algebra), which implies validity of the primitive element hypothesis. The rank of an element f of a free Lie algebra L is the minimal possible number of primitive elements of L such that f is expressible by means of them.

Theorem (Mikhalev–Zolotykh [61]). *The rank of an element f of a free Lie algebra L over a field of characteristic zero is equal to the rank of the left $U(L)$-module generated by $\frac{\partial f}{\partial x_1}, \frac{\partial f}{\partial x_2}, \dots, \frac{\partial f}{\partial x_n}$.*

Apparently, the method of the proof of the theorem allows to prove also the corresponding statements for free groups, (absolutely) free algebras, ε-algebras. For Lie algebras over a field of characteristic $p > 0$, the primitive element hypothesis is false. A counterexample [61] is the element

$$f = x + [y, z] + (ad(x))^p(z)$$

of the free Lie algebra $L = \text{Lie}\ \langle x, y, z\rangle$.

The following interesting properties of $L_1 = L/(f)$ are given in [62]:

(1) L_1 is not a free Lie algebra;

(2) $U(L_1)$ is a free associative algebra;

(3) L_1 has cohomological dimension 1.

Thus, the analog of Stallings [11] theorem for Lie algebras of a positive characteristic is false.

Question 9. Is the primitive element hypothesis for free associative algebras true?

Question 10. Is an element f of the polynomial ring $F[x_1, x_2, \dots, x_n]$ over a field F of characteristic zero primitive in the case of unimodular $\partial(f)$?

Of course, the positive answer to the question 10 solves Kähler and Zariski problems (the last one is known as the cancellation problem). The fact that a positive answer to Kähler problem solves also the cancellation problem to affirmative is stated in [63]. But the arguments presented there are not convincing.

References

[1] Schreier, O. J., Die Untergruppen der freien Gruppen. Abh. Math. Semin. Univ. Hamb. 5 (1927), 161–183.

[2] Shirshov, A. I., Subalgebras of free Lie algebras. Mat. Sb. N. S. 33(75):2 (1953), 441–452.

[3] Witt, E., Die Unterringe der freien Lieschen Ringe. Math. Z. 64 (1956), 195–216.

[4] Cohn, P. M., Subalgebras of free associative algebras. Proc. Lond. Math. Soc., III. Ser. 14:56 (1964), 618–632.

[5] — Free Rings and Their Relations. Mir, Moscow 1975. Original: Lond. Math. Soc. Monogr. 2. Academic Press, London–New York 1971.

[6] Magnus, W., Karrass, A., Solitar, D., Combinatorial group theory: Presentations of groups in terms of generators and relations (Pure Appl. Math. 13). Interscience Publ., New York–London–Sydney 1966.

[7] Lewin, J., On Schreier varieties of linear algebras. Trans. Am. Math. Soc. 132:3 (1968), 553–562.

[8] Mikhailova, K. A., The occurrence problem for direct products of groups. Dokl. Akad. Nauk SSSR 119:6 (1958), 1103–1105.

[9] Fox, R. H., Free differential calculas, I. Derivation in the free group ring. Ann. Math., II. Ser. 57 (1953), 547–560.

[10] — Free differential calculas, III. Subgroups. Ann. Math., II. Ser. 64 (1956), 407–419.

[11] Cohen, D. E., Groups of cohomological dimension one (Lect. Notes Math. 245). Springer-Verlag, Berlin–Heidelberg–New York 1972.

[12] Birman, J. S., An inverse function theorem for free groups. Proc. Am. Math. Soc. 41:2 (1974), 634–638.

[13] Bergman, G. M., Dicks, W., On universal derivations. J. Algebra 36:2 (1975), 193–211.

[14] Jacobson, N., Structure and representations of Jordan algebras (Am. Math. Soc. Colloq. Publ. 39). Am. Math. Soc., Providence, RI 1968.

[15] Zhevlakov, K. A., Slin'ko, A. M., Shestakov, I. P., Shirshov, A. I., Nearly associative rings (Monogr. Modern Algebra). Nauka, Moscow 1978.

[16] Umirbaev, U. U., Residuality of free Lie algebras relatively occurrence. In: Int. Conf. on Algebra Dedicated to the Memory of A. I. Mal'tsev (Novosibirsk, August 21–26, 1989), Abstracts, 137. Institut Matematiki SO AN SSSR, Novosibirsk 1989. English translation: Contemporary Math. 131. Am. Math. Soc., Providence, RI 1992.

[17] — Approximation of free Lie algebras relative to inclusion. In: Monoids, Rings and Algebras (Tartu. Riikl. Ül. Toimetised 878), 147–152. Tartu Gos. Univ., Tartu 1990.

[18] Reutenauer, C., Applications of a noncommutative Jacobian matrix. J. Pure Appl. Algebra 77:2 (1992), 169–181.

[19] Spilrain, V., On generators of L/R^2 Lie algebra (to appear).

[20] Mikhalev, A. A., On right ideals of a free associative algebra generated by free colour Lie p-superalgebras. Usp. Mat. Nauk 47:5 (1992), 187–188. English translation: Russ. Math. Surv. 47:5 (1992).

[21] Umirbaev, U. U., On Schreier varieties of algebras. Algebra Logika 33:3 (1994), 317–340. English translation: Algebra Logic 33:3 (1994).

[22] Kukin, G. P., Subalgebras of a free Lie sum of Lie algebras with an amalgamated subalgebra. Algebra Logika 11:1 (1972), 59–86. English translation: Algebra Logic

11:1 (1972), 33–50.

[23] Kurosh, A. G., Non-asociative free algebras and free products of algebras. Mat. Sb. N. S. 20:2 (1947), 239–262.

[24] Shirshov, A. I., Subalgebras of free commutative and free anticommutative algebras. Mat. Sb. N. S. 34 (76):1 (1954), 81–88.

[25] Mikhalev, A. A., Subalgebras of free colored Lie superalgebras. Mat. Zametki 37:5 (1985), 653–661. English translation: Math. Notes 37:5-6, 356–360.

[26] Shtern, A. S., Free Lie superalgebras. Sib. Mat. Zh. 27:1 (1986), 170–174. English translation: Sib. Math. J. 27:1 (1986), 136–140.

[27] Mikhalev, A. A., Subalgebras of free Lie p-superalgebras. Mat. Zametki 43:2 (1988), 178–191. English translation: Math. Notes 43:1-2, 99–106.

[28] Dnestrovsk Notebook. Institut Matematiki SO AN SSSR, Novosibirsk 1982.

[29] Bakhturin, Yu. A., Identities in Lie Algebras. Nauka, Moscow 1985.

[30] Umirbaev, U. U., The occurrence problem for Lie algebras. Algebra Logika 32:3 (1993), 326–340. English translation: Algebra Logic 32:3 (1993), 173–182.

[31] Zaitsev, M. V., On finite separability relative to free Lie algebras. In: Int. Conf. on Algebra in Memory of A. I. Shirshov (Barnaul, August 20–25, 1991), Abstracts, 43. Mathematical Institute Press, Novosibirsk 1991.

[32] Agalakov, S. A., Finite separability of groups and Lie algebras. Algebra Logika 22:4 (1983), 363–371. English translation: Algebra Logic 22:4 (1993), 261–268.

[33] Mihailova, K. A., The occurrence problem for free products of groups. Dokl. Akad. Nauk SSSR 127:4 (1959), 746–749.

[34] Agalakov, S. A., On the occurrence problem for free products of Lie algebras. In: 3rd Int. Conf. on Algebra in Memory of M. I. Kargapolov (Krasnoyarsk, August 23–28, 1993), Abstracts, 4–5. INOPROF, Krasnoyarsk 1993.

[35] Mal'tsev, A. I., Algorithms and recursive functions. IInd ed. Nauka, Moscow 1986.

[36] Noskov, G. A., Remeslennikov, V. N., Roman'kov, V. A., Infinite groups. In: Algebra, Topology, Geometry (Itogi Nauki i Tekhniki 17), 65–157. Vsesoyuznyj Institut Nauchnoj i Tekhnicheskoj Informasii, Moscow 1979.

[37] Kukin, G. P., Algorithmic problems for solvable Lie algebras. Algebra Logika 17:4 (1978), 402–415. English translation: Algebra Logic 17:4 (1978), 270–278.

[38] Mel'nichuk, I. L., Sapir, M. V., Kharlampovich, O. G. The problem of equality in varieties of semigroups, rings, and Lie algebras. Sib. Mat. Zh. 27:6 (1986), 144–156. English translation: Sib. Math. J. 27:6 (1986), 904–914.

[39] Sapir, M. V., Kharlampovich, O. G., Equality problem in varieties of associative algebras and Lie algebras. Izv. Vyssh. Uchebn. Zaved., Mat. 6 (1989), 76–84. English translation: Sov. Math. 33:6 (1990), 77–86.

[40] Kharlampovich, O. G., A finitely presented solvable group with the non-solvable word problem. Izv. Akad. Nauk SSSR, Ser. Mat. 45:4 (1981), 852–873.

[41] Talapov, V. V., Polynilpotent Lie algebras given by one defining relation. Sib. Mat. Zh. 23:5 (1982), 192–204. English translation: Sib. Math. J. 23:5 (1982), 743–752.

[42] Sokolov, V. G., Word-identity algorithm for one class of solvable groups. Sib.

Mat. Zh. 12:6 (1971), 1405–1410. English translation: Sib. Math. J. 12:6 (1971), 1016–1019.

[43] Tseitin, G. S., An associative calculus with insoluble equivalence problem. In: Tr. Mat. Inst. Steklova 52, 172–189. Mathematical Institute Press, Moscow 1958.

[44] Umirbaev, U. U., Algorithmic problems in associative algebras. Algebra Logika 32:4 (1993), 450–470. English translation: Algebra Logic 32:4 (1993), 244–256.

[45] — On the occurrence problem for free associative algebras. In: Interstate Conf. on Algebra, Abstracts, 145. Kazanskij Gos. Univ., Kazan' 1992.

[46] — The occurrence problem for free associative and Jordan algebras (to appear).

[47] Agalakov, S. A., On finite separability of some Lie group and algebras. In: 5th Siberian Colloquium on Varieties of Algebraic Systems, Abstracts, 3–4. Altajskij Gos. Univ., Barnaul 1988.

[48] Lyndon, R. C., Schupp, P. E., Combinatorial group theory (Ergeb. Math. Grenzgeb. 89). Springer-Verlag, Berlin–Heidelberg–New York 1977.

[49] Noskov, G. A., Occurrence problem for a polynomial ring. Sib. Mat. Zh. 19:6 (1978), 1413–1414. English translation: Sib. Math. J. 19:6 (1978), 997–998.

[50] Shannon, D., Sweedler, M., Using Gröbner bases to determine algebra membership, split surjective algebra homomorphisms, determine birational equivalence. Computional aspects of commutative algebra. J. Symb. Comput. 6:2–3 (1988), 267–273.

[51] Yagzhev, A. V., Endomorphisms of free algebras. Sib. Mat. Zh. 21:1 (1980), 181–192. English translation: Sib. Math. J. 21:1 (1980), 133—141.

[52] Umirbaev, U. U., On Jacobian of Lie algebras. In: 6th Siberian Colloquium on Varieties of Algebraic Systems, Abstracts, 32. Magnitogorskij Gos. Univ., Magnitogorsk 1988.

[53] Dicks, W., Lewin, J., A jacobian conjecture for free associative algebras. Commun. Algebra 10:12 (1982), 1285–1306.

[54] Schofield, A. H., Representation of rings over skew fields (Lond. Math. Soc. Lect. Note Ser. 92). Cambridge University Press, Cambridge–New York 1985.

[55] Umirbaev, U. U., Partial derivatives and endomorphisms of some relatively free Lie algebras. Sib. Mat. Zh. 34:6 (1993), 179–188. English translation: Sib. Math. J. 34:6 (1993), 1161–1170.

[56] Krasnikov, A. F., Generating elements of the group $F/[N,N]$. Mat. Zametki 24:2 (1978), 167–174. English translation: Math. Notes 24:2 (1978), 591–594.

[57] Umirbaev, U. U., Primitive elements of free groups. Usp. Mat. Nauk (to appear.)

[58] Roman'kov, V. A., Criteria for the primitivity of a system of elements of a free metabelian group. Ukr. Mat. Zh. 43:7–8 (1991), 996–1002.

[59] Timoshenko, E. I., About occurrence of some elements in a base of a free metabelian group. Deposited in VINITI, No 2699. Moscow 1988.

[60] Kharlampovich, O. G., The Lyndon condition for solvable Lie algebras. Izv. Vyssh. Uchebn. Zaved., Mat. 9 (1984), 50–59. English translation: Sov. Math. 28:9 (1984), 67–80.

[61] Mikhalev, A. A., Zolotykh, A. A., The rank of an element of a free colour Lie (p-) superalgebra. In: 3rd Int. Conf. on Algebra in Memory of M. I. Kargapolov (Kras-

noyarsk, August 23–28, 1993), Abstracts, 418. INOPROF, Krasnoyarsk 1993.

[62] Mikhalev, A. A., Zolotykh, A. A., Umirbaev, U. U., A nonfree Lie algebra of cohomological rank 1 over a field of characteristic $p > 0$. Usp. Mat. Nauk (to appear).

[63] Miyanishi, M., Some remarks on polynomial rings. Csaka J. Math. 10 (1973), 617–624.

Stable groups (a survey)

Frank O. Wagner *

Abstract. The paper is a survey of results on stable groups. The contemporary investigations on the subject are caused by the problem on a simple group of finite Morley rank to be an algebraic group over an algebraically closed field. Methods adapted from algebraic geometry, techniques arising from the analysis of finite groups, in particular those used in the classification of finite simple groups, play an important role here.

1991 Mathematics Subject Classification: 03C45, 03C60, 20G15.

Introduction

Stable groups first arose in the context of uncountably categorical theories, when A. Macintyre [1, 2] classified the uncountably categorical abelian groups and fields. In 1965 M. Morley [3] had published his famous theorem on theories categorical in higher powers and defined ω-stability on the way, and in the late 60's Shelah began his work on classifying the models of complete first-order theories using the notion of stability. Now, these logical constraints have being applied to algebraic structures like deferentially closed fields [4] and minimal groups [5]. Simple groups and fields of finite rank were examined in the deep and famous paper by Zil'ber [6], which is one of the cornerstones of the subject. In 1979 Cherlin wrote an influential paper [7] classifying groups of Morley rank 2, and partially also those of rank 3. There he formulated his well-known conjecture:

Conjecture 1. A simple group of finite Morley rank is an algebraic group over an algebraically closed field.

It is an analogue for groups of Zil'ber's conjecture, which states (roughly) that the geometry induced by (model-theoretic) algebraic closure on a strongly minimal set should be disintegrated or come from linear dependency in a vector space or algebraic dependency in an agebraically closed field. It also is a programme rather than a prediction: one should try to describe algebraic geometry, and in particular the Zariski topology, by model-theoretic means. And in order to do that, one

* The author would like to thank the Deutsche Forschungsgmenmeinschaft for support.

should generalize the tools from algebraic geometry to the context of structures of finite Morley rank. And indeed there is a great similarity between algebraic groups over an algebraically closed field and groups of finite Morley rank: one has chain conditions, connected components, generic elements, and dimension (called rank). However, there are two main differences: First, in the finite rank case there is no distinction between open and closed sets as we have it in the Zariski topology, and second, we lack the notion of a complete variety. Although the original Zil'ber conjecture was refuted by E. Hrushovski [8], in a way the general programme of axiomatizing algebraic geometry (over an algebraically closed field) could be successfully completed in [9] and already has important applications outside of model theory [10].

Cherlin's conjecture, however, has remained open and been studied by various people, most notably by A. Borovik, A. Nesin, and B. Poizat [11, 12]. Increasingly, besides methods adapted from algebraic geometry, techniques arising from the analysis of finite groups played an important role, in particular those used in the classification of finite simple groups.

Meanwhile, people began to look at more general stability classes. G. Cherlin and S. Shelah were looking at superstable division rings [13], fields and groups [14]; C. Berline and D. Lascar generalized many theorems from ω-stable groups to the superstable case [15, 16] and J. Baldwin and J. Saxl analyzed stable groups [17]. U. Felgner [18] and W. Baur, G. Cherlin, A. Macintyre [19] proved the nilpotency of $\aleph_0$-categorical connected stable groups, and the latter also showed commutativity in the ω-stable case. Finally B. Poizat in [20] and [21] lay the foundations for a general study of stable groups by generalizing the method of generic types to stable groups.

In this survey we shall concentrate on general stable groups. They are interesting not only in their own right, as a common generalization of finite and algebraic groups, but they also arise definably under certain circumstances in general stability theory as automorphism groups [22]. Knowledge about these groups may then successfully be used to derive further information about the general superstructure. In another direction, stable groups may be used as a kind of experimental playground for stability theory: due to the multiplicative structure and the ensuring homogeneity, many more general techniques and results have a particularly nice form for groups. Thus, groups may be used to develop and test new ideas.

Most of the further mentioned results go through several stages. First proved for groups of finite Morley rank, they may have been generalized to ω-stable, superstable, and finally p-simple groups, before settling into their present form. However, proceeding in such generality we have to face a major difficulty: The loss of dimension or rank. While on one hand this doubtlessly leads to an increased technical machinery, on the other hand we are forced to develop truly new methods, besides those borrowed from finite or algebraic group theory. It may also reveal more about the structures in question, as it is no longer possible just to use induction on rank.

The remark on the implications of all this on the ongoing work on Cherlin's conjecture should be done. In author's opinion it shows how far we still have to go if we are to succeed: If most things proved for groups of finite Morley rank can be generalized to stable groups, we cannot hope to prove the conjecture unless we are prepared to believe it to hold in the stable case. This, however, seems most unlikely indeed.

Finally, a word or two on examples and non-examples: As an algebraically closed field has Morley rank one, any algebraic group over an algebraically closed field has finite Morley rank. Separably closed fields are merely stable [23], so algebraic groups over separably closed fields are stable. All modules and hence all abelian-by-finite groups are stable; the exact stability class depends on the possible definable subgroups. As for nilpotent groups, some of them (of small class and finite exponent) have been proved to be stable by Baudisch and Mekler. On the other hand, groups interpreting the field of real numbers are unstable. As for the free group, even in only two generators, it is not known whether it is stable or not (although Gibone has proved that it is not superstable). And a lot of open problems would be solved (mostly negatively) if one could construct a stable Tarski monster.

In terms of classification, we have that a group of Morley rank one is abelian-by-finite [5], a group of rank two is soluble-by-finite of class at most two [7], and a simple group of Morley rank three which possesses a subgroup of rank two is isomorphic to $\mathrm{PSL}_2(K)$ for an algebraically closed field K [7]. And E. Hrushovski has shown that any group definable in an algebraically closed field is algebraic over that field.

1. Chain conditions

Chain conditions are the first of the two pillars on which the study of stable groups rests. Special cases, in particular periodic groups with chain conditions on centralizers, had previously been studied by algebraists [24–26]. We call a subgroup of G *definable,* if it is of the form $\varphi(G, \bar{a})$, for some formula $\varphi(x, \bar{a})$ with parameters $\bar{a}$. A family of subgroups is *uniformly definable,* if all subgroups in the family are definable using the same formula and only varying the parameters. A subgroup is *type-definable,* if it is given (in a sufficiently saturated model) by the intersection of definable groups; it is *relatively definable with respect to some subgroup H,* if it is the intersection with H of some definable subgroup. The following fundamental result is due to J. Baldwin and J. Saxl [17]:

Theorem 1. *A stable group has the uniform chain condition for intersections of uniformly definable subgroups.*

That is, for any family of uniformly definable subgroups, there is a finite bound on the length of chains of intersections of these subgroups, which depends only on the formula used for the family. In particular we have both the ascending and descending chain conditions. It follows immediately that any chain of type-definable subgroups has cardinality at most that of the language. An important consequence is that the intersection of all definable subgroups of finite index is invariant under elementary extensions and has index at most $2^{|T|}$; it is called the *connected component.* It has been noted in [27] that the Baldwin–Saxl theorem implies also the uniform chain condition for intersections of uniformly relatively definable subgroups; this becomes important when dealing with possibly undefinable objects (like maximal p-subgroups).

Under stronger stability assumptions we get stronger chain conditions; e.g. in a superstable group there is no infinite descending chain of connected type-definable groups, in an ω-stable group there is no infinite descending chain of definable groups, and in a group of finite Morley rank there is no infinite ascending chain of connected definable groups. However, chain conditions also hold in different contexts (e.g. 0-minimality), whence results merely using them are valid in greater generality than just stable groups.

An easy consequence of the uniform chain condition above is the relative definability of the ascending central series and, more generally, of the normalizer of a relatively definable subgroup. (Note that for definable groups the corresponding subgroups are obviously definable in any case.) It also follows that every soluble or nilpotent group is contained in a definable soluble or nilpotent group of the same class. Slightly more effort is involved in the following [28]:

Proposition 1 *Let $\mathfrak{N} = \{N_i : i \in I\}$ be a family of normalizing k-nilpotent subgroups. Then $\mathfrak{N}$ generates a nilpotent subgroup. The result also holds with soluble instead of nilpotent.*

Proof. As any k-nilpotent normal subgroup is contained in a uniformly (in k) definable k-nilpotent normal subgroup, we may assume that the N_i are actually uniformly definable. (This is also true for soluble instead of nilpotent.)

We first treat the nilpotent case. We put $N_i^j := Z_{k-j+1}(N_i)$, and $N_i^0 := G$. Then N_i^j is $(k-j)$-nilpotent, normal and uniformly definable. Let n be a bound for the length of a descending chain of intersections of the N_i^j, and consider a sequence $a_s \in N_{i(s)}$. Let $b_0 := a_0$ and $b_{s+1} := [b_s, a_{s+1}]$.

If t is such that b_s lies in $N^t_{i(s+1)} - N^{t+1}_{i(s+1)}$, then b_{s+1} lies in $N^{t+1}_{i(s+1)}$, and because of normality b_{s+1} lies in all N_i^j's which contain b_s. Hence

$$\bigcap\{N_i^j : b_s \in N_i^j\}$$

properly contains

$$\bigcap\{N_i^j : b_{s+1} \in N_i^j\}.$$

But the sequence can descend at most n times, whence b_n lies in $Z(N_{i(n+1)})$ and b_{n+1} must be trivial. It follows that $\langle\mathfrak{N}\rangle$ is nilpotent of class $n+1$.

In the soluble case, we use induction on the solubility class, the case of abelian groups having just been dealt with. So we may assume that the derived subgroups

$\{N_i' : i \in I\}$ generate a soluble normal subgroup N, which we may assume to be definable. But $\{N_iN/N : i \in I\}$ is a family of normal abelian subgroups of G/N and generates a nilpotent group; the result follows. □

Hence we get ascending sequences N_n and S_n of normal nilpotent (resp. soluble) subgroups containing every nilpotent (resp. soluble) normal subgroup of class at most n. The immediate question is whether these sequences become stationary, i.e. whether the Fitting subgroup (resp. the soluble radical) exist. For periodic groups with descending chain condition on centralizers, this was proved by R. M. Bryant and B. Hartley [24, 25]. However, the general case seems to require stronger assumptions.

Chain conditions work particularly well in connection with nilpotency. As an illustration of the techniques used we give a sketch of the proof that maximal 2-subgroups are locally finite:

Theorem 2 [27]. *Let S be a 2-subgroup of a stable group. Then S is nilpotent-by-finite.*

Proof. By [24] a 2-group with chain conditions on centralizers is locally finite iff it is locally nilpotent iff it is nilpotent-by-finite, and then it is soluble of bounded class. Hence, if S is not nilpotent-by-finite, we may consider pairs of distinct maximal soluble subgroups. They are uniformly definable relative to S, hence their intersection is uniformly definable relative to S, and it follows from Baldwin–Saxl that there are soluble subgroups U and V with maximal intersection I.

U and V are nilpotent-by-finite and locally nilpotent, whence hypercentral, and I is not self-normalizing in neither U nor V. Therefore there are involutions $x \in N_U(I)/I$ and $y \in N_V(I)/I$, which generate a finite dihedral group modulo I (remember that xy is a 2-element). But now $I \cup \{x, y\}$ is contained in a maximal soluble subgroup W, and $W \cap U$ and $W \cap V$ both properly contain I. By maximality of I, we get $U = W = V$, contradiction. □

The prime 2 plays a special role in this result because of the finiteness of the periodic dihedral groups. For $p \neq 2$, one can show that a sufficient condition for maximal p-subgroups to be locally finite is to be 2-finite (i.e. for any two elements to generate a finite subgroup). Using the same idea in a technically more elaborate setting, one may also prove the conjugacy of the maximal p-subgroups under certain conditions (see [27]). There also is a variant of the Baer–Suzuki Theorem:

Theorem 3. *If X is a G-invariant subset of G such that any two elements of X generate a finite p-group, then X generates a locally finite p-group.*

In [29] nilpotency is achieved not through finite p-groups, but via the Engel condition $[x, y, \ldots, y] = 1$. It is shown that any stable group satisfying such a condition must be nilpotent. In particular, a saturated stable group is nilpotent iff any two elements generate a nilpotent subgroup. It is not known, however, whether a similar result also holds with soluble instead of nilpotent.

2. Generics

Suppose H is a subgroup of a stable group G, and $\varphi(G, \bar{a})$ is a definable subset. (We shall frequently switch between formulae and their sets of realizations, and use whichever is more convenient.) We call $\varphi(x, \bar{a})$ *generic for* H if H is covered by finitely many left translates $h\varphi(G, \bar{a})$ by elements h from H. It turns out that if we had taken right translates instead, we would have obtained the same notion, and that the complement of a nongeneric set is generic. In particular there are generic types, which intuitively correspond to the most general types possible. In a saturated model they are in 1-1 correspondence with the cosets of the connected component. In particular there are only boundedly many of them, and any element is product of two generics (i.e. elements realizing a generic type). The terminology is borrowed (once again) from algebraic geometry; a generic type should contain instances of as few formulae $\varphi(x, \bar{y})$ as possible. We should note, however, that in the general stable case genericity depends on the group law: If we have a structure with two group laws, the generic types with respect to each law need not be the same. (In the case of a field, though, they coincide.) This is different in superstable groups, where generic types are just those of maximal rank.

One vexing question in this area comes from the observation that in an algebraic group an equation that holds for generics is true for the whole group. In particular a group of generic exponent n has exponent n. This is not known even for groups of finite Morley rank, except for some small values of n (where it holds in the stable case). Other equations studied include those for solubility, nilpotency, and Engel conditions (see [30] and [29, 31]).

If H is any subgroup of a stable group G, we may consider the intersection $\mathrm{dc}(H)$ of all definable subgroups of G containing H. By the chain condition this is invariant under elementary extensions, and it turns out that H and $\mathrm{dc}(H)$ have the same generic types [32]. This can be used to prove the existence of the Fitting subgroup:

Theorem 4 [33]. *Let G be a stable group. Then G has a definable Fitting subgroup $F(G)$. If G is $|T|^+$-saturated, then $F(G)$ is also the maximal locally nilpotent normal subgroup.*

Proof. As there is an ascending sequence of $\emptyset$-definable normal subgroups N_n such that N_n contains every normal subgroup of nilpotency class at most n, we only have to show that this sequence eventually becomes stationary. Hence it is enough to prove the result for $|T|^+$-saturated groups.

Let F be the maximal locally nilpotent normal subgroup of G. It is proved in [24] that F is soluble, and we may use induction on the solubility class of F. If F is abelian, it clearly is definable as centre of its centralizer $C_G(F)$. For the sake of simplicity, we shall only treat the case of solubility class two. So we assume that there is a definable normal abelian subgroup H of G containing G'.

We first show that any element of F is contained in a normal nilpotent subgroup. So fix $f \in F$. For any $h \in H$ the group generated by f and h is nilpotent, so by saturation there must be a bound n on its nilpotency class. Hence $(f-1)^n H = 0$. But $[f, f^G]$ is a subset of G' and hence centralizes H, so f and any of its G-conjugates have the same action on H. But this implies $(f^{g_1}-1)\dots(f^{g_n}-1)H = 0$ for all $g_1, \dots, g_n \in G$ and H is contained in the nth iterated centre $Z_n(\langle H, f^G\rangle)$. Therefore $F_f := \langle f^G \rangle$ is $(n+1)$-nilpotent.

Now suppose that F is not nilpotent. Put

$$Z_\omega = H \cap \bigcup_{j<\omega} Z_j(F),$$

the hypercentre of F in H, which equals the hypercentre of $\mathrm{dc}(F)$ in H. We consider the action of G on H. As G' is contained in H, the ring of endomorphisms of H generated by G is commutative.

Assume that there are some sequence f_i for $i < \omega$ of elements of F such that no image $\prod_{i<j}(f_i - 1)H$ were contained in Z_ω. The sequence of groups $\bigcap_{i\geqslant j}(f_i-1)H$ is uniformly definable and increasing, hence stationary from some j_0 onwards by the ascending chain condition. So for all $j < \omega$ there is some finite set J of indices $> j$ such that

$$\bigcap_{i\in J}(f_i - 1)H = \bigcap_{i\geqslant j_0}(f_i - 1)H.$$

In particular we get

$$\prod_{i\in J}(f_i - 1)H \leq (f_{j_0} - 1)H. \tag{$*$}$$

On the other hand, $\langle H, f_{j_0}\rangle$ is contained in $HF_{f_{j_0}}$ and thus nilpotent, so there is $n < \omega$ with $(f_{j_0} - 1)^n H = 0$. But there are n disjoint sets $J_k \subset \omega$ satisfying $(*)$, and

$$\prod_{i\in\bigcup J_k}(f_i - 1)H \leq (f_{j_0} - 1)^n H = 0,$$

contradicting our choice of the f_i.

Hence there is a finite sequence $f_1, \dots, f_n$ of elements of F such that $A = \prod_{i\leqslant n}(f_i - 1)H$ is not contained in Z_ω, but for any $f \in F$ the image $(f-1)A$ is contained there. Note that A is still G-invariant.

Suppose $(g-1)A$ is contained in Z_ω for all generic elements of $\mathrm{dc}(F)$. Then by compactness there is $j_0 < \omega$ such that $(g-1)A$ is contained in $Z_{j_0}(F)$ for any generic elements g in $\mathrm{dc}(F)$. But any element of $\mathrm{dc}(F)$ is the product of two generic elements, and $(gh-1) = (g-1)(h-1) + (g-1) + (h-1)$, so $(g-1)A$ is contained in Z_{j_0} for any g in $\mathrm{dc}(F)$. As $\mathrm{dc}(F)$ contains F, this implies that A lies in the (j_0+1)th iterated centre of F, contradicting our choice of A.

As this argument applies to any definable G-invariant subgroup of H not contained in Z_ω, we find inductively an infinite sequence of generic elements g_i for $i<\omega$ of $\mathrm{dc}(F)$, such that no image $\prod_{i<j}(g_i-1)A$ is contained in Z_ω. In particular the intersection

$$S=\bigcap_{i<\omega}(g_i-1)A,$$

equal to a finite subintersection, is not contained in the hypercentre, and there is a generic g in $\mathrm{dc}(F)$ with $(g-1)A\cap S=S$.

On the other hand for $f\in F$ the image $(f-1)A$ is contained in the hypercentre, whence $(f-1)A\cap S<S$ for all f in F. It follows that this inequality also holds for all generic elements of $\mathrm{dc}(F)$, contradiction. Hence F is nilpotent and the theorem is proved. □

One irritating feature of this proof is that it does not work for type-definable groups. It is not known whether or not they have a Fitting subgroup. Another open problem concerns the non-saturated case: Is the maximal locally nilpotent normal subgroup always nilpotent-by-finite? (This is certainly true for periodic groups with chain condition on centralizers.) And again, the soluble case seems to be more elusive, as the following is still open:

Question 1. Is the soluble radical of a (saturated) stable group definable?

This would be settled positively, if the following conjecture proposed by Haug were true:

Conjecture 2. A soluble stable group is nilpotent-by-abelian-by-finite.

This has been proven by A. Nesin [34] for groups of finite Morley rank; it also holds by results of C. Berline, D. Lascar and E. Hrushovski for other suitably connected groups (see [35] for a general statement).

3. Model theory

This section is more model-theoretic in content, as we shall try to distinguish between "small" and "large" sets using some technical machinery developed by Hrushovski. This will be done in such a way that all finite sets will be small, and the whole group will be large.

Definition 1. A type p is called *internal* in some partial type π, if there is a definable (using parameters) surjective function from some power of π to a superset of p. We say that p is *analyzable* in a set Σ of partial types, if it is internal in types from Σ in several steps. On the opposite side, p is *foreign* to π, if p is orthogonal to any completion of π (over any superset of parameters).

In the context of groups there is another particularly useful characterization:

Theorem 5 [22]. *The generic type p of a stable group G is not foreign to some π iff there is a relatively definable normal subgroup of infinite index such that the quotient is π-internal. The type p is analyzable in Σ if there is a sequence $G = G_0 > G_1 > \cdots > G_\alpha = \{1\}$ of relatively definable normal subgroups such that every quotient G_i/G_{i+1} is π_i-internal for some π_i in Σ.*

Roughly speaking, we consider a definable set X to be smaller than a definable set Y if X is Y-analyzable, but Y is not X-analyzable (i.e. there is a type containing Y which is foreign to X). As analyzability is transitive, this give us a preorder on the definable sets. In the case of a unidimensional group of finite Morley rank (e.g. a minimal counterexample to Cherlin's conjecture), this preorder collapses: Any two infinite definable sets (and, indeed, types) are analyzable in one another.

Returning to the context of groups, we note that the group is internal in any generic type (so a generic subset has the same "size" as the whole group), and that the notion is well-behaved under commutators:

Lemma 1. *Let G and H be type-definable subgroups, where G is connected, and let K be the smallest type-definable supergroup of $[G, H]$. Then K is connected, and if the generic type of G is foreign to some π, so is the generic type of K.*

If Σ is a set of formulae, then Σ-analyzability has a kind of compactness property: For any Σ-analyzable p there is a finite subset Σ_p of Σ, such that p is Σ_p-analyzable in finitely many steps. This allows us to define

Definition 2. A formula $\varphi(x)$ is *Frattini* for G, if for every set Σ of formulae such that G is analyzable in $\Sigma \cup \{\varphi\}$, G is already Σ-analyzable. The *Frattini-free component* G^Φ of G is the intersection of all relatively definable subgroups H such that the quotient G/H is analyzable in the Frattini formulae.

Intuitively, a Frattini formula is a small formula, and the Frattini-free component is some kind of strongly connected component of G. It is well a component (i.e. $(G^\Phi)^\Phi = G^\Phi$), and it generalizes notions of Berline–Lascar and Hrushovski. That is, in [16] it is shown that a superstable group G of U-rank $\omega^\alpha \cdot n + \beta$ (with $\beta < \omega^\alpha$) contains a definable normal subgroup H of rank $\omega^\alpha \cdot n$. Similarly, if G is a p-simple group for some regular type p, then there is a normal type-definable

p-connected subgroup H of the same p-weight as G. It turns out that in both cases G^{Φ} is a supergroup of H.

The main purpose of the Frattini-free component is to ensure the comparability of various sets, as exemplified by the following Lemma:

Lemma 2. *Let H be a minimal type-definable supergroup of some commutator subgroup involving G^{Φ}, N a normal subgroup of H, and A/N an M-minimal normal subgroup of H/N with respect to some group M of automorphisms of H/N. Then either H stabilizes A/N, or the generic type of A/N is not foreign to $H/C_H(A/N)$.*

Proof. Suppose neither conclusion holds. As $C_H(A/N) < H$ and H is connected, the generic type of H is not foreign to $H/C_H(A/N)$ and the generic type of G^{Φ} is not foreign to $H/C_H(A/N)$. Hence there is a definable normal subgroup $K \ln G$ with $G^{\Phi} \nleq K$ and G^{Φ}/K internal in $H/C_H(A/N)$. We claim that G/K is a Frattini formula, thus contradicting $G^{\Phi} \nleq K$.

So let Σ be a set of formulas such that G is $\Sigma \cup \{G/K\}$-analyzable. Then A/N is so analyzable; as its generic is foreign to $H/C_H(A/N)$, it is also foreign to G^{Φ}/K. Hence it cannot be foreign to $\Sigma \cup \Phi(G)$. By minimality it is already almost $\Sigma \cup \Phi(G)$-internal. Hence G^{Φ}/K is almost $\Sigma \cup \Phi(G)$-analyzable, and so is G. Hence G is already Σ-analyzable, and G/K is Frattini. □

Call a type *primary* if it is analyzable in any type it is not foreign to. Examples for such types are the generic types of a definably simple group (no relatively definable normal subgroups) or a field. A puzzling question in this context is whether a definably simple stable group is actually simple, a result proved for superstable groups using the generalized Zil'ber Indecomposability Theorem [6, 16]. Primary types ensure the comparability of sets — and as seen in the above Lemma, central subgroups may get in the way: the generic types of $G/Z(G)$ and $Z(G)$ might be pairwise foreign.

Theorem 6 [35]. *Let G be a stable group without normal abelian subgroups. Then G^{Φ} decomposes as finite direct product of normal connected subgroups with primary generic.*

However, in order to obtain the best structure theory, we need a stronger assumption than mere stability, namely

Definition 3. We say that a group is $\mathfrak{R}$ if every generic can only be algebraic over another generic.

The symbol $\mathfrak{R}$ stands for "rank-like" and allows a very rudimentary dimension theory. Recall that we had preordered sets by analyzability. Now, roughly speaking, a type is measured by how many realizations of it are needed to make a generic element algebraic over them, i.e. the arity of a surjective function from the type to the generic. This, of course, can only work if there is no definable surjection, say, from the generic p to p^2, whence the importance of condition $\mathfrak{R}$. (We view p^2 as generic of G^2, and p is in bijection with $p(x_1) \cup \{x_2 = 1\}$, a nongeneric type in G^2.)

Examples for $\mathfrak{R}$-groups are superstable groups and small stable groups (i.e. those with only countably many pure types). As an easy application we shall prove:

Theorem 7. *Let G be an infinite stable $\mathfrak{R}$-group. Then G contains an infinite abelian subgroup.*

Proof. Let H be a minimal infinite centralizer, which exists by the chain condition. We claim that H is abelian. So suppose not and consider a noncentral element $h \in H$. By minimality, $C_H(h)$ is finite, and a generic element g of H is algebraic over $\{h, h^g\}$. Now $\mathfrak{R}$ implies that h^g is generic, hence every conjugacy class h^H of a noncentral element is generic. By compactness there are only finitely many of these orbits (as the number of generic types is bounded) and the connected component H^0 is definable. Note that $H^0 - Z(H^0)$ is a single conjugacy class, as a connected group has a unique generic type.

As $Z(H^0)$ must be finite, $H^0/Z(H^0)$ is an infinite definable stable group with only one nontrivial conjugacy class. But such groups have infinite chains of centralizers, contradiction. □

In a similar way we can prove that a stable $\mathfrak{R}$-field is algebraically closed. More remarkably, though, we can retrieve the theorems about linear operations: An infinite abelian $\mathfrak{R}$-group M acting on a minimal abelian group A, whose generic type is not foreign to $M/C_M(A)$, allows the definition of an algebraically closed field K, where A is definably isomorphic to K^+ and $M/C_M(A)$ embeds into and generates $K^\times$.

The methods used for the proof, modulo considerable technical machinations, also allow to show the definability of the hypercentre of the Frattini-free component of a stable $\mathfrak{R}$-group. In case this component is soluble, we get the existence and conjugacy of Carter subgroups (self-normalizing nilpotent subgroups), and the nilpotency of the commutator subgroup. As remarked earlier, this implies the definability of the soluble radical. However, the question of the existence of the soluble radical for $\mathfrak{R}$-groups which are not Frattini-free remains open.

4. Final remarks

In this survey we have mentioned some of the characteristic subgroups well-known from finite group theory. We should note that there are also attempts to define the Frattini subgroup, in terms of the group Φ generated by all relatively definable non-supplements (normal subgroups N such that there is no proper relatively definable subgroup H with NH equal to the whole group). This has the property that if the quotient by Φ is nilpotent, then so is the whole group (modulo a non-supplement). It also seems that the theory of Carter subgroups might be put to good use in this context, replacing Sylow-subgroup arguments from finite group theory. However, there are difficulties, as the definition need not be invariant under elementary extensions.

As for further investigations, I should propose attacking the existence (and consequently the definability) of the soluble radical of a saturated stable group. This ought to be possible without dealing with Haug's conjecture. Other promising areas include generalizing classical theorems about involutions in finite groups. Much has been done for finite Morley rank groups by Borovik, Nesin and collaborators, and I should expect most of it to generalize to a wider context. However, in the absence of involutions — and if there is a minimal counterexample to Cherlin's conjecture, it has none — very little indeed is known.

References

[1] Macintyre, A., On ω_1-categorical theories of abelian groups. Fundamenta Math. 70 (1971), 253–270.

[2] Macintyre, A., On ω_1-categorical theories of fields. Fundamenta Math. 71 (1971), 1–25.

[3] Morley, M., Categoricity in power. Trans. Amer. Math. Soc. 114 (1965), 514–538.

[4] Shelah, S., Differentially closed fields. Israel J. Math. 16 (1973), 314–328.

[5] Reineke, J., Minimale Gruppen. Z. Math. Logic 21 (1975), 357–359.

[6] Zil'ber, B., Groups and rings whose theory is categorical (in Russian). Fundamenta Math. 95 (1977), 173–188.

[7] Cherlin, G.,. Groups of small Morley rank. Ann. Math. Logic 17 (1979), 1–28.

[8] Hrushovski, E., A new strongly minimal set. Ann. Pure Appl. Logic 62 (1993), 147–166.

[9] Hrushovski, E., Zil'ber, B., Zariski geometries. Bull. Amer. Math. Soc. 28 (1993), 315–323.

[10] Hrushovski, E., The Mordell-Lang conjecture for function fields. Preprint, 1993.

[11] Poizat, B., Groupes Stables. Nur Al-Mantiq Wal-Ma'rifah, Villeurbanne 1987.

[12] Borovik, A., Nesin, A., Groups of Finite Morley Rank. Oxford University Press, Oxford, to appear.

[13] Cherlin, G., Superstable division rings. In: Logic Colloquium '77, 99–111. North Holland 1978.

[14] Cherlin, G., Shelah, S., Superstable fields and groups. Ann. Math. Logic 18 (1980), 227–270.

[15] Berline, C., Superstable groups; a partial answer to conjectures of Cherlin and Zil'ber. Ann. Pure Appl. Logic 30 (1986), 45–61.

[16] Berline, C., Lascar, D., Superstable groups. Ann. Pure Appl. Logic 30 (1986), 1–43.

[17] Baldwin, J., Saxl, J., Logical stability in group theory. J. Austral. Math. Soc. 21 (1976), 267–276.

[18] Felgner, U., $\aleph_0$-categorical stable groups. Math. Z. 160 (1978), 27–49.

[19] Baur, W., Cherlin, G., Macintyre, A., Totally categorical groups and rings. J. Algebra 57 (1979), 407–440.

[20] Poizat, B. Sous-groupes définissables d'un groupe stable. J. Symb. Logic 46 (1981), 137–146.

[21] Poizat, B. Groupes stables, avec types génériques réguliers. J. Symb. Logic 48 (1983), 339–355.

[22] Hrushovski, E., Contributions to Stable Model Theory. PhD thesis, Berkeley 1986.

[23] Wood, C., Notes on the stability of separably closed fields. J. Symb. Logic 44 (1979), 412–416.

[24] Bryant, R. M., Groups with minimal condition on centralizers. J. Algebra 60 (1979), 371–383.

[25] Bryant, R. M., Hartley, B., Periodic locally soluble groups with the minimal condition on centralizers. J. Algebra 61 (1979), 328–334.

[26] Kegel, O., Four lectures on Sylow theory. In: Proc. Singapore Group Theory Conf. 1987, 3–27. Walter de Gruyter, 1989.

[27] Poizat, B. P., Wagner, F. O., Sous-groupes périodiques d'un groupe stable. J. Symb. Logic 58 (1993), 385–400.

[28] Wagner, F. O., Stable groups, mostly of finite exponent. Notre Dame J. Formal Logic 34 (1993), 183–192.

[29] Wagner. F. O., Commutator conditions and splitting automorphisms for stable groups. Arch. Math. Logic 32 (1993), 223–228.

[30] Poizat, B., Équations génériques. In: Proceedings of the 8th Easter Conference on Model Theory, Berlin 1990. Humboldt Universität.

[31] Wagner, F. O., À propos d'équations génériques. J. Symb. Logic 57 (1992), 548–554.

[32] Wagner, F. O., Subgroups of stable groups. J. Symb. Logic 55 (1990), 151–156.

[33] Wagner, F. O., The Fitting subgroup of a stable group. J. Algebra, submitted.

[34] Nesin, A., Soluble groups of finite Morley rank. J. Algebra 121 (1989), 26–39.

[35] Wagner, F. O., More on $\mathfrak{R}$. Notre Dame J. Formal Logic 33 (1992), 159–174.

Finitely presented soluble groups, Hilbert–Serre dimension, and integer polynomials

John S. Wilson

Abstract. The paper is a survey of results on finitely presented soluble groups, Hilbert–Serre dimension and integer polynomials.

1991 Mathematics Subject Classification: 20F16, 20F05.

1. A generalization of the Golod–Shafarevich theorem

Let G be a finitely presented group. We write $d(G)$ for the smallest number of elements that can generate G and G^{ab} for the abelianization of G; thus $G^{ab} = G/G'$, where G' is the derived group of G. Suppose that G has a presentation with n generators and r relations. The Golod–Shafarevich theorem [1], as improved by E. B. Vinberg [2] and Gaschütz, asserts that if G is a finite p-group, where p is a prime, then

$$r - n + d(G) \geqslant \frac{1}{4}(d(G))^2.$$

It follows from this that if G is an arbitrary finite group then

$$r - n + d(G^{ab}) \geqslant \frac{1}{4}(d(G^{ab}))^2. \tag{1}$$

This inequality breaks down as soon as one considers infinite groups. For example, if G is taken to be the infinite cyclic group, then G has a presentation for which $n = d(G^{ab}) = 1$ and $r = 0$, so that the right-hand side in (1) exceeds the left-hand side by $\frac{1}{4}$. However, for soluble groups, this is the worst that can happen:

Theorem 1 (Wilson [3]). *If G is a soluble group then*

$$r - n + d(G^{ab}) \geqslant \frac{1}{4}((d(G^{ab}))^2 - 1) \tag{2}$$

Let me state an immediate consequence of this theorem:

Corollary 1. *Suppose that G is a finitely presented soluble group. Then there is a $\kappa > 0$ such that $d(H^{ab}) \leqslant \kappa|G:H|^{\frac{1}{2}}$ for all subgroups H of finite index in G.*

The significance of this result lies in the exponent $\frac{1}{2}$. Of course, the Reidemeister–Schreier theorem shows that for any finitely generated group G one has $d(H^{ab}) \leqslant \kappa|G:H|$ for all subgroups H of finite index in G, where $\kappa = 1$.

The proof of Corollary 1 follows by applying Theorem 1 to the subgroups of finite index in G and using the Reidermeister–Schreier theorem. Let G again have a presentation with n generators and r relations, and let H have index h in G. Then H has a presentation with $hn - (h-1)$ generators and hr relations, and so we have

$$hr - (hn - (h-1)) + d(H^{ab}) \geqslant \frac{1}{4}((d(H^{ab}))^2 - 1).$$

Hence

$$(d(H^{ab}) - 2)^2 \leqslant 4h(r - n + 1) + 1,$$

and the result follows on taking square roots.

2. Examples and questions

To motivate our discussion of finitely presented soluble groups it will be helpful to have a supply of examples illustrating the diversity of these groups, and some specific test questions. We begin with some examples.

Example 1. It was proved by P. Hall [4] that if K is a normal subgroup of G and both $K, G/K$ are finitely presented then G is finitely presented. Consequently, polycyclic groups, and, in particular, finitely generated nilpotent groups, are finitely presented.

Example 2. The group D with presentation $\langle a, t \mid t^{-1}at = a^2\rangle$ is metabelian but not polycyclic; indeed, the normal subgroup generated by a is isomorphic to the group of dyadic rationals.

Fix $\theta \in \mathbb{Q}$ and define

$$G_\theta = \left\langle \begin{pmatrix} \theta & 0 \\ 0 & 1, \end{pmatrix} \begin{pmatrix} 1 & 1 \\ 0 & 1 \end{pmatrix} \right\rangle \leqslant \mathrm{GL}_2(\mathbb{Q}).$$

It is easy to see that G_θ is a metabelian group, and that $G_2 \cong D$. If fact G_θ is finitely presented iff $\theta \in \mathbb{Z}$ or $\theta^{-1} \in \mathbb{Z}$.

Example 3. Let U be a finitely presented soluble group and let $\phi : U \to U$ be an injective homomorphism. Then the group with "presentation" $\langle U, t \mid u\phi = t^{-1}ut \text{ for all } u \in U\rangle$ is finitely presented and soluble.

Example 4. Some soluble groups of matrices over fields (especially arithmetic and S-arithmetic groups) are finitely presented. All such groups are virtually nilpotent-by-abelian, i.e. they have a nilpotent-by-abelian normal subgroup of finite index.

Example 5. There are some perplexing groups constructed by D. J. S. Robinson and R. Strebel [5] which are finitely presented and abelian-by-(nilpotent of class 2) and are not virtually nilpotent-by-abelian.

We mention that the finitely presented groups in Example 2 above are particular cases of both Examples 3 and 4.

Now we list three problems. The first two of these come from Baumslag's survey article [6] of 1973.

Problem 1. Is there any way of discerning finitely presented metabelian groups from the other finitely generated metabelian groups?

Problem 2. Can every finitely generated abelian-by-polycyclic group be embedded in a finitely presented abelian-by-polycyclic group?

Problem 3. Is every finitely presented soluble group virtually nilpotent-by-(nilpotent of class at most 2)?

Later we shall discuss very briefly the solution to the first problem by R. Bieri and R. Strebel [7] in 1980 and the solution to the second problem this year by C. J. B. Brookes, J. E. Roseblade and J. S. Wilson [8]. The third problem remains open.

3. Hilbert–Serre dimension

The considerations in this section are an attempt to find a useful and natural context for the information given by Corollary 1.

Let R be a ring. We write $\lambda(M)$ to denote the composition length of a right R-module M. Choose an ideal I of R with $\lambda(R/I)$ finite.

Suppose that M is an R-module with $\lambda(MI^k/MI^{k+1})$ finite for all $k \in \mathbb{N}$. We define the *Hilbert–Serre dimension of M with respect to I*, denoted $\operatorname{hs}(M, I)$, as follows:

$$\mathrm{hs}(M,I) = \limsup_{k\to\infty} \frac{\log\lambda(M/MI^k)}{\log k}.$$

Thus if $\mathrm{hs}(M,I) = d$ is finite, then for all $\epsilon > 0$ one has

$$\lambda(M/MI^k) \leqslant k^{d+\epsilon} \quad \text{for } k \text{ large enough;}$$

and d is the smallest number with this property. Equivalently, $\mathrm{hs}(M,I)$ is the Gelfand–Kirillov dimension of the associated graded module $\bigoplus_{k\geqslant 0}(MI^k/MI^{k+1})$.

Lemma 1. *Suppose that either* (i) *R is a commutative noetherian ring or* (ii) *$R = \mathbb{F}_pH$ where p is a prime, H is a finitely generated nilpotent group and I is the augmentation ideal of R. Let M be a finitely generated R-module and form the Poincaré series*

$$P(M,I,t) = \sum_{k\geqslant 0} \lambda(MI^k/MI^{k+1})t^k.$$

Then $P(M,I,t)$ has the form

$$\frac{f(t)}{\prod_{i=1}^{s}(1-t^{l_i}),}$$

where $f(t) \in \mathbb{Z}[t]$ and each l_i is a positive integer. Moreover, $\mathrm{hs}(M,I)$ *is the degree of the pole of $P(M,I,t)$ at 1. In particular,* $\mathrm{hs}(M,I) \in \mathbb{N}$.

For the case when (i) holds this is the classical Hilbert–Serre theorem (see, for example, [9, Chapter 11]), and for the case when (ii) holds it is a theorem of P. F. Pickel [10].

For the rest of this section, we suppose that $R = \mathbb{F}_pH$, where H is a polycyclic group. We need information which is stable under passage to subgroups of finite index. It is easy to prove that $\mathrm{hs}(M,I) \leqslant r_0(H)$ for every finitely generated R-module M, where $r_0(H)$ is the torsion-free rank (or Hirsch number) of H. We define the *total Hilbert–Serre dimension* of an R-module M, denoted $\mathrm{ths}(M)$, to be the supremum of the numbers $\mathrm{hs}(M,I)$ as I runs through the ideals of R with R/I finite. If H_1 is a subgroup of H, we write $\mathrm{ths}_{H_1}(M)$ for the total Hilbert–Serre dimension of M regarded as an $\mathbb{F}_pH_1$-module. It is easy to prove

Lemma 2. *Let H_1 be a subgroup of H and write $d = \mathrm{ths}_H(M)$ and $d_1 = \mathrm{ths}_{H_1}(M)$. The following assertions hold:*

(i) *$d \leqslant d_1$;*

(ii) *if $|H : H_1|$ is finite then $d = d_1$;*

(iii) *if H_1 is normal in H and d_1 is finite, then $d = d_1$.*

The connection with finitely presented groups comes from the next result, whose proof depends on Corollary 1.

Corollary 2. *Let G be an extension of an elementary abelian normal p-subgroup M by H and suppose that G is an image of a finitely presented soluble group. Then there is a constant κ such that*

$$\dim_{\mathbb{F}_p}(M/[M,G_1]) \leqslant \kappa |G:G_1|^{\frac{1}{2}}$$

for each subgroup G_1 of finite index in G. Moreover, regarding M as an $\mathbb{F}_pH$-module, we have $\operatorname{ths}(M) \leqslant \frac{1}{2}r_0(H/C)$, *where C is the centralizer of M in H.*

For the next three results we assume that H is nilpotent. We recall that if R is a ring and M is a non-zero R-module then the Krull dimension of M, denoted k-dim(M), is the greatest ordinal d such that M has a descending chain of submodules of order-type ω^d. We have

Lemma 3 (Groves and Wilson [11]). *If M is a non-zero $\mathbb{F}_pH$-module then*

$$\operatorname{ths}(M) = k - \dim(M).$$

Using Lemma 3 and a considerable amount of representation theory for finitely generated nilpotent groups one can prove the following theorem.

Theorem 2 (Groves and Wilson [11]). *Let H be a torsion-free nilpotent group and suppose that there is an $\mathbb{F}_pH$-module M satisfying* $\operatorname{ths}(M) \leqslant \frac{1}{2}r_0(H)$ *on which H acts faithfully.*

(i) *If $d(H) \leqslant 2$ then H is abelian.*

(ii) *If $d(H) \leqslant 4$ then H is nilpotent of class at most two.*

(iii) *The centre of H cannot be equal to H' and be infinite cyclic.*

Taken together with Corollary 2, assertion (i) above leads to the following result.

Theorem 3 (Groves and Wilson [11]). *If G is a finitely presented nilpotent-by-nilpotent group and $d(G) \leqslant 2$ then G is virtually nilpotent-by-(nilpotent of class at most two).*

A result valid for arbitrary finitely presented nilpotent-by-nilpotent groups but with a weaker conclusion has been proved by C. J. B. Brookes and J. R. J. Groves [12]: if G is such a group then G is virtually nilpotent-by-(nilpotent of class at most $d(G)$).

Now we return to the case where H is a general polycyclic group. For such groups the behaviour of $\mathrm{ths}(M)$ is rather harder to understand; indeed, it is not yet clear whether this dimension function is integer-valued. We shall illustrate that $\mathrm{ths}(M)$ and $k\text{-dim}(M)$ do not in general coincide. We begin by describing some important examples of polycyclic groups.

Let A be a free abelian group of finite rank and let T be an free abelian subgroup of the automorphism group of A such that $A \otimes_{\mathbb{Z}} \mathbb{Q}$ is irreducible as a $\mathbb{Q}T_1$-module for all subgroups T_1 of T of finite index. Set $H = A \rtimes T$. Slightly more specifically, let F be an algebraic number field, with ring of algebraic integers R, and let V be a subgroup of the group of units of R such that F is generated by V_1 over $\mathbb{Q}$ for all subgroups V_1 of V of finite index. Define

$$H = \left\{ \begin{pmatrix} v & r \\ 0 & 1 \end{pmatrix} ; v \in V, r \in R \right\},$$

$$A = \left\{ \begin{pmatrix} 1 & r \\ 0 & 1 \end{pmatrix} ; r \in R \right\},$$

$$T = \left\{ \begin{pmatrix} v & 0 \\ 0 & 1 \end{pmatrix} ; v \in V \right\}.$$

It is apparent that some assertions from number theory have consequences for the structure of these groups. For example, Dirichlet's unit theorem has the consequence that $r_0(T) < r_0(A)$.

With H as above, define $M = \mathbb{F}_p A$. Then M becomes an $\mathbb{F}_p H$-module if we let A act on M by multiplication and T act by conjugation. From the commutative Hilbert–Serre theory we have $\mathrm{ths}_A(M) = r_0(A) > 1$ (see for example M. F. Atiyah and I. G. Macdonald [9, Chapter 11]). From Lemma 2 we have $\mathrm{ths}_A(M) = \mathrm{ths}_H(M)$. On the other hand, from the Bergman theorem [13], all non-zero $\mathbb{F}_p H$-submodules of M have finite index in M, so that the Krull dimension of M is 1. It seems probable that $\mathrm{ths}(M)$, which extracts information of a delicate kind about the finite images of M, is the "correct" dimension function for modules for polycyclic groups.

Some ideas related to those above play an important part in the proof of the following result.

Theorem 4 (Brookes, Roseblade and Wilson [8]). *Suppose that H is a polycyclic group and M is an $\mathbb{F}_p H$-module. If some extension of M by H is finitely presented, then H has a normal subgroup L such that H/L is virtually nilpotent and such that L acts nilpotently on M. Finitely presented abelian-by-polycyclic groups are virtually nilpotent-by-nilpotent.*

Since there are many examples of finitely generated abelian-by-polycyclic groups which are not virtually nilpotent-by-nilpotent (for example, standard wreath products of $\mathbb{Z}$ with polycyclic groups which are not virtually nilpotent), this result gives a negative answer to the second problem in Section 2 above.

The proof of this theorem is rather difficult. The first step in the proof is to reduce to the case when H is a split extension $A \rtimes T$ of the type described above. The hypothesis of the theorem has consequences not dissimilar to the assertion $\mathrm{ths}(M) \leqslant \frac{1}{2} r_0(H)$. Suppose that this assertion holds, and that in addition M is finitely generated and torsion-free regarded as an $\mathbb{F}_p A$-module: we need to obtain a contradiction. However from above we have

$$r_0(A) + r_0(T) = r_0(H) \geqslant 2\,\mathrm{ths}_H(M) = 2\,\mathrm{ths}_A(M) = 2r_0(A),$$

so that $r_0(T) \geqslant r_0(A)$, in contradiction to Dirichlet's unit theorem.

4. Degenerate spheres and integer polynomials

We now discuss briefly the invariant introduced by R. Bieri and R. Strebel [7] in connection with their work on the first problem in Section 2 and we analyse the finitely presented soluble groups for which this invariant cannot provide much useful information.

Let Q be a finitely generated abelian group and let $V(Q) = \mathrm{Hom}(Q, \mathbb{R})$. Thus $V(Q)$ is a real vector space of dimension n where $n = r_0(Q)$. Choose a norm $\| \ \|$ in $V(Q)$ and let $S(Q) = \{v \in V(Q) \mid \|v\| = 1\}$, so that $S(Q)$ is an $(n-1)$-sphere. For $v \in S(Q)$ let Q_v be the submonoid $\{q \in Q \mid v(q) \geqslant 0\}$. With each $\mathbb{Z}Q$-module M, Bieri and Strebel associated the open subset

$$\Sigma_M = \{v \in S(Q) \mid M \text{ is a finitely generated module for the ring } \mathbb{Z}Q_v\}$$

of $S(Q)$, and they proved the following result.

Theorem 5 (Bieri and Strebel [7]). *Let G be a finitely generated soluble group, and write $K = G'$, $Q = G/K$ and $M = K^{ab}$. Thus M can be regarded as a $\mathbb{Z}Q$-module.*

(i) *If G is finitely presented then $\Sigma_M \cup -\Sigma_M = S(Q)$.*

(ii) *Suppose that $K' = 1$. If $\Sigma_M \cup -\Sigma_M = S(Q)$ then G is finitely presented.*

Thus, to determine whether a finitely generated metabelian group is finitely presented, it is sufficient to examine the invariant associated with its derived group. Indeed, if G is an arbitrary finitely generated group then important structural information will frequently be recoverable from the set Σ_M where $M = (G')^{ab}$. Of course this will not be the case if G^{ab} is finite or if G^{ab} has torsion-free rank 1; in the former case $S(V)$ is empty and in the latter it consists of just two points. Since we are frequently free to pass to a subgroup of finite index, the groups for which the Bieri–Strebel invariant is ineffective are those having no subgroup of finite index which maps surjectively to $\mathbb{Z}^2$. Our final theorem describes the finitely presented such groups.

Theorem 6 (Wilson [14]). *Suppose that G is a finitely presented soluble group and that G has no subgroup of finite index which maps surjectively to $\mathbb{Z}^2$. Then G has a subgroup G_0 of finite index which has the structure of the group in Example 3 in Section 2, with U a finitely generated nilpotent group.*

The structure of the groups G_0 in the above conclusion is well understood from work of R. Bieri and R. Strebel [15]; in particular, each such group G_0 is an extension of a nilpotent group of finite rank by an infinite cyclic group. The proof of Theorem 6 begins with some representation theory, to reduce to the case in which G has a rather special form. Then one would like to apply Corollary 2; however it appears that the information on Hilbert–Serre dimension given by Corollary 2 is not decisive in this situation. Instead, some delicate calculations with indices of subgroups and generating numbers have to be done, and in the course of these a problem in elementary number theory arises.

Let f be a monic polynomial over $\mathbb{Z}$ and let $n \in \mathbb{Z}$, with $|n| > 1$. We say that f is *n-generating* if it satisfies one of the following equivalent conditions:

(i) infinitely many rational primes are divisors of numbers of the form

$$hcf(f(n^l), n^{lm} - n), \qquad l, m \in \mathbb{N};$$

(ii) f has a root $c \in \mathbb{C}$ such that the ring $\mathbb{Z}[c]$ has infinitely many prime ideals I for which $n + I, c + I$ generate the same multiplicative group in $(R/I)^*$.

Conjecture. *f is n-generating iff f has a root in $\mathbb{C}$ which is not 0 or a root of 1.*

Let us consider the case when f has degree 1, so that $f = t - m$ for some $m \in \mathbb{Z}$. We may assume that $m \neq 0, \pm 1$. In this case the conjecture asserts essentially that there are infinitely many primes q such that the images of m, n in $\mathbb{Z}/q\mathbb{Z}$ generate the same multiplicative group. Even this case is not easy. For the case when $mn < 0$, I was able to give a short proof using Dirichlet's theorem on primes in arithmetic progression and quadratic reciprocity. The validity of the conjecture when $mn > 0$ was not needed for the group-theoretic application, but this has now been affirmed by A. Schinzel and J. Wójcik [16]; the proof requires eight journal pages.

For the group-theoretic application it was also necessary to settle the case when f has arbitrary degree and where n is a prime which divides $f(0)$. This is not difficult and indeed the restriction that n is a prime can be lifted. Another more substantial result concerned with the conjecture has been proved by J. W. S. Cassels (unpublished). Let $f = \Pi_{i=1}^{d}(t - c_i)$ be the factorization of f in $\mathbb{C}[t]$, and define $f_{(r)}(t) = \Pi_{i=1}^{d}(t - c_i^r)$ for $r = 2, 3, \ldots$. Thus each $f_{(r)}$ is also a monic polynomial over $\mathbb{Z}$. Cassels proved that for each f there is an integer r for which $f_{(r)}$ behaves in accordance with the conjecture. These are the only results that I know of concerning the conjecture, and the general case seems likely to be difficult. It is possible that other results related to the conjecture will have applications to the study of finitely presented groups. Of course, it is a pleasing and by no means unusual phenomenon in mathematics that attempts to prove results in one area, such as infinite group theory, lead to interesting and apparently difficult questions in another area, such as number theory.

References

[1] Golod, E. S., Shafarevich, I. R., On the class field tower. Izv. Akad. Nauk USSR, Ser. Mat. 28 (1964), 261–272.

[2] Vinberg, E. B., On the theorem concerning the infinite dimension of an associative algebra. Izv. Akad. Nauk USSR, Ser. Mat. 29 (1965), 209–214.

[3] Wilson, J. S., Finite presentations of pro-p groups and discrete groups. Invent. Math. 105 (1991), 177–183.

[4] Hall, P., Finiteness conditions for soluble groups. Proc. London Math. Soc. 3:4 (1954), 419–436.

[5] Robinson, D. J. S., Strebel, R., Some finitely presented soluble groups which are not nilpotent by abelian by finite. J. London Math. Soc. 2:26 (1982), 435–440.

[6] Baumslag, G., Finitely presented metabelian groups. Proceedings of the Second International Conference on the Theory of Groups. Lecture Notes in Mat. 372, 65–74, Springer-Verlag 1974

[7] Bieri, R., Strebel, R., Valuations and finitely presented metabelian groups. Proc. London Math. Soc. 3:41 (1980), 439–464.

[8] Brookes, C. J. B., Roseblade, J. E., Wilson, J. S., Finitely presented abelian by polycyclic groups and exterior powers. Preprint 1993.

[9] Atiyah, M. F., Macdonald, I. G., Introduction to Commutative Algebra. Addison Wesley, Reading, Massachusetts.

[10] Pickel, P. F., Rational cohomology of nilpotent groups and Lie algebras. Comm. Algebra 6 (1979), 409–419.

[11] Groves, J. R. J., Wilson, J. S., Finitely presented metanilpotent groups. J. London Math. Soc. (2) 50 (1994), 87–104.

[12] Brookes, C. J. B., Groves, J. R. J., Modules over nilpotent group rings. Preprint 1993.

[13] Bergman, G. M., The logarithmic limit set of an algebraic variety. Trans. Amer. Math. Soc. 157 (1971), 459–469.

[14] Wilson, J. S., On finitely presented soluble groups with small abelian-by-finite images. J. London Math. Soc. (2) 48 (1993), 229–248.

[15] Bieri, R., Strebel, R., Almost finitely presented soluble groups. Comment. Math. Helv.53 (1978), 258–278.

[16] Schinzes, A., Wojcik, J., On a problem in elementary number theory. Math. Proc. Cambridge Philos. Soc. 112 (1992), 225–232.

Solution of Mazurov's problem

Viktor I. Zenkov

Abstract. A partial answer to the problem on a subgroup generated by some special elements to be a p-group posed by V. D. Mazurov in the "Kourovka Notebook" is given.

1991 Mathematics Subject Classification: 20D15, 20E07, 20F05.

In [1, Question 5.32] V. D. Mazurov formulated the following

Question. Let p be a prime number, C a class of conjugate p-elements of a finite group G and let, for any elements x, y from C, the product xy^{-1} be a p-element. Is the order of the subgroup generated by C a power of p?

Here we answer to this question in the affirmative for the case $p = 2$; namely, the following theorem is proved.

Theorem. *Let G be a finite group, C a class of conjugate 2-elements of G. If, for any elements x, y of C, the product xy^{-1} is a 2-element then $\langle C \rangle \leqslant O_2(G)$.*

This theorem was announced by the author in [2]. The case $p > 2$ was considered by the author with the use of the classification of finite simple groups; this will be published elsewhere.

1. Preliminary results

Lemma 1. *Let L be a finite group, $L_1 \leqslant L$, p a prime number, $1 \neq u$ a p-element of L, and $C = C_L(u)$. If F is a subgroup of L_1 such that $N_{L_1}(F)$ is a p'-subgroup, L/L_1 is an abelian group, and $t = |L : C| > |L : N_L(F)|$, then there is an element v in L conjugate to u such that uv^{-1} is not a p-element.*

Proof. The inequality $|L : C| > |L : N_L(F)|$ means that the number of elements conjugate to the element u in L is strictly greater than the number of subgroups conjugate to F in L. Then there are at least two equal subgroups among $F^u, F^{u_1}, \dots, F^{u_t}$ where $\{u, u_1, \dots, u_t\} = u^L$. Thus, $F^{u_k} = F^{u_m}$ for some k and m, $1 \leqslant k < m \leqslant t$, whence $F^{u_k u_m^{-1}} = F$. Therefore $1 \neq u_k u_m^{-1} \in N_{L_1}(F)$ since L/L_1 is an abelian group. By the condition of the lemma, $u_k u_m^{-1}$ is a p'-element. Since $u_k = u^f$ for some $f \in L$, we have

$$(u_k u_m^{-1})^{f^{-1}} = u_k^{f^{-1}} (u_m^{-1})^{f^{-1}} = u(u_m^{f^{-1}})^{-1} = uv^{-1}$$

where $v = u_m^{f^{-1}} = u^g$ for some $g \in L$. Lemma 1 is proved.

Lemma 2. *Let $G = K\langle x\rangle$ where $K = F^*(G)$ is a finite simple nonabelian group and x is an involution from G. Then there is an involution y in G conjugate to x such that xy^{-1} is not a 2-element.*

Proof. $\langle x, x^g\rangle$ is not a 2-group for a certain element g of G by Theorem 1.1 of [3]. However $\langle x, x^g\rangle \simeq D_{2t}$ whence the involution x inverts a certain odd order nonidentity element r from $\langle x, x^g\rangle$. Therefore the order of $1 \neq xrx^{-1}r^{-1} = xy^{-1}$, where $y = rx^{-1}r^{-1} = rxr^{-1}$, is odd. Lemma 2 is proved.

Lemma 3. *Let $G \simeq Sz(2^n)$ or $U_3(2^n)$, $n > 1$, and let x be a 2-element of G. Then there is an element $y \in G$ conjugate to x such that xy^{-1} is not a 2-element.*

Proof. By Lemma 2, we may assume that $|x| > 2$. By [3, p. 155] the exponent of the Sylow 2-subgroup T of G is equal to 4. Hence $|x| = 4$.

If $G \simeq Sz(q)$ then by [4] G has one class of subgroups of order 4 and contains a cyclic subgroup A_1 of odd order such that $N(A_1)/A_1 \simeq Z_4$ and $N(A_1)$ is a Frobenius subgroups of G. Hence $1 \neq x(x^f)^{-1} \in A_1$ for a certain element f of A_1 and Lemma 3 is proved in this case.

If $G \simeq U_3(q)$ and $3 \mid (q+1)$ then by [5, Chapter 11] and [1] G contains a subgroup $Q \simeq E_9 \lambda Q_8$ which is a Frobenius group. By [7, p. 493] all subgroups of order 4 are conjugate in $\mathrm{Aut}\,(G)$. Hence $1 \neq x(x^f)^{-1}$ is a 3-element for a certain element f of Q and Lemma 3 is proved in this case.

If $3 \mid (q+1)$, then by [6; 3, Chapter 11] or 1.10 of [8, EII] G contains a subgroup P of order $q^2 - q + 1$ and $|N(P)| = 3|P|$. By [7, p. 493] $|C(x)| = q^2$. Since $|N(P)| > |C(x)|$ we can apply Lemma 1 with $L_1 = G$, $F = P$, and $u = x$. Lemma 3 is proved.

2. Proof of the theorem

Let G be the minimal order counterexample for the theorem, x the element of G for which the conclusion of the theorem fails. Let $S(G)$ be the solvable radical of the group G. The condition of the theorem is evidently inherited by the quotient groups and the subgroups of G.

Lemma 4. *We have $S(G) = 1$.*

Proof. Indeed, if $S(G) \neq 1$, then, applying the induction assumption to the group $\overline{G} = G/O(G)$, we have $\bar{x} \in O_2(\overline{G})$.If $[x^{-1}, f] \neq 1$ for a certain element f from $O(G)$, then $[x^{-1}, f] = xf^{-1}x^{-1}f = x(x^f)^{-1}$ is a $2'$-element from G contradicting the condition of the theorem. Therefore $[\langle x \rangle, O(G)] = 1$.

By the Hall—Higman theorem $x \in O_2(G)$, a contradiction. Lemma 4 is proved.

Lemma 5. *We have $G = K\langle x \rangle$, where K is a simple nonabelian group.*

Proof. Assume that the element x^{-1} does not normalize a certain component K_t from $E(G)$, $1 \leqslant t \leqslant n$. Let $K_{t'} = K_1^{x^{-1}} \neq K_t$, $1 \leqslant t' \leqslant n$ and $t \neq t'$, and let f be an arbitrary nonidentity $2'$-element of K. As the subgroup K is simple, such an element exists. If $1 \neq [x^{-1}, f] = xf^{-1}x^{-1}f = x(x^f)^{-1} = (f^{x^{-1}})^{-1}f$, then it is a $2'$-element of G. Since $[f, (f^{x^{-1}})^{-1}] = 1$, we obtain a contradiction. Hence the element x normalizes each component of $E(G)$. If $n > 1$, then by the induction assumption we have $x \in O_2(\langle x, K_i \rangle)$ for $1 \leqslant i \leqslant n$. Hence $[\langle x \rangle, K_i] = 1$ for $1 \leqslant i \leqslant n$ and $[\langle x \rangle, E(G)] = 1$. By [3, 1.27] $\langle x \rangle \leqslant F^*(G) = E(G)$, a contradiction. Consequently, $n = 1$. Lemma 5 is proved.

Lemma 6. *The element x lies in a unique maximal subgroup M of G and, moreover, in $O_2(M)$.*

Proof. Let M be an arbitrary maximal subgroup of G, and let x belong to M. By the induction assumption $x \in O_2(M)$. By Lemma 4 $O_2(G) = 1$, and Lemma 6 follows from Theorem 1 of [9].

Lemma 7. *Let $W(x)$ be the weak closure of the element x in the subgroup M relative to G. Then $G = \langle W(x), W(x)^g \rangle$ whenever $W(x) \neq W(x)^g$.*

Proof. By Lemma 6, $W(x) \leqslant O_2(M)$. If $W(x) \neq W(x)^g$, then there is an element x^{g_1} in the subgroup $W(x)^g$ which does not belong to $W(x)$. If $\langle W(x), W(x)^g \rangle = G_1 \neq G$, then by Lemma 6 $G_1 \leqslant M$ and hence $W(x) = W(x)^g$, a contradiction. Lemma 7 is proved.

Lemma 8. *We have either $K \simeq L_2(q)$, $q = p^\alpha > 3$ is odd, or $K \simeq L_3(4)$ and $|W(x) : W(x) \cap K| = 2$.*

Proof. By the main theorem of [10] the B-hypothesis holds in each section of G. By Lemma 8 $G = \langle W(x), W(x)^g \rangle$ for $W(x) \neq W(x)^g$. Hence by Theorem 1 of [11] and Lemma 4 either $G \simeq L_2(2^n)$, $U_3(2^n)$, $Sz(2^n)$ or $K \simeq L_2(q)$, $q = p^\alpha > 3$ is odd, or $K \simeq L_3(4)$. In the latter case we have $|W(x) : W(x) \cap K| = 2$.

If $G \simeq L_2(2^n)$, $U_3(2^n)$, $Sz(2^n)$, then G is not a counterexample by Lemmas 2 and 3. Lemma 8 is proved.

Lemma 9. *$K \not\simeq L_2(q)$, $q = p^\alpha$ is odd.*

Proof. Let $K \simeq L_2(q)$, $q = p^\alpha > 3$ is odd. By [12, Section 9] $\mathrm{Aut}(K) \simeq GL_2(q)\lambda\langle f\rangle$ and f acts on K also as an automorphism of the field. Identify G with its image in $\mathrm{Aut}(K)$. Let $T \in \mathrm{Syl}_2(G)$, t be an involution from $Z(T) \cap K$, $T_1 \in \mathrm{Syl}_2(\mathrm{Inn}(K))$, $T_1 \leqslant T_2 \in \mathrm{Syl}_2(\mathrm{Inndiag}(K))$, $T_2 \leqslant T_3 \in \mathrm{Syl}_2(\mathrm{Aut}(K))$, $T_3 \geqslant T$. We may assume that $T_3 = T_2\lambda\langle f_1\rangle$ where $\langle f_1\rangle \in \mathrm{Syl}_2(\langle f\rangle)$ and $[T_1, \langle f_3\rangle] = 1$ for an involution f_3 from $\langle f_1\rangle$ in view of the structure of $C_K\langle f_3\rangle$ [12, Section 9]. Then $x = d_1 f_2$ where $d_1 \in T_2$ and $f_2 \in \langle f_1\rangle$. By Lemma 2, $|x| > 2$.

If $4 \mid (q+1)$ then $f_1 = 1$ and by [6, p. 285, 286], $K \geqslant H \simeq E_q\lambda Z_{\frac{q-1}{2}}$. Since $|H| = \frac{q(q-1)}{2} > q+1 \geqslant |C(x)|$ and H is a maximal subgroup in K, we have a contradiction by applying Lemma 1 with $L = G$, $L_1 = K$, $F = H$, $u = x$. Hence $4 \mid (q-1)$.

If $f_2 = 1$ then $G \leqslant \mathrm{Inndiag}(K) \simeq PGL_2(q)$. From [6, p. 285, 286] and the structure of the centralizer of involutions in $GL_2(q)$ it follows that $\mathrm{Inndiag}(K)$ contains a Frobenius subgroup $H \simeq E_q\lambda Z_{q-1}$. Hence $x \in H \cap G$, a contradiction by Lemma 6. Therefore $f_2 \neq 1$. If $d_1 = 1$ then from the structure of the centralizer of the element f_2 [12, Section 9] we have that p divides $|C_K(f_2)|$. Since the Sylow p-subgroup Q of K is a TI-subgroup in K, we have $f_2 \in N(Q)$ and $C_Q(f_2) \neq Q$, a contradiction by Lemma 6. Hence $d_1 \neq 1$.

If $q \notin \{5, 7, 9, 11\}$ then by [6, p. 285-286] and [12, Section 9] $C_K(y)$ is a maximal subgroup of K for any involution y from $\mathrm{Aut}(K)$. In this case, the involution x_1 from $\langle x\rangle$ is equal to t by Lemma 6. Since $q > 11$ and $f_2 \neq 1$, we have $O(C(t)) = R \neq 1$. If $|d_1| > 2$ then $[\langle d_1\rangle, R] = 1$ by the structure of $C_{\mathrm{Inndiag}(K)}(t)$ and $[\langle x\rangle, R] = 1$ by Lemma 6. Hence $[\langle f_2\rangle, R] = 1$. Then $C_K(f_3) \geqslant \langle T_1, R\rangle$ by [12, Section 9] contradicting maximality of $C_K(t)$. Hence $|d_1| = 2$ and d_1 inverts R by the structure of $C_{\mathrm{Inndiag}}(K)$. Since $[\langle x\rangle, R]=1$, we have that f_2 inverts R. Hence $[\langle f_2^2\rangle, R] = 1$ and if $f_2^2 \neq 1$ then $[\langle f_3\rangle, R] = 1$ contradicting maximality of $C_K(f_3)$. Therefore $f_2^2 = 1$ and $G \simeq PGL_2^*(q)$ with T is semidihedral. The case $G \simeq PGL_2^*(9)$ is also included in the reasoning below. Let $z \in C_K(f_2)$. Then $y = x^z = (d_1 f_2)^z = z^{-1} d_1 z f_2$ and $y^{-1} = f_2 d_1^z$. Hence $xy^{-1} = d_1 f_2 f_2 d_1^z = d_1 d_1^z$. Since $D = C_K(f_2) \simeq PGL_2(\sqrt{q})$, we have $D \geqslant T_1$ and therefore $d_1 \in T \backslash T_1$. Then $C_K(d_1) \simeq D_{q+1}$, $|C_D(d_1)| = 2$ and we have

$$|d_1^D| = |D : C_D(d_1)| = \frac{\sqrt{q}\,(q-1)}{2}, \quad |D_1^K| = \frac{q(q-1)}{2}.$$

Since $|f_2| = 2$, we have that the Sylow p -subgroup Q_1 from $C_K(f_2)$, which is contained in Q, has order $\sqrt{q}$. Hence $Q = C_Q(f_2) \times [\langle f_2\rangle, Q]$ where $|[\langle f_2\rangle, Q]| = \sqrt{q}$. Let $X = d_1^D$. Then $X \cap X^r = \emptyset$ for any element $r \in [\langle f_2\rangle, Q]$. Indeed, if $X \cap X^r \neq \emptyset$, then $C(d_1^g) \geqslant \langle f_2, f_2^r\rangle$ for $g \in D$ and p divides $|C_K(d_1)|$, a contradiction. Hence $|X^{[\langle f_2\rangle, Q]}| = \sqrt{g}$. If $X^{r_1} \cap X^{r_2} \ni d_1^g$ for $g \in K$ and $r_1 \neq r_2$ then we have $C(d_1^g) \geqslant \langle f_2^{r_1}, f_2^{r_2}\rangle$ and p divides $|C_K(d)|$, a contradiction. Hence the set d_1^K divides into $\sqrt{q}$ subsets which are conjugate to X by elements of $[\langle f_2\rangle, Q]$. Since $C_K(d_1) \simeq D_{q+1}$, $L = O(C_K(D_1))$ is a Hall π-subgroup in K. By [6, p. 285, 286] L is a TI-subgroup in K, $N_K(L) \simeq D_{q+1}$ and then 2 divides $|K : N_K(L)|$.

Hence d_1 inverts L^g for a certain $g \in K$ and $L^g\lambda\langle d_1\rangle$ is a dihedral group. Since $|L^g| = \frac{q+1}{2} > \sqrt{q}$, we have that a subset X^r for a certain element $r \in [\langle f_2\rangle, Q]$ contains two different involutions from the subgroup $L^g\lambda\langle d_1\rangle$. Hence X contains two different involutions from the subgroup $(L^g\lambda\langle d_1\rangle)^{r_3}$ for $r_3 \in [\langle f_2\rangle, Q]$ and $|xy^{-1}| = |d_1 d_1^z| \neq 1$ and is odd for a certain element z from D, a contradiction.

If $q \in \{5, 7, 9, 11\}$, then we have $q = 9$ and $G = K\lambda\langle f_2\rangle$, $x = d_1 f_2$, $|x| = 4$. There are two classes of elements of order 4 in G. It follows from the Sylow theorem that $G \geqslant K_2 \simeq Z_5\lambda Z_4$ and since $f_2 \neq 1$, $x \in K_2$, a contradiction. Lemma 9 is proved.

Lemma 10. *We have $K \not\simeq L_3(4)$, $|W(x) : W(x) \cap K| = 2$.*

Proof. Let $K \simeq L_3(4)$, $|W(x) : W(x) \cap K| = 2$. We preserve the notation of Lemma 9. By the structure of $L_3(4)$ and Lemma 6 we have $x \notin K$ and $x_1 \in Z(T)$. Since $N_G(T_1)$ contains x, we have from Lemma 6 and the Frattini argument that $[\langle g\rangle, S] = 1$ for a certain Sylow 3-subgroup from $N_G(T_1)$ and a 2-element g from T. By the structure of the centralizers of involutions in K [7, p. 492] we have $\langle g\rangle \cap T_1 = 1$. Since $|W(x) : W(x) \cap K| = 2$, we have $|g| = 2$. From the structure of centralizers of involutions in Aut (K) [12, Section 9] we have that g is a graph automorphism and $C_K(g) \simeq L_2(4)$. Hence $Z(T_1) = Z(T) \simeq E_4$ and the involution g maps a subgroup Y_1 from T_1, Y_1 is isomorphic to E_{16}, onto a subgroup Y_2. From Lemma 2 and the structure of T, we have $8 \geqslant |x| > 2$. If $|x| = 8$, then we have $|x^2| = 4$ and $x^2 \in K$. By [7, p. 492] $|C_K(x^2)| \leqslant 16$. If $|x| = 4$, then we have $x^2 \in Z(T)$. Hence $\bar{x}$ is an involution in $\overline{T} = T/Z(T)$.

Since $x = gi$ for a certain element $i \in T_1$, $\bar{x}$ maps $\overline{Y}_1$ onto $\overline{Y}_2$ in $\overline{T} \simeq E_4 wr Z_2$. Hence $|C_{\overline{T}_1}(\bar{x})| = 4$. Since $\overline{C_{T_1}(x)} \leqslant C_{\overline{T}_1}(\bar{x})$ we have $|C_{T_1}(x)| \leqslant 16$. In all the cases we have a contradiction by applying Lemma 1 for $F \simeq Z_7\lambda Z_3$, $L_1 = K$, $x = u$. Lemma 10 is proved.

Lemma 10 completes the proof of the theorem.

References

[1] Unsolved problems in group theory. The Kourovka Notebook. (12th edn). Novosibirsk 1992.

[2] Zenkov, V. I., On p-elements of a finite group. In IIIrd International Conference on Algebra, Theses of talks, INOPROF, Krasnoyarsk 1993, 127.

[3] To the Theory of Finite Groups. Collection of papers, Mir, Moscow 1979.

[4] Suzuki, M., On a class of double transitive groups. Ann. Math. 75:1 (1962), 105–145.

[5] Hartley, R. W., Determination of the ternary collineation groups whose coefficients lie in the $GF(2^n)$. Ann. Math. 27:2 (1925–1926), 140–158.

[6] Dickson, L. E., Linear Groups with an Exposition of the Galois Field Theory. Teubner, Leipzig 1901; reprinted by Dover, N.Y., 1958.

[7] Simpson, W., Frame, J., Character tables for $SL(3,q)$, $SU(3,q^2)$, $PSL(3,q)$, $PSU(3,q^2)$. Can. J. Math. 25:3 (1973), 484–494.

[8] Seminar on Algebraic Groups and Related Finite Groups. Lect. Notes Math., 131, Springer-Verlag, Berlin e.a. 1970.

[9] Robinson, J., Of maximal subgroups of finite groups. Proc. Amer. Math. Soc. 92:3 (1984), 325–326.

[10] Walter, J., The B-conjecture, characterization of Chevalley groups. Mem. Amer. Math. Soc. 61, AMS, Providence, 1986.

[11] Stellmacher, B., Über endliche Gruppen mit 2-lokalen maximalen Untergruppen. J. Algebra 70:1 (1981), 48–77.

[12] Gorenstein, D., Lyons, R., The Local Structure of Finite Groups of Characteristic 2 Type. Mem. Amer. Math. Soc. 42, AMS, Providence, 1984.

List of Plenary and Sectional One-Hour Lectures

Boris APANASOV
Varieties of discrete representations of groups of hyperbolic 3-manifolds and low-dimensional topology

Konstantin I. BEIDAR, W. MARTINDALE, Aleksandr A. MIKHALEV
Lie isomorphisms of associative rings

Vissarion V. BELYAEV
Locally finite simple groups from the point of view of the classification of finite simple groups

Oleg V. BOGOPOLSKII, Dmitriĭ G. KHRAMTSOV
Fundamental groups of graphs of groups

Zenon I. BOREVICH, Vladimir A. KOIBAJEV
Garlands in linear groups

Roger BRYANT
Partially ordered sets, groups of permutations, and linear groups

Nikolaĭ S. CHERNIKOV
Generalized soluble groups with factorizations

Georgiĭ P. EGORYCHEV, Vladimir M. LEVCHUK
Enumeration problems for nilpotent groups and their associated Lie algebras

Yuriĭ L. ERSHOV
Relatively regularly closed fields

J. R. J. GROVES
Finitely presented soluble groups

Chander-Kanta GUPTA
Automorphisms and primitivity of relatively free groups

Narain D. GUPTA
Dimension and Lie-dimension subgroups

Anatoliĭ S. KONDRAT'EV
Decomposition matrices of finite quasisimple groups

Andreĭ S. MOROZOV
Groups of computable automorphisms

Yakov N. NUZHIN
Generating elements of simple groups and their applications

Evgeniĭ A. PALYUTIN
Coverings by classes of commuting equivalencies and elementary theories close to modules

Nikolaĭ D. PODUFALOV
On solubility of collineation groups of semi-field planes

Vladimir N. REMESLENNIKOV
Results of M. I. Kargapolov and his school

Vitaliĭ A. ROMAN'KOV
Infinite generation of the automorphism groups of free pro-p-groups

Vladimir V. RYBAKOV
Algebraic and model-theoretical methods in the theory of non-classical logics

Vladimir P. SHUNKOV
T_0*-groups*

Ualbaĭ U. UMIRBAEV
Universal derivations

Frank O. WAGNER
On the structure of stable groups

John S. WILSON
Finitely presented soluble groups, Hilbert–Serre dimension and integer polynomials

Mikhail V. ZAITSEV
Special Lie algebras

List of Contributors

Vladimir A. ANTONOV, Troitskoe shosse, 50, apt. 44, Chelyabinsk, 454053, Russia

Boris APANASOV, Department of Mathematics, University of Oklahoma, 601 Elm ave., Norman, OK 73019, USA

Serikzhan A. BADAEV, Mametovoi st., 25, apt. 42, Almaty, 480016, Kazakhstan

Vissarion V. BELYAEV, Kirenskogo st., 5, apt. 79, Krasnoyarsk, 660074, Russia

Andreĭ A. BULATOV, Ural University, Dept. of Algebra and Analysis, Turgeneva st., 4, Ekaterinburg, 620083, Russia

Vyacheslav P. DOBRITSA, Pushkin st., 75, apt. 63, Almaty, 480100, Kazakhstan

Georgiĭ P. EGORYCHEV, Krasnomoskovskaya st., 1a, apt. 149, Krasnoyarsk, 660028, Russia

Yuriĭ L. ERSHOV, Institute of Mathematics, Universitetskiĭ pr., 4, Novosibirsk 90, 630090, Russia

Sergeĭ S. GONCHAROV, Institute of Mathematics, Universitetskiĭ pr., 4, Novosibirsk 90, 630090, Russia

Aleksandr V. GRISHIN, Kashirskoje shosse, 84–1–327, Moscow, 115569, Russia

Peter HILTON, Dept. of Mathematical Sciences, State University of New York, Binghampton, NY 13902, USA

Hans-Jürgen HOENKE, Mendelstr., 4, Berlin, D-1100, BRD

Anatoliĭ S. KONDRAT'EV, Institute of Mathematics and Mechanics, Sophia Kovalevskaya st., 16, Ekaterinburg, 620219, Russia

Alekseĭ A. KOROBOV, Institute of Mathematics, Universitetskiĭ pr., 4, Novosibirsk 90, 630090, Russia

Andreĭ A. KROKHIN, Department of Algebra and Geometry, Ural State University, Lenina pr., 51, Ekaterinburg, Russia

Vladimir M. LEVCHUK, Akademgorodok, 17Б, apt. 45, Krasnoyarsk, 660036, Russia

E. K. LOGINOV, Department of Higher Mathematics, Ivanovo Energetic University, Rabfacovskaya st., 34, Ivanovo, 153548, Russia

Aleksandr A. MAKHNEV, Ural State Technical University, Ekaterinburg, 620002, Russia

David I. MOLDAVANSKII, Ivanovo State University, Ivanovo, 153377, Russia

Yakov N. NUZHIN, P.O. Box 16849, Krasnoyarsk, 660074, Russia

Konstantin A. PAVLOV, Department of Mathematics, University of Wisconsin, Madison, Wisconsin 53706, USA

Vitaliĭ A. ROMAN'KOV, Omsk State University, Mira pr., 55-A, Omsk, 644077, Russia

Vladimir V. RYBAKOV, Mathematical Department, Krasnoyarsk State University, Svobodnyĭ pr., Krasnoyarsk, 660062, Russia

Iskhak I. SAKHAEV, Galeeva st., 8, apt. 81, Kazan', 420072, Russia

Anatoliĭ M. SEBEL'DIN, Kovalikhanskaya st., 30, apt. 11, Nizhniĭ Novgorod, 603155, Russia

Yuriĭ V. SELIVANOV, Moscow Aircraft Technological Institute, Petrovka st., 27, Moscow K-31, 103767, Russia

Vladimir I. SENASHOV, Uritskogo st., 98, apt. 2, Krasnoyarsk, 660049, Russia

Sergeĭ I. SENASHOV, Uritskogo st., 98, apt. 2, Krasnoyarsk, 660049, Russia

Christopher SCHUCK, Dept. of Mathematical Sciences, State University of New York, Binghampton, NY 13902, USA

Vladimir P. SHUNKOV, Akademgorodok, 1a, apt. 5, Krasnoyarsk, 660036, Russia

Ualbaĭ U. UMIRBAEV, Timiryazeva st., 46/12, apt. 304, Almaty, 480121, Kazakhstan

Frank O. WAGNER, Mathematical Institute, 24–29 St. Giles, Oxford OX1 3LB, UK

John S. WILSON, School of Mathematics and Statistics, University of Birmingham, Birmingham B15 2TT, UK

Victor I. ZENKOV, Vostochnaya st., 174, apt. 28, Ekaterinburg, 624100, Russia